The Farmer's Calendar

*Containing the Business Necessary to be
Performed on Various Kinds of Farms
during every month of the year*

ARTHUR YOUNG

CAMBRIDGE
UNIVERSITY PRESS

CAMBRIDGE UNIVERSITY PRESS

Cambridge, New York, Melbourne, Madrid, Cape Town,
Singapore, São Paolo, Delhi, Tokyo, Mexico City

Published in the United States of America by Cambridge University Press, New York

www.cambridge.org
Information on this title: www.cambridge.org/9781108037167

© in this compilation Cambridge University Press 2011

This edition first published 1804
This digitally printed version 2011

ISBN 978-1-108-03716-7 Paperback

THE

FARMER'S

CALENDAR.

———

[*Price Half-a-Guinea.*]

Printed by B. M'Millan,
Bow-Street, Covent-Garden.

THE

FARMER'S CALENDAR:

CONTAINING THE

BUSINESS NECESSARY TO BE PERFORMED

ON

VARIOUS KINDS OF FARMS

DURING

EVERY MONTH OF THE YEAR.

BY ARTHUR YOUNG, ESQ. F. R. S.

SECRETARY TO THE BOARD OF AGRICULTURE,

Honorary Member of the Societies of Dublin, Bath, York, Salford, Odiham,
South Hants, Kent, Essex, and Norfolk; the Philosophical and Literary
Society of Manchester; the Veterinary College of London; the Economi-
cal Society of Berne; the Physical Society of Zurich; the American
Society of Massachusetts; the Palatine Academy of Agriculture at
Manheim; the Imperial Economical Society established at Peters-
burgh; the Royal and Electoral Economical Society of Celle;
Member of the Society of Agriculture for the Department of
the Seine, France; and Corresponding Member of the Royal
Academy of Agriculture at Florence; of the Patriotic
Society at Milan; and of the Economical
Society at Copenhagen.

A NEW EDITION,

GREATLY ENLARGED AND IMPROVED.

LONDON:

PRINTED FOR RICHARD PHILLIPS, ST. PAUL'S CHURCH-YARD;
AND TO BE HAD OF ALL BOOKSELLERS.

1804.

ADVERTISEMENT.

GARDENERS have found great use in Calendars of their necessary work for every month in the year; and, if the two employments of the Farmer and the Gardener be well considered, it will appear that the former wants such a remembrancer, at least as much as his brethren in the garden.

At the beginning of every month, a good Farmer, whether he has or has not a book of this sort, is obliged to reflect on the work he has to perform in that month: he ought to foresee the whole at once, or it is impossible he should make a proper provision for its due performance. I leave it to any one to judge, if such an estimate of monthly business can be gained so easily, completely, or systematically, without such an assistance to the memory as is afforded by this Work; and even if a book of this sort but once in a year gives intimation of some important work, which might otherwise have been forgotten, its worth must be acknowledged.

In

In respect to the Calendars which had appeared previously to this Publication, they were very slight and imperfect sketches, generally nothing but additions to other books ; and their authors omitted at least as many useful articles as they inserted.

TO THE NEW EDITION.

In various parts of the Correspondence published during the last fifteen years in the Annals *of Agriculture, I have been called upon for a New Edition of this Calendar, and have as often resolved to give it; but the new improvements which have taken place, made so many and such great alterations necessary, that other and more pressing employments have prevented the undertaking. It is at last completed; and I hope the Reader will find it, in the present form, worthy of his attention.*

A. Y.

Bradfield-Hall,
February 3, 1804.

CONTENTS.

CONTENTS.

JANUARY.

Potatoes,

Cabbages,

Sow

FARMER'S CALENDAR.

JANUARY.

SHEEP.

IN this month ewes of some breeds of sheep will lamb. Great care ought to be taken of them: till then they seldom want turnips; most farmers having grass either in whole fields, kept walks, or in borders, &c. sufficient for lean-stock till they are near lambing, when they should have turnips regularly given them. If the land be not dry, the best method is to draw the turnips, and cart them to a dry pasture, and there bait the sheep on them twice a day, observing well, that they eat clean, and make no waste; which is not a bad rule for ascertaining the quantity necessary. In this way, the turnip crop goes the farthest. On dry soils, the best way, for the sake of manuring for barley, is to eat the crop on the land, hurdling off a certain quantity for the flock; and, as fast as they eat pretty clean, to remove the hurdles farther. This method saves much trouble, and is highly improving to the land; but it should be practised

B   only

only on lands that are dry, otherwise the sheep poach, and do mischief. The crop will not go quite so far as if drawn and laid in a grass field ; for the sheep dung, and stale, and trample on many of the roots after they are begun, which occasions some waste : nor is there any loss of manure in carting them, only it is left, in one instance, on the arable field, and, in the other, on the grass one. No improvement can be greater than this of feeding the sheep with turnips. On whatever land they are given, the benefit is always very great.

It is further to be observed, that many sheep are fattened on turnips, particularly wethers ; in which system of husbandry many of the turnips are wasted, if you have not two flocks, one lean, to follow the fatting sheep, and to eat up their leavings ; for sheep will not eat clean in fattening.

In very wet weather, storms, or deep snows, the sheep and lambs should be baited on hay. Some farmers drive them to hay-stacks, which shelter and feed them, but it is wasteful. Others give the hay in moveable racks ; and allow a certain quantity every day. It is an excellent method, to allow them in their racks a small quantity of hay daily while on turnips, let the weather be good or bad : but this is not absolutely necessary.

In some parts of the kingdom, the best farmers give their ewes and lambs in this month bran and oats, or oil-cake, in troughs, while they are feeding

on

on turnips ; but it must be a good breed, for such a practice to repay.

FOLDING SHEEP.

In respect of folding, a very great change has taken place on inclosed farms in the practice of the best farmers, especially in Norfolk. They are now fully convinced, that it is an unprofitable practice, except where the openness of downs and common fields renders it necessary for the purpose of confinement. The number of sheep that may be kept on a farm without folding, is much greater than can be supported with it. This is a very essential point. There is a deduction from the farmer's profit, in the injury done by folding to both ewe and lamb, which has been estimated by experienced judges, at from 2 s. 6 d. to 4 s. per ewe ; so that a farmer should consider well, before he determines to follow a practice, which, from a multitude of observations, is pronounced unprofitable. Mr. Bakewell used to call it robbing Peter to pay Paul. The arguments now used in its defence are not satisfactory : it is contended, that if sheep be not folded, they will draw under hedges, &c. for shelter in bad weather ; if so, they ought to be allowed to do it, for more would be lost in such cases by forcing the sheep from shelter, than the value of their fold. Where this practice takes place, good shepherds will, in case of rain, get up in the night and let their flocks out of fold, knowing the consequence of confinement on arable land in wet weather. The instinct of these animals will

B 2 conduct

conduct them much better than our reason, not
only where to fly for shelter, but also for choosing
their own time to go to rest, and to feed in the
morning. These they vary according to seasons
and weather; but folding prevents it, and forces
them to a regularity never called for by the weather.

When I began first to entertain doubts of the
propriety of folding sheep on any farms in which
they can be kept to certain fields in the night without that practice, I desired earnestly to try some
experiments that might throw more light on the
question than it was possible for reason to do ; but
to effect this comparatively, was very difficult, as
the trial I wished for was such, as should carry
some positive conviction with it. I have not been
able to effect it fully ; but the trials I have made,
may not be found destitute of power to throw some
light on this interesting question. I am perfectly
persuaded, that it would have been impossible for
me to have kept on the same land, nearly such a
stock in one parcel with folding. I do not conceive that the fields would have carried three-
fourths so managed. Four drivings in a day make
them trample much food, disquiet the sheep, and
transfer the choice of their hours of feeding and rest
from themselves to the shepherd and his boy. While
lambs are young they are injured by this, and the
ewes are liable to be hurried and heated ; all which
are objects that should weigh in the question.
When sheep are kept in numerous parcels, it is not
only

only driving to and from fold that affects them, but they are, in fact, driving about in a sort of march all day long, when the strongest have too great an advantage, and the flock divides into the head and the tail of it, by which means one part of them must trample the food to be eaten by another. All this points the very reverse of their remaining perfectly quiet in small parcels.

But the question turns on the benefit to be reaped by the fold ; for if that be great enough to compensate for the loss by such circumstances, the practice may not be condemned.

I believe the reason why farmers are such warm advocates for folding, arises from the power it gives them of sacrificing the grass lands of a farm to the arable part of it. Their object is corn, by which they can carry off a farm whatever improvement they bring to it. Grass improved is profit to the landlord in future; and tenants are too apt to think, that this is done at their expence. They do not at all regard impoverishing a grass field in order to improve a ploughed one ; and I need not observe, that every sort of sheep-walk is thus impoverished ; so that ancient walks, which have been sheep-pastured perhaps for five centuries, are no better at present than they ever were before; whereas most fields sheep-fed, without folding from them, are in a constant state of amelioration : this leads me to remark the effect I observed on several of my own fields.

I attended, through the course of a summer,

B 3　　　　　　　　many

many gentlemen over my fields, with a view to examine whether the sheep had seemed to have rested
only on spots, to the too great manuring of such; or,
on the contrary, to have distributed themselves
more equally; and it was a pleasure to find, that
they seemed generally to have spread in every part,
if not quite equally, at least nearly so. The improved countenance of several old lays fed in the
same manner, when examined in autumn, convinced me as well as my bailiff, that the ground
had been unquestionably improved considerably.
Those fields had carried a very bad appearance for
some years, but they were, after sheep-feeding, of
a rich verdure, and as full of worm-casts as if they
had been dunged. I rolled them heavily in November, but they soon became rough again by
worms, and demanded much rolling in spring.
They had afterwards a greener and more fertile
appearance by far than ever they wore before.

The whole of this circumstance, the value of
which I shall be able to appreciate in the trials of
future years, belongs to this method of dividing
flocks, to the exclusion of folding. The fold is
valuable, but so is the improvement of the grass
land, and may, for what I know, nearly equal it:
when, in addition, we include the greater number
of sheep that can be kept, and the favour done to
them by letting them alone, there remains, in my
mind, no further doubt of the fact.

It is common to hear flock-farmers in open countries say, they have not the power to manage so.
This

This may be very true, upon the major part of the farms, but such have often many inclosures, in which this management might be applied without difficulty.

But if we suppose folding to be the system pursued, I may remark, that the farmers in those parts of the kingdom which understand it best, do not extend it so far as they might ; they give over folding in November or December, whereas it may certainly be carried on through the whole winter with profit; even supposing that the practice is *necessary :* on those farms which have a perfectly dry gravelly pasture or two, it is advisable to fold all winter on such dry grass land. It must not be attempted on moist arable land, nor on moist grass land ; but on dry pastures. The safety to the sheep is greater, and the benefit to the grass an object. There is another method of gaining all the benefit of folding, quite through the winter, and on all soils ; this is, to confine them at night in a sheep-yard, well and regularly littered with straw, stubble, or fern ; by which means you keep your flock warm and healthy in bad seasons ; and at the same time raise a surprizing quantity of dung : so great a quantity, if you have plenty of litter, that the profit will be better than folding on the land. A great improvement in this method, would be giving the sheep all their food (except their pasture) in such yard ; viz. hay and turnips ; for which purpose they may be brought up not only at night, but also at noon, to be baited ; but if their pasture be

at

at a distance, they should then, instead of baiting at
noon, come to the yard earlier in the evening, and
go out later in the morning. This is a practice
which cannot be too much recommended ; for so
warm a lodging is a great matter to young lambs,
and will tend much to forward their growth ; the
sheep will also be kept in good health ; and, what
is a point of consequence to all farms, the quantity of dung raised will be very great. If this
method is pursued through the months of December, January, February, March, and April, with
plenty of litter, 100 sheep will make a dunghill of
at least 60 loads of excellent stuff, which will amply manure two acres of land : whereas 100 sheep
folded (supposing the grass dry enough) will not
in that time equally manure one acre.

SHEEP IN ROUEN.

Such ewes as have lambed before, and in this
month, should be drawn off from the flock, and
put into rouen in inclosed farms, to give early
lamb ; but this remark is not applicable to flock-
farms, where the grand support is the turnip crop.
On such, the rouen should be preserved till the period of distress arrives, when turnips are done, and
forward grasses not ready.

FARM-YARD.

In this month a strict attention should be given
to the cattle in the yard or yards ; those I mean
which run loose there. Care should be taken to
have them regularly supplied with straw, if that be
the food, and that they have always water at command.

mand. The threshers should be so proportioned to the stock of lean cattle, as to make the straw last just through the winter. Take good care also to keep the yard well littered from the stacks of straw, stubble, fern, &c. raised in autumn, so that the cattle may always lye perfectly dry and clean. Their health requires this attention; which should, at any rate, be given, were it merely for raising large quantities of manure.

STRAW.

While it is noted, that if the cattle are fed with straw, it should be done with certain necessary attentions, it would be an omission not to remark, that the best farmers in Norfolk are generally agreed that cattle should eat no straw, unless it be cut into chaff mixed with hay; but, on the contrary, that they should be fed with something better, and have the straw thrown under them, to be trodden into dung: and I am much inclined to believe, that in most, if not in all cases, this maxim will prove a just one. The common cases of straw-feeding are, of cows, young cattle, or black cattle just bought in, and not yet put to fatting. With regard to cows, the food is certainly insufficient, and lets them down so much in flesh, that when they calve, and are expected to yield productively, they lose a considerable time, and that, perhaps, the most valuable, in getting again into flesh, before they give their usual quantity of milk; but if they have been well and sufficiently wintered, they

are

are half summered, and yield at once adequately. For young cattle, it is still worse management; for their growth is stunted, and they never recover it. Black cattle from poor mountains had better be put to straw than any other stock; but here again care must be taken that the system be not deranged by it. If well fed, and the beasts be not large, they may be cleared off between harvest and the end of November; but if they are wintered on straw, this may not be effected, and the farmer may be forced to put himself to the expence of corn or oil-cake, to feed beasts not of a size to pay well enough for those articles. The evil is less if he has plenty of turnip or cabbage, but for these he may have other applications. In so far as regards the quality of the farm-yard dung, all this reasoning becomes still more forcible; for from straw-fed cattle, the farmer will, at the end of winter, find perhaps a large heap, of so poor a quality, that it will go but a little way in manuring his fields; whereas, one load of dung made by fat or well fed cattle, will be equal to two or three of it.

The proper food for cows in this month is cut chaff, one half hay and the other half straw, with a good bait of turnips or cabbages. For young cattle, the same chaff, and as much cabbage as they will eat; and the same, or turnips, for black cattle.

BEAN-STRAW.

" Bean-straw, if well harvested, forms a very
" hearty and nutritious diet for cattle in the winter
" time,

" time, and both oxen and horses, when not worked,
" will thrive on it: sheep, also, are very fond of
" browzing on the pods, and the caving is a very
" nutritious manger meat."—*Bannister.*

The importance of putting beans in early, and reaping soon enough, will appear clearly, when I observe that the straw, well harvested, is worth from two to three pounds per acre. Mr. Arbuthnot's teams, which were always hard worked, never had a truss of hay while his bean-straw lasted.

CUT-CHAFF.

The number of engines which have of late years been invented for cutting hay and straw into chaff (most of which execute their work sufficiently well), leaves no farmer in the kingdom under the necessity of using the common chaff-box, worked by those only who have acquired the art of using it, and who usually made much greater earnings than the common pay per diem. Of these machines, I believe Salmon's has the preference ; the price is 12l. 12s.; but a very good one is made at Thetford for eight guineas. The practice of cutting both hay and straw for all sorts of stock, is one that has been found very important by many practical and intelligent cultivators of great experience. General observations are not so satisfactory as comparative experience ; but there are not many persons who have opportunity, time, and power, to compare the food and labour of two different teams, the one fed in the common way, with hay, and the other with cut chaff, half or one-third straw. The opinion

nion of the best informed persons is decidedly in favour of the latter. However, if racks are permitted
in a stable, it is not an easy matter to prevent
horse keepers from cramming them full of hay, and
especially at night. The best contrivance I have
heard of to supply the place of racks, was that of Mr.
Vancouver, who made a sort of hopper the whole
length of the manger, which delivered chaff from
a loft above it gradually, as the horses moved the
lower lip of the hopper with their noses, in this
manner supplying themselves ; but a very intelligent nobleman trying it, found that it would not
deliver regularly : this might arise from the dimen-
sions not having been sufficiently attended to ; for
if the hopper be not of a due breadth, the
chaff might arch above the moveable board, and
not come down : the aperture in the manger
through which it passes, must necessarily be of a
certain size, neither too wide nor too narrow. It
certainly seems to be a practical idea, and very capable, after some trials and regulations, of being
fully applicable to common practice. It well deserves attention, especially as the expence of an experiment for one stall could not be considerable.
I have often determined to try it myself, but have
always been prevented by some journey or excursion taking me from home at the moment when I
could otherwise have given the requisite attention.
I conceive that it would demand a manger from
four to six inches wider than common ones.

The practice, however, of giving hay cut with a
mixture

mixture of straw, instead of feeding in the common way with hay, is to be recommended, at all events, to as great a degree as can be effected; for the saving is unquestionable. Nor is it to be practised for the teams only, but also for all other stock that eat hay. Mr. Page, of Cobham, in feeding his stock, gives no hay or straw but what is cut into chaff. At the expence of only 5l. he added a mill-wheel to his chaff-cutter, by which means a boy and a little poney cut 20 bushels per hour. This practice he finds so profitable, that he earnestly recommends it *.

For sheep, attention must be paid to the troughs in which it is given, to see that they be so boarded as to prevent the wind from blowing the chaff out: this is effected in Lord Clarendon's sheep-yard, in Hertfordshire, by a semicircular boarding, which covers the sheep's heads while feeding in the troughs.

COWS.

Several cows will probably calve in this month; about a month before which, they should be taken into the cow-house from the straw-yard, and be baited twice a day with green food; turnips, cabbages, carrots, potatoes, or whatever is the field winter food. After they calve they should be kept quite separate from the lean stock, either in the house or in another yard, and be fed upon those articles and hay, or *very good* straw. Cabbages will maintain cows in the cheapest manner, and make the butter perfectly sweet; but the decayed and yellow

* Annals of Agriculture, vol. xxviii. p. 107.

leaves

leaves must be picked off, giving the cows nothing but the heart of the cabbage : the refuse leaves will be eat clean up by the lean cattle. The great expence of winter feeding cows with hay alone, eats up half the profit of the dairy, even if none be given till they calve; for supposing them to calve in January or February, there remains three or four months for that food.

If the dairy consists of more than one or two cows kept for the use of the farmer's family, the system of feeding them becomes an object of considerable importance, and should be well digested. This subject demands most attention at Michaelmas, when all arrangements of stock take place. In the Calendar for October, it will be particularly considered ; but as in January the cows are probably calving, their food demands a careful attention. The preceding remarks suppose them taken from the straw-yard ; but let it be remembered, that superior managers, especially about Epping, are cautious of letting their cows at any time depend on straw.

If no other food be provided, they have hay only; it is not necessary to consider whether it will answer to give it to them when at certain high prices, as many farmers are, by their leases, deprived of the power of selling hay : where this is the case, the hay must be valued at what it costs, and not at what it would sell for : this estimate is easily made.

Rent

Rent of an acre,	-	-	-	-	-	-	£.1	0	0
Tithe,	-	-	-	-	-	-	0	3	0
Rates,	-	-	-	-	-	-	0	5	0
Mowing, making, carting, and stacking,							0	10	0
Manuring once in four years,			-	-	-		0	15	0
Fences,	-	-	-	-	-	-	0	1	0
							£.2	14	0
Interest and profits, ten per cent.			-	-	-		0	5	4
							2	19	4
After-grass, if sold,	-	-	-	-	-		0	10	0
Expence of Hay,	-	-	-	-	-		£.2	9	4

Such land, therefore, if it produces one ton of hay, ascertains the cost of the hay to be 49s. 4d. a ton—say 50s. Supposing then a cow to be fed at the rate of 56lb. per diem, and that only during 120 days, it is exactly three tons, which, at 50s. is 7l. 10s. No other calculation is necessary to prove that feeding cows with hay is ruinous. I have fed cows with my own hands, that have ate 56lb. per diem; but supposing only 30lb. per diem, it is above one ton and a half, at 50s. above 3l. 15s. which, for four months only, is much too high; and ought to convince the young farmer how necessary it is for him to provide green winter food.

In the Annals, vol. xvi. p. 361, is an experiment of mine, in feeding smaller cows, such as would fatten to about 45 stone (14lb.) Three milch ones ate, in October, 96lb. each, of cabbages, per diem: and, in another trial, 39lb. of cabbages, and 10¾lb. of hay each per diem; or, in the proportion
tion

tion of 2 tons, 18 cwt. of cabbages, and 15 cwt. of hay, in six months.

In the weaning of calves there are many different methods.—In Suffolk they do not wean till after Christmas, letting them suck six or seven weeks ; then they give bran and oats, with flet milk and water, and some very sweet hay by them, continuing this till grass is ready : but, if the farmer has carrots, they make an excellent article of their food, and save oats. The Duke of Northumberland's method succeeded with his Grace repeatedly, and I tried it with equal success. His account is this : " I have for some time entertained an idea " that skimmed milk might be prepared with pro- " per ingredients, effectually to answer the purpose, " where the practice is to give new milk from the " cow, and at about a third of the expence. The " articles are treacle, and the common lint-seed " oil-cake, ground very fine, almost to an impal- " pable powder, and the quantities so small, that to " make thirty-two gallons would cost no more, " exclusive of the milk, than about sixpence. It " mixes very readily, and almost intimately with the " milk, making it more rich and mucilaginous, " without giving it any disagreeable taste.—Take " one gallon of skimmed milk, and in about a pint " of it add half an ounce of common treacle, " stirring it till it is well mixed ; then take one " ounce of lint-seed oil-cake finely pulverized, and " with the hand, let it fall gradually in very small " quantities into the milk, stirring it in the mean " time,

" time, with a spoon or ladle, until it be tho-
" roughly incorporated, then let the mixture be
" put into the other part of the milk, and the
" whole made nearly as warm as new milk from
" the cow. After a time, the quantity of oil-cake
" may be increased."

THE DAIRY.

Mrs. Chevallier, a lady very attentive to a very
successful dairy, remarks, that in winter, it is a
good way to add hot water to milk, directly as it
comes from the cow; it makes it yield the cream
better. The trays in which it is set, should also be
scalded with hot water, or else warmed by the fire,
before the milk is set in them. All trays should
be of deal, about three inches and a half deep:
they are preferable to leaden ones, which not only
blister when hot water is poured into them, but
are also said to be unwholesome. About twelve
square yards of tray, with some spare bowls, will
do for twenty cows. The churn for such a dairy
should contain about 50 gallons beer measure.
The copper should hold 100 gallons. Chaffing
dishes of charcoal are kept in dairies in frost, but
then the cream does not rise so well. The best
dairy-maids never put the butter in layers in the
firkin; but leave the surface every day rough and
broken, in order to unite better with that of the succeeding churning. In Suffolk, from three and a half
to four pints of salt are commonly used to a firkin of
butter; but two, with good management, are better.
The milk, after the first skimming, is left twelve

hours more in the farm-houses, to make a second butter, which is sold to the poor at an inferior price. A dairy-maid commonly milks seven or eight cows in an hour.

YEARLING CALVES.

" These are very subject to the garget, *sup-* " *posed* to resemble the rheumatism in the human " body : lying wet either in yards or in fields will " give it. To be kept perfectly dry, is an almost " sure preventative."—*Mrs. Chevallier*.

YOUNG CATTLE.

The last year's calves should now be fed with hay, and roots, either turnips, carrots, or potatoes ; and they should be thoroughly well fed, and kept perfectly clean by means of litter : at this age it is a matter of great consequence to keep such young cattle as well as possible, for the contrary practice will inevitably stop their growth, which cannot be recovered by the best summer food. If hay is not to be had, good straw must be substituted ; but then the roots should be given in greater plenty, and with more attention. To steers and heifers two years old, the proper food is hay, if cheap ; or straw, with baits of turnips, cabbages, &c. It is not right to keep yearling calves and two year olds together ; because, in general, the younger the cattle are, the better they should be fed.

FATTENING BEASTS.

At this time, the farmer who makes it his business to winter fatten, is in the height of his work. There are three methods of fattening cattle,

tle, viz. carrying their turnips, &c. to a dry grass-
field, to a farm-yard, or to the house where the
beasts are tied up ; the two latter methods are the
best. Not many pastures are dry and sound enough
to bear the tread of an ox in winter ; and great
numbers are fattened in the field, in Norfolk,
eating the turnips where they grow. If fattened in
a yard, the food, viz. turnips, cabbages, or carrots,
must be given in mangers under open sheds, with
good cut chaff always in them, if hay is not plentiful ; but they will pay well for the best hay. The
same rule is to be followed in stall-feeding ; but
they must be littered well, or else they will presently
have a bound hide, and not thrive. In either
of these methods, plenty of litter must be provided. I would advise the use of three waggon-
loads of straw, stubble, or fern, to every beast, for
so much they will make into dung, which ought to
be the guide, and not the expence of the litter,
as the dung will repay that with great profit. I am
sensible that less will do ; but it should always be
remembered, that raising dung is the grand pillar
of husbandry.

OIL-CAKE AND CORN-FED BEASTS.

Feeding beasts in this manner is so very expensive, that if the greatest attention be not paid to
them, the loss will be considerable. The points
to be constantly attended to, are ; 1st, cleanness of
lodging, by constantly removing dung, sweeping
the pavement clean, and giving plenty of clean
litter, to prevent all filth from sticking to their
hides ; 2d, clean mangers often washed ; 3d, and

the

the most material article, to give very little food at
a time, and to vary it properly. Beasts which are
carelessly fed in this respect never thrive well. The
master's eye is constantly necessary. 4th, To keep
them warm enough by shelter, for warmth fattens
almost as much as food. These attentions are necessary for all cattle stalled, whatever their food
may be, but if neglected with that, which is very
expensive, the mischief is more felt.

SWINE.

This is a principal season with swine, both for
fattening, rearing, and bringing forth. As the
two first are mentioned largely under other months,
I shall at present speak only of the management
of sows and pigs. They must be kept each litter
in a sty, and fed with dairy-wash out of cisterns,
and with the food stored for them in autumn, such
as carrots, parsnips, potatoes, and cabbages; all
these do excellently for them. To substitute barley or pease, or even purchased bran or pollard, is
therefore unprofitable. The sows should always
have as much as they will eat, or the pigs will suffer; and what is of as much consequence, is keeping them them well littered. Let them be always
perfectly clean; it ensures the health of the pigs,
and at the same time raises a large quantity of
the best manure on a farm.

The breeding of swine being one of the most
profitable articles in the whole business of a farm,
the husbandman cannot pay too much attention to
it. I shall, in as few words as the subject will admit, give an account of the best system to be pursued

sued in this branch of his business. The farmer
who would make a considerable profit by hogs, must
determine to keep a proper number of sows, in order to breed many pigs , but this resolution ought
to be preceded by the most careful determination to
prepare crops proper for supporting this stock. The
proper ones for that purpose are barley, buck, beans,
pease, clover, and potatoes, or carrots. In the
common management, a farmer keeps only a sow or
two, because his dairy will do no more ; but in the
system of planting crops purposely for swine, a different conduct must necessarily be pursued. Potatoes,
carrots, Swedish turnip, and cabbages, must be provided for the sows and stores from October till the
end of May, by which time clover, chicory, or lucerne, should be ready to receive them, which will
carry them till the stubbles are cleared ; so that the
whole year is filled up with these plants, the common offal of the barn-door and the corn-fields.
When the sows pig, meal must be provided to make
wash, by mixing it with water. This in summer
will be good enough for their support, and in winter it must be mixed with boiled roots, oats and
pea-soup, for the young pigs. If cows are kept,
then the dairy wash is to be used in the above mixtures.

Upon this system, a farmer may proportion his
swine to his crops, or his crops to his swine ; and
he will find that for the whole year he should have
about an equal quantity of roots and grass, and half
as much corn as potatoes. For carrying the profit to
the highest advantage, the sows should pig but

twice a year; that is, in April and August; by
which means there will never be a long and expensive season for rearing the pigs before they are put
to the staple food of clover or potatoes, &c.; but
this circumstance is much removed by the provision of crops raised expressly for swine.

Upon this plan the annual sale of lean hogs
should be in October, the litters of April sold then
as stores, and those of August kept till October
twelvemonth, to sell for baconers, if the farmer fats
none himself. The stock upon hand this month will
therefore be the sows, and the pigs littered in the
preceding August, all which should have roots
from the store, and run at the same time in the
farm-yard, for shacking the straw of the barn
doors. In proportion to what they find in this,
you must supply them with roots, giving enough to
keep them to their growth.

WEANED PIGS.

It has been often remarked, that winter pigs are
unprofitable; and it is certainly true, if they are
not kept with great care and attention. Where
there is a dairy, the milk and whey may be so profitably applied to their use, that it should be preserved carefully for that purpose. The best addition,
or which alone will wean them incomparably, is
pea soup. Six pecks of pease boiled in a hogshead
of water till well broken and dissolved, and then
mixed in a tub or cistern with dairy wash, or given
alone, will wean them well. If dry meat be given
in addition, or alone, it should be oats, which do
for young swine far better than other sorts of
grain

grain. Barley does not agree nearly so well with them.

HORSES.

One of the most useful general lessons that can be given to an arable farmer, is to keep his horses always at work. The expence of a team is so great, that, if he does not pursue this rule, he must lose by them. January is a month in which all business of tillage ought to be at a stop. If the weather is a hard frost, care should be taken to make use of it in carting manures on the farm. If there are composts ready, a frost should not be let slip ; or, if there is faggot carting to be performed, or the earth of borders under hedges to be carried, the carts should be kept close to work of that kind, as long as the frost lasts. But, if the weather is open, road-work must be done. Carting out the corn may not nearly employ the teams : on other days the carts should go to the nearest town for manure. There certainly are situations precluded from this advantage, but not many. How well it would answer to keep a team on purpose for the employment, depends on various circumstances; but we may be assured, that it must answer to employ the teams about it, when they would otherwise stand still ; for then the expence is little more than labour and wear and tear.

The same observations are partly applicable to the ox teams ; and the farmer should have a strict eye, that both horses and oxen have plenty of litter ; otherwise his farm will suffer from a deficiency of manure.

THRESHING.

I before remarked, that the farmer, in threshing his crops of corn and pulse, should be attentive to proportion his threshers to his stock of lean
cattle, that neither more nor less straw may arise,
than is regularly consumed. Relative to the management of the threshers, the farmer should be
very clear-sighted to their motions, both as to the
cleanness of their work, and as to their honesty.
He may lose immensely if his straw is not threshed
clean ; and, as it is a work generally performed by
measure, the men are too apt to turn it over
too quickly, and to thresh out only that corn
which comes the easiest from the ear. In respect
to pilfering, the work gives them greater opportunities for it than any other ; for which reason, he
should have a sharp look out, and take care now and
then to meet the men of an evening in their way home,
and to come upon them in the barn, at various times,
and unawares. Such a conduct will keep the men
honest, if they are so already, and will prevent
many knaves from practising their roguery ;
whereas an indolent, inattentive master, will make
pilferers.

THRESHING-MILL.

If the farmer has one of these most useful implements, he is safe from the two evils mentioned in
the preceding article. The expence of a fixed mill,
is from 60 to 100 guineas for one that requires two
or three horses. It will thresh about 15 quarters of
wheat in eight or nine hours, and from 15 to 20
quarters

quarters of barley, oats, pease, or beans. Barley is
the grain that threshes worse with them than any
other; but I have seen several that thresh it as well
as other corn, such as Mr. Asbey's, at Blyborough,
Suffolk. His price, for a fixed one, 100 guineas,
and for a moveable one 160 guineas. The granary
should always be over the fixed mill, that the corn
may be drawn up at once and lodged safe under the
farmer's key. For *feeding* cattle, fresh threshed
straw is better than old; for littering (the proper
application) they are equal; but the best management for eating straw clearly is, to cut it into chaff
by the power of the mill, and to have the chaff-
house adjoining, so as to receive the cut straw at
once, without any carriage. This house should
have brick walls, in order that fermentation may
not set fire to any thing, and then if water be
thrown on the chaff it ferments, and is much more
nutritious * than when used in the common way.

FENCES.

This is a principal season for hedging and ditching. A farmer cannot give too much attention
to the fences of his farm; for, without good ones,
he might as well cultivate open fields: he cannot manage them as he pleases, but is for ever
crampt, for fear that his own or other people's cattle should break into his corn or hay fields. In
fencing, he should determine to execute the work
in the best manner, which is the plashing method.

---

* Annals, vol. iii. page 480.

It

It is done in the following manner : the men first clear the old hedge of all the dead wood, brambles, and other irregular growing rubbish, leaving along the top of the bank the straightest and best-growing stems of thorns, hazel, elm, oak, ash, sallow, beach, &c. about five or six in a yard; but, if there are any gaps or places thin of live wood, on each side of such places they leave the more. When this work is done, they repair the ditch, which I should never advise to be less than three feet by two and a half, and six inches wide at bottom, in the driest soils; but in all wet or moist ones, never less than four by three, and one at bottom. All the earth that arises from the ditch is to be thrown on the bank. The men, if no bargain is made with them before-hand, will lay some of it on the brow of the ditch ; but this must not be allowed, unless the ditch-earth happens to be extraordinarily rich, and to pay well for carrying it to the land, otherwise the grass of the border is spoiled, and the farmer is at the expence of carting earth which may be worth but little. When the ditch is finished, the men begin the hedge. Such of the stems left in cutting the old hedge, as they find growing in the line where the new hedge is to run, they cut off three feet from the top of the bank, to serve for hedge-stakes to the new hedge. This practice cannot be too much commended ; for these stakes being immovable, and never rotting, keep up the new hedge, so that it never falls, or leans either way. In the next

next place, they drive in their dead hedge-stakes where wanted, chusing sallows or willows, that they may grow. The hedgers then plash down the remainder of the live wood left standing. They cut the stick twice, one stroke near the ground, and the other about 10 or 12 inches higher, and just deep enough to slit out a part of the wood between the two, leaving the stem supported by little more than the bark, or about a quarter of its first size. It is then laid along the top of the bank, and weaved among the hedge-stakes. All are served thus; and, where they are not thick enough to finish the hedge, dead thorns are wove among them; then the top of the hedge is eddered in the common manner.

The fence thus made, consists of a good ditch and a hedge, most parts of which are alive ; that is, the stakes, and much of the wood that is weaved between them. The importance of having as much as possible of the hedge alive, cannot be too strongly impressed. This management ensures a lasting fence ; whereas, the hedges that are all dead, presently rot, and fall into the ditch. Those farmers, who live in countries that know nothing of the plashing method, cannot give too much attention to teaching it to their men. The best way is, to send for labourers from the plashing countries, who, in one season, will easily instruct their regular men in the business, which they may afterwards perform without difficulty.

DITCHES.

DITCHES.

In very wet soils, where draining is an essential improvement, and where the soil is a poor, loose loam, and not sufficiently consolidated, an evil that demands clay or marle, it is the custom of many farmers to dig ditches of much greater depth and width, for the double purpose of making better drains, and of raising clay or marle, wherewith to manure the fields. But I have heard other, and very practical farmers, object to this ; urging, that the expence, when compared with marling from a pit, is more than doubled ; for it costs more to throw it out of a ditch than into a cart, and, when removed, only two men can stand to fill : and, further, that for want of the greater depth to which pits are dug, the marle is neither so good nor so pure. These objections are powerful ones, and seem to authorize the farmer to reject any greater size of ditch, than the two purposes of draining and fencing demand.

DRAINING.

January is a proper season for draining. There are several sorts of drains ; but I shall confine myself at present to the covered ones. There are two methods of making them ; one by ploughs, which cut them either at one, or various furrows, according to their merit ; another by digging with a system of spades, which work one after the other, so as to dig a drain about two inches wide at bottom, and of various depths and breadths at top.

If

If a farmer occupies land that has no stones in it large enough to obstruct a plough, that implement is by all means eligible ; for the expence of cutting the drains with a plough, is not so great as with spades. But it should be observed, that draining-ploughs can only cut the small drains ; spades must be used for the main ones : their various courses and superior depths require manual work. Suppose a large field drained by parallel cuts of a plough, still the water must be carried out of those cuts by deeper drains dug, unless the land has a regular descent ; but, whether the operation be performed by a plough, at a small expence, or by spades at a large one, still the necessity of the improvement for wet soils remains the same, and those who have had experience of their nature, will not regret the expence of performing the work effectually. Wet grass-lands are for ever over-run with rushes, and other aquatic rubbish ; the hay of little value, and small in quantity. Arable land that is wet, can never be applied to a profitable purpose. It is too adhesive to be ploughed, when kindlier soils have received their tillage, and are sown. In wet seasons, the crops are too trifling to pay expences. Whatever attention is given to water-furrow them, still the land will not have that mellow, favourable nature, that enables it to yield advantageous crops. The expence of covered drains may be estimated, on an average, at 3l. an acre, when done with spades. Now this expenditure will, in moderate

cases,

cases, be repaid by the mere saving of the extra expence of water-furrowing, exclusive of all the superior benefits of it. Covered drains, dug 32 inches deep, two inches wide at bottom, and 12 at top, and filled about 10 inches deep, may be completely executed at 3 d. a perch, where labour is 18 d. per day in winter. In respect to filling up these drains, the farmer must be guided by the circumstances of situation. If stones are to be had in great plenty, he should fill with them. Bushes, common faggot-wood, bricks, horns, and bones, turf laid in like a wedge, straw, fern, ling, stubble, &c. are all used in various places; and in *Essex*, where these drains have been made almost time immemorial, the farmers insist, that the great object is not durability of materials, but the arching of the earth, when the materials are rotten and gone. In many parts of that county, drains run well to this day, that were filled with nothing but straw, more than thirty years ago. The extending such a practice should, however, depend absolutely on soil; for most certainly there are soils, in which such a practice would be totally inexpedient.

A very economical way of doing this work is the following : first, the farmer ploughs four or five times in a place with his common plough, and then shovels out the loose mould ; after which, on that smooth bottom, he takes one spit with a very long spade, about four inches wide at top, and narrowing to two at bottom ; then with a scraper cleans

out

out the moulds, and fills them up. Digging this
spit is three halfpence a rod, and filling up a halfpenny. The newly-invented mole-plough will be
mentioned hereafter.

BEANS.

If the autumnal sowing of Mazagan beans has
been prevented by the weather being uncommonly
wet, it ought to be done this month, for the earlier
the beans are planted, the better will be the crop.
And as the season must regulate the article of tillage in all cases, the farmer ought to sow his beans
the first month his land is dry : some seasons will
be dry in January, that were wet in December.

Beans are a crop that will pay very well for manuring ; and if there are not many turnips, potatoes, &c. all the dung of the farm should be laid
on for them, by way of a preparation for wheat ;
in which case the manure may be laid on at any
time when it can be done previously to the ploughing.

CARROTS.

The best culture of carrots is to let the barley
or wheat stubble lie till you plough and sow, putting them in on one earth. If much previous tillage is given, the second earth will probably fall in
this month. I mention this circumstance, in case
the farmer is determined on much tillage, which
for carrots I do not think advisable : possibly on
very running sand, winter tillage might be beneficial.

POTATOES.

The above observation is also applicable to this
root. There may be cases in which a ploughing in
January

January may be advisable ; but in general the land
should either lie till the planting, or at least have
only autumnal tillage. In dry soils, upon which
weeds have come up since that time, a ploughing
now may be right. If the dung designed for the
potatoe land is laid in the field ready for it, and the
weather happens to be frosty, the first opportunity
may be taken for carting it on, especially if the
land is wet enough to make a frost necessary.

WOOD.

There are not many districts in which woods
are profitable to a farmer to hire ; but when he
finds them a part of a farm, it is not always that he
can have a choice whether to take or leave them,
and must therefore apply himself to convert them
to the best profit. This month is generally a busy
one in felling : the men who do the work are commonly paid by measure, or tale. In some countries the falls are only cut and laid in rows, and
sold in that manner by the rood ; in others, the
farmer converts the stuff to the proper use, and
sorts it into faggots, poles, hoop-stuff, or hurdles ;
and this, I believe, will generally prove the most
profitable way.

In cutting woods there is one point much disputed, which is, the number of years growth at
which to cut. Customs vary from nine years to
twenty-seven, but generally about twelve or fourteen. I have seen many woods, in cutting which,
one stem on a stool was left to be of a double age
at next cutting, in order to have some large wood

in

in each fall. The question is, whether such stems draw from the root so much nourishment as to lessen the young growth as much as the large shoots amount to. From viewing such woods, I have observed, that the part of twelve years growth, among which were some of one stem, twenty-four years old, was as good as others, where the whole was only twelve. If so, the additional growth is nearly all profit; but if not, it certainly makes the wood when cut more saleable, and applicable to more purposes.

One great point to be attended to in the management of woods, and to which too much attention cannot be paid, is to keep the fences in the very best order possible; for a farmer or landlord had better let cattle into their wheat, than into their underwood; because their biting and mangling one year's shoot, is doing mischief to the amount of at least three years growth. But if woods are so ill fenced and so extensive, as to be too great an expence for the person to afford repairing; in that case, the longer the growth is, the greater will be the profit; for supposing that cattle, upon an average, eat three years growth, then there are three in twelve or fourteen destroyed; whereas, if the term is twenty-four years growth, still there are but three destroyed; which is doubly advantageous. These are points which should be well considered; and also what is the age at which the various sorts of underwood attain the greatest weight, having

D always

always in view the variations of soil. For instance, it should be tried, what weight twenty single stems of sallow, ash, oak, hazel, hornbeam, &c. come to at six, twelve, and twenty-four years growth, to see whether the produce is proportioned to the age. This would be a very easy experiment in every respect, but that of the time it would demand.

In the beech woods of Buckinghamshire, this system is carried exceedingly far ; for they are not cut till thirty or forty years growth ; the consequence of which is, they are destroyed as underwood, and nothing appears but single stems, which are successions of young trees. The way of cutting them is not by falls, as in common woods, but by singling out, every year, the largest of the trees, and cutting enough of them to pay 12, 15, or 20s. per acre, per annum, according to the goodness of the wood. These trees, though some of them when cut would more than measure as timber, are all sawn into lengths of four feet, or thereabouts, and rived into billets for fire-wood, for the London market, being conveyed there by the Thames. Good beech woods, upon this system, will pay 20s. an acre, clear of expences, which is more than underwood would pay upon the same soil. I believe it will generally be found, that the older the growth the greater will be the profit. At twelve years growth of ash, the land must be very good to have a crop of hop-poles ; but at twenty years growth,

growth, you will have very fine ones, and pay yourself much better than by the younger growth.

Some woods are so very wet, that the ash, hazel,
hornbeam, and oak stubs, will not thrive ; in that
case, the sallow and willow should be multiplied, or
the wood hollow-drained ; which is a practice beginning in some parts of Essex.   There they have
so long seen the advantages attending drains of that
kind, in their corn and grass lands, that they now
think their wettest woods will pay as well for them
as an arable field.   It cannot be doubted, but the
practice must be exceedingly advantageous; and this
month is a very proper time for doing it.

THE MOLE PLOUGH.

The accounts which have for the last two or
three years been received, of the effects of this implement, are extremely contradictory. With some
farmers the use of it has been great, and the duration of the drains extremely satisfactory ;   with
others the reverse.   I have attended to these circumstances in various districts, and have employed
the tool on my own farm, and from all I could observe or hear of it, the effect seems to depend entirely on the soil.   In clay the result has given
much satisfaction; but in loose spungy loams, however wet, and where sand *gauls* (as they are called)
abound, the drains have generally stopped.   A
young farmer should therefore acquaint himself well
with his soil to the depth of from 12 to 16 inches,
for if he has not a pretty regular stratum of clay,
or stiff marle, he may expect the pipes to fail in

two years : whereas in stiff soils, those are running well at present that were made six or seven years ago. Another remark that should be made, relates to the strength of the draft. Without wheels the mole plough demands 10, 12, and even 14 horses; these, when the land is wet, do almost as much mischief in trampling as the drains can do good. The addition, therefore, which has been made to this implement, of wheels before and a roller behind, is of essential importance, and reduces the team to six horses. It is certainly an excellent tool, if well applied.

A precaution, which should be mentioned here, is, on grass land, to open a furrow with a good common plough, or better, with an open furrow drain plough, in order that the mole plough may follow; and finish the work by turning back the furrow. This defends the slit from the frost, which otherwise is apt to moulder down the earth, to the hazard of stopping the drains.

The state in which land should be for draining, admits but of one question, the right moment for applying the Essex method of hollow-draining ? With respect to this there is a diversity of opinion, and, perhaps, with some propriety, it being a point on which something may be advanced on both sides. As to all other circumstances, such as the greater exertions of large open drains; the brick soughs of Mr. Elkington, &c. they should be unexceptionably performed previously to all tillage ; an assertion which many practical reasons support.
They

They imply a degree of wetness which would be ruinous in tillage, and as that wetness commonly proceeds from powerful springs, much carting and trampling could not be permitted after ploughing, even in summer, without essential mischief. Another object in extensive works, especially on moors upon the slopes of hills and in mountains, is the union of draining and irrigation. The general mouth of many drains may, in numerous cases, be made the constant supply of a system of watering the land below it. Till this is settled, the improver is uncertain what land it will be advisable to break up, and what otherwise to improve by water, for it may be laid down almost as a maxim, that water, where it can be had, should be tried with mere levelling before any other method be attempted. Upon the high moors to the south of Paitely-bridge, in Yorkshire, there are some remarkable instances of small abandoned mill-courses having overflowed the ling moor below them, and thereby destroyed the ling, and established a large family of grasses, converting black into what they call *white land*, and which, though (like all the rest of the moors) in a state of utter neglect (this accidental circumstance excepted), would let, with a wall fence around it, at 10s. an acre, instead of 6d. or 9d. When the effect of water is so remarkable, it is a proof that not a drop should be lost, but that what comes from the drains should be converted from an enemy into a friend.

It is the practice of many farmers to chuse the

year of fallow for this work, because they think
it is better done in summer than in winter,
and while the land is in fallow than when under
clover, though that clover be fed. If this be
really the fact, it probably depends not only on
the land being dry, by which means it admits carting of the straw and bushes, but also because the
sun and air have the effect, by drying the earth or
the sides of the cuts, to avoid that plastering which
the action of the spade has in digging them in wet
weather, and by which they are apt afterwards to
bleed through the pores less freely. There may be
something in this. Other farmers do the work in
winter, partly because they have a better opportunity, on flat fields, of seeing how the drains *draw*,
as it is termed; and partly because at that season
labourers are easier to be had. But poaching the
surface upon arable land is an objection. It seems
on every account to be desirable, that such soils as
well as others, and whether the work be done in
winter or summer, should be drained while in grass,
by which means poaching is avoided, and if the
surface turf be tough, it gives an opportunity of
making sod drains, which are cheaper than filling
them either with bushes or straw.

MARLING.

The marle and clay carts may work all this
month. This is so important an object, that too
much attention cannot be given to it, nor can a
great breadth of land be thus manured, if the teams
and men assigned to it be not employed regularly

the

the year through. Upon dry soils no difficulties
occur, but upon wet ones the teams can stir in winter only, while the surface is frozen, unless it be
on *layers* of some years standing, and well drained.
Upon sand, marle or clay should be laid on, in
the proportion of 50 or 60 cubical yards per acre ;
but on loose wet loams, upon which clay or marle
works a very great improvement, it should be laid
on to the quantity of 100 yards. The cheapest
way of doing it is, to contract for the whole job
with some little farmer, or horse keeper, who works
for hire.   In Suffolk, it is not uncommon to give
8d. a cubical yard for all expences whatever, except
spreading, which accurate farmers chuse to do by
the day, as a minute attention is in nothing more important.   If this be not well executed, some spots
in the field will have in the proportion of 200 loads,
and others not more than 50.

If the young farmer wants any inducement to
undertake the work of marling, it will be best
found in the register of what his brethren have
done.

Mr. Rodwell's account of this operation is very
interesting.

" My operations at first, were to inclose with
" thorn hedges, marle or clay, and break up 300
" acres of the heath ; and in the first seven years
" of the lease I finished what I meant to improve
" in that term. I marled or clayed 600 acres, at 70
" loads an acre, being 42,000 large tumbril loads. In
" this work I employed three teams, two of my

" own, and one I hired for several years. It is se-
" vere work, and the second year I lost nine horses,
" attributed to feeding on pea-straw from the new
" broken heath, a circumstance that deserves the
" attention of improvers.

" In the eleventh year of my lease I applied to
" my landlord for a renewal ; on which the farm
" was valued again by Mr. Hare, the surveyor at
" Peterborough, and I took a fresh lease of 15
" years, to commence at the termination of my old
" one, at the rent of 400l.

" I immediately clayed and broke up 200 acres
" more, at 100 loads an acre, 40 bushels per load,
" inclosing all with quick hedges, and ditches five
" feet wide and 4 feet deep ; after this I improved
" 100 acres more in the same manner.

" In the two leases of 28 years I clayed or marled
" 820 acres ; and I have clayed or marled so much
" over the second time, at 70 loads an acre, that
" the quantity I have carried in all, is very little
" short of 140,000 loads.

" Upon taking a third lease, I was, in 1798-9,
" particularly steady to this work, and in 49 weeks
" and three days carried 11,275 cubical yards, pay-
" ing by measure of pits, and not by loads, which
" were filled and spread by four men and a boy,
" and carted by six horses and two tumbrils.

" In this business of carrying clay or marle I
" have practised hand-barrowing : the men can
" make good earnings at 10d. per yard, wheeling
" it 30 rod ; and down to 7d. a yard at shorter dis-
                                        " tances ;

" tances ; and I am much inclined to think that if
" we had workmen used to the operation, and
" handy at it, like those employed in navigations,
" that· this method would be of all others the
" cheapest, especially on heavier soils. But by
" far the greatest part I have done by tumbrils, the
" expence of which put out is 5d. per yard for
" team, and 2¼d. a yard for labour ; this, with pay-
" ing for laying picks, wedges, &c. also for stones
" that rise, increase the whole expence to 8d. per
" yard, which is at least a halfpenny per yard cheaper
" than I can do it with my own teams ; the reason
" of which is, that the man who contracts with me
" drives his own horses, and looks after them. At
" 8½d. per yard, 140,000 yards have cost me 4,958l.
" excepting the small proportion *hired* at a half-
" penny a yard lower.

" I come now to mention a few circum-
" stances, which I hope may " tend to render
" this paper useful to others not having the
" experience which I have acquired : I shall use
" but few words, but they shall be founded on
" positive experiment or attentive observation.

" Clay is much to be preferred to marle on these
" sandy soils, some of which are loose, poor, and
" even a black sand. By clay is to be understood
" a grey clayey loam, some of it brick earth, and
" all has with vinegar a small effervescence.
" Marle is a white, greasy, chalky substance, that
" effervesces strongly with acids : I make a univer-
" sal

" sal rule, on a second improvement, to lay clay
" on the fields marled before, sometimes marle
" where clay was spread before ; but this not ge-
" neral, as clay answers best on the whole.

" On 90 acres, clayed 100 loads an acre, I have
" had, after two crops, the one turnips, the other
" barley, cole-seed, and sold it on the ground for
" 1000 guineas : then turnips, a famous crop, fol-
" lowed by barley, on 75 acres, 16 coombs an
" acre (the coomb is $\frac{1}{2}$ a quarter ;) and by oats on
" 15 acres (poorer land), 10 coombs an acre.
" These crops are, for the soil, *great*; but in ge-
" neral my products have been highly to my satis-
" faction."

Since this account was written, I have heard of
9d. per yard being given in Suffolk, and even as
far as 10d. offered, to induce a little farmer to set
up a team strong enough for the work.

DRAW CHALK.

" The method pursued in Hertfordshire in chalk-
" ing land is this ; and the persons employed there-
" in follow it as a trade : a spot is fixed upon,
" nearly centrical to about six acres of land to be
" chalked. Here a pit, about four feet in diame-
" ter, is sunk to the chalk, if found within twenty
" feet from the surface ; if not, the chalkers con-
" sider that they are on an earth pillar ; fill up
" the pit, and sink in fresh places, till their labour
" is attended with better success. The pit from
" the surface to the chalk, is kept from falling in
                                             " by

" by a sort of basket-work, made with hazel or
" willow rods and brushwood, cut green, and manu-
" factured with the small boughs and leaves remain-
" ing thereon, to make the basket-work the closer.
" The earth and chalk is raised from the pit by a
" jack-rowl on a frame, generally of very simple
" and rude construction. To one end of the rowl
" is fixed a cart-wheel, which answers the double
" purpose of a fly and a stop. An inch-rope of
" sufficient length is wound round the rowl; to
" one end of which is affixed a weight, which
" nearly counterbalances the empty bucket fast-
" ened to the other end. This apology for an
" axis in peritrochio, two wheel-barrows, a spade,
" a shovel, and a pick-axe, are all the necessary
" implements in trade of a company of chalkers,
" generally three in number. The pit-man digs
" the chalk and fills the basket, and his compa-
" nions alternately wind it up, and wheel its con-
" tents upon the land : when the basket is wound
" up to the top of the pit, to stop its descent till
" emptied, the point of a wooden peg, of suffi-
" cient length and strength, is thrust by the per-
" pendicular spoke in the wheel into a hole made
" in the adjoining upright or standard of the
" frame, to receive it. The pit is sunk from twenty
" to thirty feet deep, and then chambered at the
" bottom, that is, the pit-man digs or cuts out the
" chalk horizontally, in three separate directions;
" the horizontal apertures being of a sufficient
                                        " height

" height and width to admit of the pit-man's
" working in them with ease and safety. One pit
" will chalk six acres, laying sixty loads on an
" acre. If more be laid on, and to the full ex-
" tent of chalking, viz. 100 loads, then a propor-
" tionable less extent of land than six acres is
" chalked from one pit. Eighteen barrow-fulls
" make a load, and the usual price for chalking is
" 7d. per load, all expences included ; therefore the
" expence of chalking, at sixty loads per acre, is
" 1l. 12s. 6d. ; and at 100 ditto, 2l. 18s. 4d.
" As the chalk is considered to be better the
" deeper it lies, and the top chalk particularly, if
" it lie within three or four feet of the surface very
" indifferent, and only fit for lime, or to be laid on
" roads, gateways, &c. the chalkers must be di-
" rected to lay by the chalk for the first three or
" four feet in depth, to be applied to the above
" purposes, or if not wanted, to be again thrown
" into the pit when filled up ; and also to pick
" out the flints from the chalk before it is carried
" on the land, for if they are not narrowly watched,
" they will chalk with both.

" Mr. John Hill, of Coddicot, farms upwards
" of 1200 acres in the adjoining parishes of
" Coddicot and Kimpton, a considerable part of
" which is his own estate. He has chalked many
" acres of land, and approves much of the prac-
" tice. He chalked a field of strong clay-land in
" the autumn of 1793, laid on sixty loads to an
                                                   " acre,

" acre, and the chalk where the pits were sunk lay
" about ten feet from the surface. I viewed the
" field the 7th of August 1794 ; it had borne a
" crop of pease since it was chalked, and was then
" under the plough, preparatory for a crop of wheat.
" The chalk was good, and the land appeared to
" work well, though the chalk was not then tho-
" roughly incorporated with the soil. Mr. Hill
" never lays on more than sixty loads of chalk on
" an acre : this, he finds, will not only make the
" land work much better with less strength of
" cattle, but also, with a light coat of dung, or
" spring dressings occasionally laid on to quicken
" the vegetation, produce abundant crops for ten
" years ; he then chalks again with equal success."
—*Mr. Walker.*

EXAMINE WATER-FURROWS.

At this season, if snow melts, all water-cuts made
in autumn for keeping arable fields dry, should be
carefully examined, to see that leaves, weeds,
frozen snow and ice do not impede the passage of
the water, and overflow the stitches. If this be
neglected, mischief may presently occur ; and nothing demands, in this respect, more attention than
young clover and other *seeds.*

BURN LIME.

If a farm affords the opportunity of burning
lime, and experiments made for the purpose, or
common practice, have proved its efficacy, it is a
business which may go on through all the winter.

Perpe-

Perpetual kilns are not uncommon in Ireland, which have burnt through the entire year, coal or culm being the fuel, in layers between layers of broken stone. They are best situated on the slope of a hill or mountain, for ease of conveying the stone to the kiln, and for drawing out the burnt lime at bottom. Sheds, or stores, should be conveniently situated for receiving it, that it may be kept from the weather, if preserved any time before using ; as in many cases it should be used unslacked. The common way is to contract with the burners, for quarrying and burning, by the quarter hogshead or bushel. In that case, the chief attention to be given them is, to see that the coals delivered produce a proper quantity of burnt lime.

LIMING.

Should lime burnt in January be used or kept till the spring ? There are two motives for burning stone or chalk : one is, for the sake of reducing the material to powder, for accuracy in spreading : the other is, for the application of a caustic body destructive of living vegetables. For the former purpose, the lime had better be kept ; for the latter, it is usually laid on in such large quantities, that it is not very material at what season it is spread, provided it be done fresh from the kiln. It will have a greater effect in spring and summer, but the superiority is not such as to induce delay from a time in which the teams have little to perform, to a season in which there is much work for them.

The

The grand effect of this manure is on uncultivated
waste land.   On moors, mountains, bog, and
boggy bottoms, the effect is very great, but the
quantity applied considerable.   The more the better.   In Derbyshire, as far as 600 bushels an acre
have been used ; or 20 one-horse cart loads of 30
bushels : such a dressing, when the space to be improved is large, demands the employment of regular teams to be kept continually at work.   In such
undertakings, it is idle to be nice about the season
of applying the manure ;   convenience demands
that the work should go on at all seasons, but in
the English counties where lime is most used, the
common season is summer, and on fallows.

MOUNTAIN IMPROVEMENT.

Throughout this month, if the snows be not
very deep and falling, quarrying stone, and building walls may proceed ; but the stones must have
been provided and laid ready for the latter.   In
some high districts, where there is much snow,
carting cannot be executed.   The improver will
have no difficulty in knowing what he can, and
what he cannot execute ; but every work should be
in his mind, that no days be unnecessarily lost.
He who can contrive to employ most hands through
the year, will do his work cheaper than if he were
not regular in his employment.

TOWN MANURES.

It is not very easy to give advice to a young farmer touching this article, because experiments for
ascertaining the value of these manures have been
few,

few, and not varied sufficiently to afford adequate
information. I was largely in the practice myself,
very early in life, when they were much cheaper
than they are at present, and left it off from an
idea (but not founded on experiment), that it did
not answer, induced very much by the cheap price
at which I could then buy straw. Several farmers
with whom I have conversed on the subject, have
been of opinion that it answers when the horses
have nothing else to do ; but that it will not answer
at any other time ; that for back-carriage it answers,
but not otherwise. This confines the object so
much, that it becomes no longer a matter of great
consequence. The grand question is, will it answer to set up a team for that purpose only, and
to keep it the whole year at work, upon the same
principle that teams are thus set up and employed
in marle-carting ? This is an object for the young
farmer to calculate ; and the mode of doing it is
this, supposing three one-horse carts employed:

	£	s	d
Price of three carts and harness, - - -	£.42	0	0
Price of three horses, - - - -	90	0	0
	£.132	0	0
Charge interest for this sum, at ten per cent. -	13	4	0
Keeping three horses a year, - - - -	60	0	0
One man and two boys, 17s. a week, 52 weeks,	44	4	0
Three hundred days work, 900 loads, at 2s. -	20	0	0
Allowance for beer, turnpikes on some roads, and sundries, 1s. 6d. a day, - -	22	10	0
	£.229	18	0

Sup-

Supposing them to bring each a ton weight, according to the various authorities in Scotland, 900 tons cost this sum, or about 5s. per ton.

This might probably answer; but suppose four horses in a waggon, and to bring the common load of a waggon (not nearly that of carts, proportionally to the team), the expence would be so increased, that I do not conceive it could answer.

However, in all these cases, where carriage runs dear, the more valuable manures should be brought, such as soot, ashes, malt dust, night soil, bones, rape cake, rabbit dung, &c. &c. in which the carriage bears but a small proportion to the total expence. It must be remembered by the young farmer, that all these speculations evidently demand a large capital, which he should carefully calculate before he enters into them.

WATERED MEADOWS.

Mr. Wright directs that the floater should take care in this month to keep the land sheltered by the water from the severity of frosty nights. It is necessary, however, every ten days or fortnight, to give the land air, and to lay it as dry as possible, for the space of a few days. Whenever the frost has given a complete sheet of ice to a meadow, it is advisable to discontinue floating, for the frost will sometimes take such strong hold of the land, as to draw it into heaps, and injure the evenness of the surface. Attention is also to be paid to prevent the equal distribution of the water being obstructed by the continual influx of weeds, leaves, sticks, &c.

E   CARTING

CARTING TURNIPS.

If a farmer occupies land which is not suitable to feeding sheep with turnips where they grow, it is of very material consequence so to arrange the lands, stitches, or ridges, as to be able to cart off the crop with as little damage to the field, that is, with as little treading and poaching, as possible. To avoid this evil, is a point which should be particularly attended to when turnips are drilled or sown. In the Calendar for June and July this subject will be particularly treated. As the wheels of carts are five feet four inches asunder from centre to centre of the fellies, this demands drilled turnips to be in double or treble rows on ridges of that breadth, for two horses working double, or for one horse in a quarter cart; or single rows on ridges of 32 inches. In either case, the wheels and horse move only in the furrows, and consequently do as little damage as possible, and that only at the bottom of the ridge-furrows. Throughout the winter, especially if it be a wet one, the great use of this precaution will be found.

CARTING CABBAGES.

Cabbages will of course be planted with an eye to the same circumstance: the ridges so arranged, as to admit carts to move, as for turnips, in the ridge-furrows only. In carting both turnips and cabbages, attention should be paid to the men, for preventing all movements *across* the ridges, that they turn from the gateway along the border till they come to the ridges whence the load is to be taken.

FELL

FELL ASH.

If February should prove a forward month, ash timber had better be felled in January than delayed longer : this note upon a supposition that the farmer occupies his own land, and employs a wheelwright constantly, which I have found to be a cheaper plan than employing others in the common way. If elm abounds on the farm, this is the season to fell that also ; and the aquatic timbers likewise.

WINTER EVENINGS.

Some readers may smile at this title in a Calendar of the business of a farm ; but if they do, it will be for want of due consideration. In fact, there is no part of the day can be more profitably employed. Every work for the next day is to be arranged, whether for a fine or a rainy day, and the farm-books to be made up for the transactions of the past day. Besides these, he should have another book, for miscellaneous observations, queries, speculations, and calculations, for turning and comparing different ways of effecting the same object, for estimates of the different kinds of food he may have it in his power to give to the same cattle ; with all such inquiries, doubts, or propositions, worth attention, as he may have heard in conversation. Loose pieces of paper are generally lost after a time, so that when a man wants to turn to them to examine a subject formerly estimated or discussed, he loses more time in searching for a memorandum, than would be sufficient

E 2   for

for making half a dozen new ones ; but if such matters are entered in a book, he easily finds what he wants, and his knowledge will be in a much clearer progression, by recurring to former ideas and experience.   Formerly farmers never read books of husbandry : many do read them now; and there are few that will not furnish very valuable hints.   These should be noted, that when an occasion offers, use may be made of them.   Such an employment of a winter evening, is a very different one from spending it at a public-house, in the same company over and over again, which, after a time, becomes a pump that yields no water.

TRAVELLING.

The reader may also be surprized to see such an article as this in a Calendar of the business of farming; but were I to name one circumstance which has, in the last twenty years, advanced the husbandry of this country more than any other, I should fix on the practice of farmers *taking their nags* (to use an expression of Bakewell), *to see what other people are doing.*   Men who are confined their whole lives to one spot or vicinity, necessarily contract a too limited range of thought.   Their ideas flow so much in the same channel, and dwell so much on the same objects, that new ones, however useful, make too faint an impression : nor can they know what is doing by the best farmers, on soils perhaps exactly similar to their own.   To take a ride, for a fortnight, through four or five hundred miles of
country,

country, with an eye scrutinizing every thing they
see, and calling upon noted farmers to make inquiries about such objects as appear interesting,
must necessarily give a new movement to their
minds, a new spring to thought, and remove many
prejudices. If only one journey be taken in a year,
and that at a vacant time, perhaps June would be
the best season ; but, as I propose that two should
be taken, one may as well be in January as in any
other month. This season will explain the winter
management of live stock, the important objects of
the farm-yard, fattening beasts, sheep-feeding in
many branches, winter irrigation, and many other
objects, which are truly interesting. A farmer may
take such a ride, at the expence of as many guineas as he is out days, and he will not find this
money the most unproductive that he expends.

OATS.

Early in this month, if the weather be open, the
young farmer may examine such fields as he intends sowing with oats, that he may consider if he
has any apprehension of having his hands full of
business in February and March, whether he
should not lessen the work of those more busy
seasons, by sowing some oats now. The temper
of the soil must govern him : but it is necessary
that he should know that oats sown so early succeed well, as will appear from two very important
experiments, one made by the late Mr. Macro, of
Suffolk, and the other by the present Earl of Win-
chilsea. The former observes : " Having tried
early and late sowing of barley, in the year 1784

and 1785, I had a mind, the following season, to try the same experiment with white oats ; and began by sowing one acre, in December 1785, with one coomb of seed, harrowed in upon a wheat stubble, with one earth.—Value of land 10s. an acre.

" In January 1786, I sowed exactly another acre by the side of it, with the same quantity of seed, and dressed it in the same manner. In February, another acre the same, except half a bushel *less* seed. In the beginning of March I ploughed the remaining part of the piece of land a second time ; and, about the middle of that month, sowed it, at the rate of three bushels of seed an acre, ploughing in one cast, or half the seed, and harrowing in the other half; and marked out another acre for the experiment. This last acre had three clean earths,

" Produce of the four acres :

			C.	B.	P.
On that sown in December,	-	-	8	2	0
January,	-	-	8	3	2
February,	-	-	6	2	1
March,	-	-	6	2	2

Or seed deducted :

On that sown in December,	-	-	7	2	0
January,	-	-	5	3	2
February,	-	-	5	2	3
March,	-	-	5	3	2

" That sown in December, though it came up thick enough at first, lost so much of its plant, by the winter frosts, that I expected it must have been

been ploughed up, and sown again in the spring; but observing the plants that were alive beginning to flourish very early, I gave them time; yet it never got to be a full plant, nor did I expect, though the straw was very stout, and the hawes, or ears, very fine ones, that it would have turned out so well. Those sown in January and February both lost some of their plants, so that that sown in March, *with the least seed,* was the fullest and evenest plant of any."

The Earl remarks: " I was induced to make the following trial, from having seen upon two very capital farms in Kent and Essex, great crops of oats, sown as early as Christmas week, and from being informed by the gentlemen who occupied those farms, that they always sowed their oats as early as that, if the season admitted of it; and that they thought it the best time for sowing that grain. I wished to ascertain whether this plan would answer in this more northern county. The general time of sowing oats here, is from the beginning of March to the end of April; and it is the opinion of most people here, that oats sown much earlier would be liable to be destroyed by spring frosts. Last winter was very favourable for the experiment, as the weather was open at Christmas for sowing, and the frost in the spring not severe. I divided a field of eight acres equally : one half was sown the day after Christmas day ; the other half the middle of March. Five bushels per acre were sown broad cast on each part, and the same oats : the sort, a small white oat, here called short-smalls. The early sown were ripe and cut one week before the others; they were harvested equally well, without being exposed to any bad weather. I had a rood of each set out very carefully in the middle of the field, reaped and threshed as soon as carried. The produce and weight were as follows:

		Winch. Bushels.			Qrs.	
Early sown,	-	22	per acre,		11	0
Late sown,	-	19	do.	-	9	4

" Weight per Winchester bushel as soon as threshed:

| Early sown, | - | $44\frac{1}{2}$ lb. |
| Late sown, | - | $42\frac{1}{2}$ lb. |

" The crop was, as you see, very good. The land yielded potatoes the preceding year, 450 bushels per acre, and was not manured for that or the oat crop: it had before that been in grass for six or seven years. The soil a red loam. I am inclined to think that the early sowing will answer here, as this field is very high and much exposed."

FEBRUARY.

FEBRUARY.

BEANS.

In this month, a farmer should begin to sow his bean crop, and, if the soil and the season agree, finish it if possible; for later sown crops will not succeed so well. The land ought to have been ploughed into the three-foot ridge, and well water-furrowed the autumn before; by which means his only object now will be dibbling in the seed: so that the *first* dry season may be taken. To get the bean crop in the land in February is an object of consequence, if the soil is dry enough.

As to the methods of sowing, there are many. Some farmers sow the beans over the land, and plough them in; others plough first, and harrow in the seed; and these both on ridge and flat work. A better way of sowing is, either to half plough the ridges, sow broad-cast, and afterwards finish; or to sprain them by hand before the plough, so that they may rise in rows, on the tops of the ridges. In the latter way, they are in single rows, but in the former double. In the following summer, the single rows are ploughed between, in the horse-hoeing manner, and the double ones hand-hoed. Both methods are common husbandry in several parts of the kingdom. But I shall recommend, in preference to them, other methods, and using a drill plough, as it executes that work with

much

much greater accuracy than any hand can do.
Light drills may be had to wheel along the ground,
like a wheel-barrow. The use of such an instrument will save money, at the same time that it
performs the work much better. A farmer who
has land proper for beans, should, on no account,
avoid giving a particular attention to that crop;
for it will prove one of his surest funds of profit.
By means of beans, he may be able to lessen, if
not to banish, the custom of fallowing; for a crop
of beans, rising in single rows on three-feet ridges,
or double rows at one foot, on four-feet ridges,
gives so good an opportunity for ploughing the intervals, and also admits hand-hoeing the rows, that
the land may be cleaned as well as by a fallow, and
the crop succeeded by corn. But, if the soil be
in such order that this culture is insufficient to
clean it, then a second crop of drilled beans should
succeed, which will be very profitable husbandry,
and cannot fail of bringing the land into order.
Whenever beans are cultivated with this view of
substituting them in the room of a fallow, the farmer
should absolutely determine to drill or dibble them,
so as to admit the plough between the rows; for
no hand work will clean and pulverize the land sufficiently for this purpose, at least without an expence too great for the object. If the spirited
husbandman calculates the expence of a summer
fallow, and also the account of a drilled bean crop,
he will find the necessity of this culture. Beans
do very well on loams, and on lighter ones than
                                        com-

commonly imagined ; but on light gravels, sands, &c. more profitable crops may be substituted. Let the farmer remember the general maxim, if he ploughs for beans this month, never to allow his ploughs to stir while the land is wet : if his horses poach at all, or his ploughs do not go clean through the land, he will lose, or greatly damage his crop. But improvements, and especially those which have taken place in Middlesex, but most of all in Suffolk, have opened a new field for this cultivation, which will be explained in this work. The grand basis of it is, to banish spring ploughings, by laying the land ready in autumn, for either dibbling or drilling.

BEANS AFTER BARLEY.

The barley stubbles intended for beans, or land whereon clover failed, having been ploughed into the proper stitches, and laid dry for winter, are now ready for drilling or dibbling. It will probably be the end of the month before the season is suitable for this work. The same attention must be paid to this crop, as to barley, in respect of avoiding spring ploughings, and also to effect every operation without permitting the horses to set a foot *on the land*. They are ever to move, in spring, only in the furrows. As this is the first month for putting in beans, it will be proper for the young farmer to consider, whether he shall adopt the system of drilling or of dibbling, setting, or planting, as the operation is in different districts differently termed.

Dibbling

Dibbling is an excellent method, when well performed; but the grand objection to it is, the difficulty of getting it well done. When it becomes
the common husbandry of a district, the workmen
find that great earnings are to be made by it; and
this is much too apt to make them careless, and
eager to earn still more; and if a very minute attention be not paid to them, by the constant attendance of the farmer, they strike the holes so
shallow, that the first peck of a rook's bill takes
the seed, and acres may be destroyed, if the breed
of those birds be encouraged as they ought to be.
Boys are employed for weeks together, to keep the
fields, but all works that depend on boys are
horribly neglected, and thus the farmer suffers
materially; but if the seed is deposited two and a
half, or (better) three inches deep, it is not so
easily got at. The imperfect *delivery* of beans
by all the drill-machines which I have seen, causing many gaps in the rows, is an additional motive
to dibble. But, on the contrary, the power to put
in the seed at the desired depth, with the drill, is a
great motive to use it; nor should the difference of
the expence be forgotten. To dibble beans well,
at eighteen inches equi-distant, will cost 5 s. an
acre; but drilling will not come to the half of that
sum. On layers, whether of grass or clover, I prefer dibbling, because, on such, it is easier to deposit the seed at a safe depth, by the dibble than
by the drill, unless it be on clover of one year,
ploughed with Mr. Ducket's skim-coulter before
                                        winter,

winter, and left for frosts to work upon. On such, the drill will work well. This is, however, a point that must be left in some degree of latitude. No general rule can safely be laid down : the farmer must judge according to soil, season, his dependance on dibblers, and other circumstances : both methods, when well applied, are good.

The dibbled crops demand harrowing with fine, light, short-toothed harrows, which will not displace the seed, and it should be carefully done, in order to hide the holes from rooks. The drilled crops want only one light harrowing, to smooth the land.

In putting in beans after barley or wheat, on land ploughed in autumn, the farmer must remember, that if the frosts have had full play, the surface will probably be in such friable order, in a dry February, that he *must* drill, as the mould would run in, and fill the holes before the seed is dropped. This is a circumstance that will sufficiently explain itself.

There is a practice about Coggeshal, in Essex, that should here be noted. Their course is,

1. Fallow,
2. Barley,
3. Clover,
4. Wheat,
5. Fallow,
6. Barley,
7. Beans,
8. Wheat.

Designed,

Designed, probably, to throw the return of clover to the eighth, instead of the fourth year. The barley stubble of the sixth year, is dunged in autumn, with farm-yard composts, and ploughed after wheat-sowing is finished, on ridges of three feet; two bout ones. In February, they dibble a double row of Windsor-beans, on the crown of each ridge, nine inches from row to row, which leaves an interval of twenty-seven inches for cleaning. They are exceedingly deficient, in not horse-hoeing so wide an interval, applying the hand-hoe only ; but they do this three or four times ; and, if the stubbles are in the least foul, they are very attentive to hand-hoe them for the wheat which succeeds. Their avoiding spring-tillage for the beans has much merit. This practice they carry so far, as neither to scarify, nor even harrow, putting the seed into the frost-worked surface, and their success is a justification of the system.

SORT OF BEAN.

The common little horse-beán has the advantage of all others, in being more generally marketable ; for in certain situations, it is not always easy to dispose of ticks, Windsors, long-pods, and various other sorts. They also grow higher, shade the ground in summer more from the sun, and yield a larger quantity of straw, which makes excellent manure. But some of the other sorts are generally *supposed* to yield larger products. This, however, is a point in which some well-conducted comparative experiments are wanting.

SOIL

SOIL FOR BEANS.

Every one knows, that all the sorts of strong
and heavy soils are the common ones generally applied to this crop. In Kent they wisely cultivate
them to great extent, upon rich dry sound loams ;
but it is not generally known, and very rarely practised, to venture them on light turnip loams and
middling sands. I have, however, seen them succeed
so well on such, that a note of it ought to come into
this work ; and as this is the month in which a farmer
will first turn his thoughts to beans, it deserves his
attention to consider, whether he has not land
upon his farm which would do for that crop, although he never before thought of venturing it.
The soundness of a man's farming practice may be
judged of by this cultivation, as well as by any
other criterion ; for he ought to have beans where-
ever it is possible to have them. They do not exhaust the soil—they prepare it better for wheat than
any other crop—they stand erect to harvest, admitting horse-hoeing to the last ; they shade the
ground from the sun, and the straw is valuable, if
harvested in a favourable time, or, if not so harvested, makes excellent dung. The favourable circumstances attending this crop are so many, that
every man who can have them ought to determine
on the culture. A bad crop of pease fills the
land with weeds, but a bad crop of beans may
be as clean as a garden. Some of the greatest products of this plant, which I have seen, were on a
rich sand ; but I have known beneficial ones on a
                                          sand

sand of 10s. an acre. Beans are never seen in Norfolk, on sands that let from 10s. to 15s., and even more per acre; and this is a deficiency in their husbandry.

QUANTITY OF SEED BEANS.

The quantity of seed will depend much on the distance at which the crop is drilled or dibbled. It takes about two bushels of horse-beans to an acre, the rows equi-distant at 18 inches; and it demands six bushels of Windsors, put in in the same manner.—The quantity of seed proper for other varieties, will necessarily be in proportion to the size of the grain; and the variation of distance in the rows, will demand seed in proportion to these quantities for the distance named. It is in almost every case better to put in a peck too much than half a peck too little.

THE ROWS OF BEANS.

Beans are drilled from 12 to 24 inches, equidistant. In Suffolk, by many farmers at 12 inches; but, on good land, they will then be evidently too thick, and draw themselves up, without podding below. Eighteen is a better distance, and used by the best farmers. In Kent, 14 and 16 inches is the distance adopted by many. In Essex, I have just stated double rows at nine, with intervals of 27 inches. I have had great products on layers, from double rows, at nine, with intervals of 18, and also 27, that is, two *flags* planted and two or three missed, for intervals; the former, viz. the double rows, with intervals of 9 and 18 inches, have,

have, I think, been most productive. But this point will entirely depend on the fertility of the soil ; for in proportion as the land is rich, whether from nature or from manuring, the distance should be large.

In Berkshire they have a custom, which, in this respect, varies from all other countries with which I am acquainted: it is, to plant in clusters four or five beans in a hole, and nine inches from hole to hole; the space between the rows varied according to soil. Their crops are large. This method admits effective hand-hoeing in the rows, and the intervals are horse-hoed. It may be combined with *de Chateaux vieux* well-known experiment on planting barley in clusters, which seems to have been very carefully made, and in which four, five, or six grains in a hole, produced more than the same number of grains singly, in as many holes as grains. It is in vain to reason about such results ; but it appears as if the germination of the grains, in such close contact, caused a fermentation in the soil around, that was beneficial, even in the produce at harvest. In the case of the Berkshire beans, something is certainly to be attributed to the hoeing being more effective than in common rows.

BEANS AFTER CLOVER, &c.

To put in beans after clover and other *seeds*, is most excellent husbandry, and preferable to sowing wheat, which does better after beans, and also enables the farmer to get two profitable crops in-

F   stead

stead of one, with the land preserved at the same
time in good heart, and clean.

    1. Fallow, turnip, cabbage, winter tares, or
       potatoes ;
    2. Barley ;
    3. Clover, &c. ;
    4. Beans ;
    5. Wheat.

Here is a much more profitable course than that
of four years ending with wheat ; or of five years,
by taking barley or oats after the wheat.

The clover lay should be dunged before wheat
sowing, if the time should be too dry for that
operation, or after it, and then ploughed into such
stitches as suit the drill-plough or scarifiers, and
planted in this month without more ploughing.
This is an excellent system, that cannot be too
much commended. The layer affords a good opportunity for carting the manure, which is wanting
in some courses.

BEANS AFTER WHEAT.

There are some rich soils, upon which the most
profitable husbandry that can be practised is, to
take beans and wheat alternately ; others on which
the same husbandry may be repeated twice in five
years, or thrice in seven. There may be one or more
such fields on a farm ; but wherever found, this
management should not be neglected. In all eases
the land ought to be ploughed in autumn ; no
spring ploughing to be given ; and the stitches drilled

                              or

or dibbled this month, if the weather be favourable ;
if not, in March.

BEANS AFTER TURNIPS.

From the wetness of the soil or season, turnip-
land, after sheep-feeding, will sometimes be found
in very bad order for barley. The general practice is, to persist in the intention for barley, and to
effect a partial pulverization, by much tillage and
much patience : but if land is found in such order,
it is much better to give one deep earth, and to
dibble in beans. For this grain, it is no objection
that the land breaks up a whole and *clung* furrow,
as the farmers term it. The beans succeed well,
and the horse and hand-hoeings, with the effect of
the seasons through summer, bring the land into
proper order for scarifying for wheat. I have found
this husbandry successful, and every one knows
how easily a crop of barley is lost in such a case.

BARLEY AFTER TURNIPS.

Towards the end of this month, part of the turnip-land will be ready for being tilled for barley ;
and, as this is the first time of mentioning the
sowing of that grain, it will be necessary to explain
a system that has, for a few years past, and since
the former editions of this work were published,
been making a greatly rapid progress in Suffolk :
it is that of putting in barley on turnip-lands, by
means of drilling, without any ploughing.

For this purpsoe, and for many others, the surface of the field is thrown on to lands *(stitches,* as

they

they are called in Suffolk, and *ridges* in some
counties), of such breadth, as shall very exactly
suit for one stroke or going of the drill-machine,
or for two ; a *bout*, as it is termed. The shafts
of the drill are fixed, like those of a cart for one
horse, that quarters. This will be more particularly explained elsewhere, but the horse-hoeing
implements, and scarifiers, and scufflers, whatever
may be used, must be prepared according to the
drill-machine, to fit the stitches exactly. We shall
suppose the turnips to have been drilled, or sown,
on stitches sixty-six inches wide, which will admit
seven rows of barley to be drilled, at nine inches
asunder, besides leaving twelve inches for each
furrow. These lands being cleared of turnips,
either by sheep-feeding, if the soil be dry, or by
carting off with double breast carts (the horses and
wheels moving only in the furrows), and the soil
on the surface being pulverized and opened in
some degree by frosts, the question will be, how
to prepare it for barley or oats. The husbandry
universal in the kingdom, till very lately, was that
of ploughing such land once, twice, or thrice, for
spring corn ; the better farmers thrice, others
once, and a few twice. Upon very dry soils, the
evil was little more than that of a useless expence,
except, probably, a greater dissipation of the volatile particles of the urine of the sheep that had fed
on the turnips ; but upon all other soils more stiff
and unmanageable, the surface, which had been
                                        rendered

rendered friable by the frosts, being turned down, and the more stiff and *clung* bottom not influenced in the same manner by those natural agents being brought up ; it might also, if very favourable weather ensued, be brought into good order, but if the season proved the least unfavourable, the farmer could have no chance of obtaining so fine and safe a tilth as the surface was capable of, without any such reversal of it by ploughing. The new system is, to apply the scarifiers instead of such ploughing. Mr. Cook's, with his cast-iron beam, or any other heavy enough, is used, the horses walking only in the furrows, and consequently without any trampling of the land. These scarifiers are of different breadths, but all narrow, usually about three inches, or, at most four, and they will go as deeply as may be thought proper. They ought to stir to the depth to which it would have been ploughed, whether four, five, or six inches. They completely loosen the soil, let down the air, to dry it at bottom, give a very good tilth, with the material advantage of not burying that pulverized surface which frosts have given, and which, if once lost, may not be regained in time for barley. In some cases, one scarifying, and two or three harrowings, will effect the preparation ; in others two. Three operations may be wanted in others, that is, two scarifyings and one scuffling, with broader triangular shares. These variations will depend entirely on the degree in which the soil is tenacious, and to ascertain which, the far-

mer's

mer's eye and foot can alone enable him to judge.
These operations go off very quickly, and leave the
lands, or stitches, in excellent order for the drill-
machine to follow, and deposit the barley-seed ;
the farmer, during the whole of these operations,
being as little liable to be thrown out by unfavourable weather, as it is possible he should be, and
much less so, than if he had ploughed the land.
Those who are used to attend to the effects of tillage on different soils, know well, that loams and
clays of various degrees of tenacity, if they have
been properly formed into lands for winter, and
not poached by horses trampling, receive the frosts
to advantage, and are found with a friable surface in
the spring.   If rain comes, it dries, and leaves the
surface still in good order, and ready for any operation : but plough such land, and turn up the
more adhesive bottom, not acted upon by frost,
and let rain fall on such fresh turned furrows ; it
remains stiff and saddened ; it does not become
porous again ; the air cannot get into it ; and if
drying sharp winds at north-east follow, the furrows become longitudinal slices of clod, very difficult to be acted upon by any instrument, and the
farmer finds himself in a most unpleasant situation.
He no more recovers a fine friable surface, and it
becomes twenty to one whether he has a good
crop.   His only chance is, to have abundance of
patience, to wait for favourable weather, and lay
his account to sow very late.   The motive for ad-
vising him to avoid such spring ploughings, is not
derived

derived from the practice of a few individuals, but from that of a considerable district, occupied by numerous and intelligent farmers.

BROAD-CAST BARLEY.

The preceding directions are not confined to the drill-husbandry, but are applicable to the preparation of the land for broad-casting. In this method, the same attention must be given to the breadth of the lands, because the operations must be effected by horses that walk only in the furrows ; and when the seed is covered by harrowing, the same attention must be paid to that circumstance.

BARLEY ON FALLOW.

In some very well-cultivated districts of heavy land, it is the common practice to sow barley on a summer fallow ; it is particularly so in Essex. There the farmers plough their fallows in August or September, on two bout ridges, of three feet breadth ; if in August, some will reverse the ridges immediately after wheat-sowing, others before it. They water-grip the field well, and in February plough and sow, still on the same ridge, but harrowed nearly flat, by harrows made for the purpose. If they have a dry season to plough and sow, they get good crops, but much ever depends on this in spring tillage. To lay their lands in such form as to admit the scarifier and drill, the horses walking only in the furrows, and avoiding any spring ploughing (now the common practice on the strong lands in Suffolk, where they also

F 4   fallow

fallow for barley), is a very superior practice, and attended with great success.

SOW CABBAGE-SEED.

The seed of cabbage intended to be planted in June, may now be sown upon land which has been pared and burnt in August (see the Calendar of that month), and carefully manured and dug in October; but it must be well raked before sowing. Before the farmer determines on this matter, he is to consider another mode of cultivation, which is upon the whole preferable, and will preclude his trusting principally to the transplanting method. This is, drilling the seed where the plants are to remain, and for which April is the proper time, under which month it will be particularly described. Transplanting cabbages demands a very wet time of at least two or three days ; and, if hands are not to be procured plentifully, of a longer duration, such a time may not occur when wanted : it must then be waited for, perhaps while the plants are drawing themselves up to long shanks in the seed-bed, and thereby much damaged. This is a great objection to the method, and often causes a light crop on land, which, from soil or preparation, is equal to giving the largest. This inconvenience is prevented by drilling the seed where the plants are to remain. It will be the safer way to practice both methods, and it is consequently necessary to describe both in this work. Three ounces of seed should be sown on each square perch of the prepared nursery, well raked in, and then a peck of soot

soot sown over each rod. A cabbage-nursery cannot be too rich, nor too much care taken to have fine strong plants, by afterwards thinning carefully. If this crop is meant to be cultivated on a large scale, an acre of land should be well inclosed for a nursery, kept highly manured, and the seed drilled at nine inches, for the purpose of weeding and hoeing.

SORT OF CABBAGE.

The great American cabbage, which thirty years ago was to be had, and which came to 50, 60, and even 80lb. weight, is, I fear, lost at present. The great cattle cabbage, the great Scotch, the Drumhead, the Dutch, and other sorts, are not distinct varieties, and little dependence is to be placed on the manner in which orders to seedsmen are executed. A farmer should, at first, get the best stock he can, and then trust only to the seed he raises himself. At present, I am inclined to believe, that the best sort to be procured, is the large red cabbage. It comes to a good size, and is hardier than most others, green boor cole, brown cole, rape, chou de vache, &c. may now be also sown for transplantation.

REYNOLD'S CABBAGE-TURNIP.

The latter end of the month is the proper time, if the weather be open, for sowing the seed of this plant, if it be intended for transplantation. There are some objections to it, on comparison with the ruta baga, particularly its being still harder, and growing more with fangs, whereby it is more difficult

cult to get it up clean, but its impenetrability by
frost will always render it a valuable article, and
more so still, if complaints should continue to be
heard, of the latter plant degenerating here, and
rotting with frosts. The preparation of the nursery should be the same as for the preceding articles.

PARE AND BURN GRASS-LAND.

This husbandry is mentioned in the present
month, merely that if the north-east winds should
happen to set in the last week of it, the farmer
may not lose the opportunity. Those are the most
evaporating and drying winds of the year, so that
this operation never goes on better ; and it is to
be borne in mind, that the land should be ready
pared, to receive their influence when they blow.
In the Calendar for March, when this work should
be in full operation, I shall examine the question
in relation to the expediency of this husbandry,
and endeavour to shew, that there is no other way
of breaking up old grass, and all waste lands,
heaths, commons, downs, moors, fens, mountains,
&c. that is comparable to it. I have pared and
burnt layers, of my own sowing, which were only
ten years old, that did well, and yielded plenty of
ashes ; and this husbandry is so very valuable, that
it is no inconsiderable motive for sowing seeds to
last nine or ten years, expressly with a view to it.

BLACK OATS.

This month is the proper season for sowing
black oats. The land should have been ploughed
in

in autumn, and the seed harrowed in. Four or
five bushels per acre is a proper portion of seed, in
rich soils; but six does better on poorer ones.
They suit best on turf land ploughed up before
the winter *, and left till this time for dibbling in,
which is a profitable husbandry. The farmers
usually sow them after other crops of corn, but
that practice is always to be condemned. They
likewise plough for them at the time of sowing.
On the contrary, I suppose the land to have been
ploughed in the preceding autumn. They follow
beans or pease properly, or any ameliorating crop
of roots, &c. Supposing the land too wet for
dibbling, they cannot be sown this month; but,
if the soil and the season will allow, there should
be no delay in getting them into the ground; for
early sowing of all hardy crops, when the land is
dry enough, is of great importance, and many
times more than sufficient to balance other very
expensive circumstances.

SORT OF OATS.

The sorts of oat which chiefly demand attention are,

1. *Poland*, which produce, on dry warm lands,
a very large plump and beautiful grain.

2. Essex *short smalls*. This is remarkably short,

---

* The reader will have the goodness to remark, that the direction to plough the land, at a former season, previous to sowing, was given (however imperfectly) in the first edition of this
work, printed in 1771.

and

and weighs better than most other kinds. It does
on any land that is tolerably dry.

3. *Black.* These are well known; they bear a
wet harvest well, and do on the wettest soils.

4. *Churche's oat*, yield well; are white, and
come into ear more early than any other oat.

5. *Potatoe oat*, lately introduced; is very heavy,
and yields largely.

PEASE.

This is also the season for sowing the hardy
sorts of pease. The land should have been ploughed in autumn. A farmer, desirous of keeping his
land always in good and clean order, should, in the
arrangement of his crops, take great care not to
be too free with wheat, barley, and oats, which
are all exhausting plants. He should sow beans
and pease enough, because they are ameliorating
ones, and admit hand-hoeing, to kill the weeds.
In those fields, in which the common husbandmen
sow oats or barley, after wheat, or after each other,
let the good cultivator substitute pease or beans,
or some other ameliorating crop, which will pay
him better than white corn, under such circumstances, and at the same time keep his land clean.
Pease may be sown, like black oats, on turf ploughed up before winter, and now harrowed in.

I must in general remark, on the culture of
pease, that bad farmers are too apt to sow this
pulse, when the land will yield nothing else. They
have a proverb among them, which signifies, that
the season does as much for pease as good husbandry;

bandry; and they from thence take care, that good crops shall be owing to season alone. Hence arises the general idea of pease being the most uncertain crop of all others. This is owing to their being scarcely ever sown on land that is in good order. Let the good husbandman lay it down as a maxim, that he should sow no crop on land that is not in good order; not merely in respect of fine tilth at the time of sowing, but also of the soil's being in good heart, and clear of weeds. I would not, however, here be understood to rank all these crops together; because beans and pease will admit of cleaning while they grow. On that account, if a farmer comes to a field which his predecessor has filled with weeds, a horse-hoed crop of beans will be expedient, when a barley crop would be utterly improper, and, after land has yielded one crop of barley, certainly another should not be sown, but one of pulse substituted. If these ideas are well executed, the pease and beans, in every course, will find the land in heart enough for barley, the soil will always be clean, and the crop good. Pease, when managed in a spirited manner, will not have the reputation of being so very uncertain a crop, which character has in some measure been owing to ill conduct.

PEASE ON LAYERS.

The white boiling pea, of many sorts and under various names, is more tender than the greys, and various kinds of hog pease; but I have many times put them into the ground in February, and though

very

very smart frosts followed, they received no injury. I have uniformly found, that the earlier they were sown the better. There is also a particular motive for being as early as possible ; that is, to get them off in time for turnips. This is most profitable husbandry, and should never be neglected. If they are sown in this month, and a right sort chosen, they will be off the land in June, so that turnips may follow, at the common time of sowing that crop. All the sorts of early pease should be cultivated on dry soils only. They will grow on moist, and even wet ones, but the crop is seldom beneficial. Upon sands, dry sandy loams, gravels and chalks, they succeed well. They are not, however, to be much recommended on land in tilth. Great success is rarely commanded, especially in a wet season, if they be not on a layer. The following courses do well for them :

    1. Turnips,
    2. Barley,
    3. Clover,
    4. Pease and turnips,
    5. Barley,
    6. Carrots,
    7. Barley,
    8. Clover,
    9. Wheat.

Also,

    1. Turnips,
    2. Barley,
    3. Seeds for 2, 3, 4, or more years ;

                    4. Pease

4. Pease and turnips,
5. Barley,
6. Beans,
7. Wheat.

Broad-cast pease are to be utterly rejected in every case. The only question that can arise in their culture, is between drilling and dibbling. If the former is determined on, the land should have been ploughed late in autumn, with Ducket's skim-coulter, into lands adapted to the drill machine, and scarifiers. The surface being then worked shallow, in this month drilling should directly follow. If dibbling is determined on, the land should not be ploughed till the time of planting, and a heavy roller follow the plough. On lighter soils, if the frost has worked the surface, after ploughing, the ground will be in too friable a state, the holes will moulder in, and the seed will be laid too shallow. Dibbling pease on a layer cannot be too much commended. It is an excellent practice.

There is a remarkable circumstance observed by that excellent husbandman, Mr. Overman, of Norfolk, relative to pease, which should be in the mind of every farmer fond of a pea crop. It is, that they do not succeed well, if sown oftener than once in 10 or 11 years; and I have heard it more generally observed by others, that pease should not be sown *too often.*

SHOULD PEASE BE MANURED?

It is the practice of some farmers to manure for pease; I must confess that I have been always so much

much against it, that I never did it myself, and
therefore can only state my reasons for avoiding it.
If land is in heart, and they are put in on a layer,
they do not want manure. A very good crop may
be gained without it. I have had 5, and even $5\frac{1}{2}$
quarters an acre, without any manure applied for
this crop. Dung makes them run to long straw,
and that is not favourable for podding productively.
Dung encourages weeds; and pease, except in the
early stage of their growth, do not admit of such
hoeing as a farmer would wish to give. Beans
cannot have too much dung, but with pease the
case is different. There are very few situations in
which the farmer can have such a command of manure as to give him a sufficiency. It is therefore
of much consequence to him, never to spread a load
but where it will be sure to answer best. Every
man complains of a want of dung; how very careful therefore ought all to be, to give it to the crops
that will pay best for the expence.

ROWS OF PEASE.

The practice of various farmers differs exceedingly on this point. Equi-distant rows from 9 to
18 inches are commmon. I have seen them at 2
feet and even at 3. In dibbling, it is common in
Norfolk and Suffolk, to put in two rows, on every
flag of 9 or 10 inches breadth; some farmers one
row on such a furrow: and I have known very
good crops in most of these distances. If horse-
hoeing, or much hand-hoeing, is intended, double
rows at 9 inches, with intervals of 18, do well;
but

but the greatest crops I have known, have been
from planting every flag.

QUANTITY OF SEED PEASE.

From two to two and a half bushels an acre is
the usual quantity, in planting every flag. If they
are drilled at greater distances, 6 or 7 pecks will do.
Some have trusted to one bushel per acre, but that
quantity is too small.

BORDERS.

This is a proper season for bringing the borders
of the inclosures into good order. They are generally found to be high, irregular ridges of land,
from earth thrown out of ditches, and not carted
away, and from the turning of the ploughs and
harrows. They are often over-run with bushes and
wood, and much land is thereby lost. The best
method to be used with them, is first to cut all the
wood, and make it into faggots, and then to grub
up all the roots, and make them into stacks, for
which work labourers are generally paid by the piece.
It is proper to agree with them for raising the earth
into a high ridge, in the middle of the border. In
most countries, this will be done for 6s. to 10s. a stack,
of the roots 16 feet long, 3 high, and 3 broad; but in
others it may cost more. The earth then lies ready,
and without any obstruction, for carting away, either
to the field, to the farm-yard to make a compost, or
for dung to be brought to it. But, in case one spit
deep is not sufficient to make the border lower than
the surface of the field, which it should always be, or,
at the least, on a level with it, if it is grass land :

G   then

then it will be advisable to let the men who stub
up the roots, leave it level, and set others to dig it
to the proper depth. I have seen many farms so over-
run with rubbish, that the borders occupy a considerable part of the whole. They then yield a very
contemptible profit ; for the product by wood that
is spontaneously planted, and open to all cattle, is
(consisting three parts in four of brambles and rubbish) of little value ; upon the whole, no object,
compared with the land lost. When cleared, and dug
away to a proper depth, they are ready to be laid down
for grass, so as to pay rent as well as the rest of the
farm. In arable fields, the plough will advance much
nearer the hedges than before, and yet leave space
enough for a grass border. Such an object as this
may appear trifling to some farmers, who have not
attended to the great loss of land from this slovenly
practice, but to good husbandmen, desirous of
making the most of every part of their farms, it
will not appear in such a light.

WOODS.

This month, as well as the preceding, is a good
season for felling underwood, in which work, and
the converting of the product to the best profit,
lies much judgment. When a farmer has taken a
farm that has a wood in it, he should consider well
which is the most advantageous use to put it to.
In some countries hoop stuff pays best ; in some,
hop-poles are, of all other articles, the most profitable ; in others, faggot wood of various sorts. In
some situations, bushes, loose or tied in faggots, are
particularly valuable. In many parts, nothing in a
wood

wood pays so well as hurdles.   Whatever answers
best, the farmer should apply his wood to, and
subject his management of it to such changes as
a variation in demand may occasion.   This may
appear superfluous advice to old farmers, but there
are many young ones that want reminding now and
then of such circumstances.

CARROTS.

This crop is of vast importance to the farmers
who have spirit enough to cultivate it.   It is common husbandry in some parts of this kingdom.
March is the proper season for sowing it; but, on
some soils, part of the preparation may be made
this month.   I suppose the land ploughed as deep
as possible in October.   If, on examining it, there
is any reason to expect that it will be deficient in
fineness at sowing, let it receive a common ploughing in dry weather this month, which can scarcely
fail of ensuring a good tilth the ensuing one.
The soil proper for carrots being dry gravels, sands,
or loams, it may probably be practicable to plough
them this month.   This tillage will not be necessary, if the land bids fair to work fine in March.
Let me here remark further, that in case the land
is very mellow, and in good enough order for being
harrowed on this month's tillage, it should by no
means be omitted to sow upon this ploughing, and
harrow in the seed; for although March is the
common season, yet the uncertainties of weather
are such, that the state of the land, in most cases,
requires a greater attention than the name of the
month ; and carrot seed, let the weather be ever

so severe, will take no harm. It may be sown without danger in November. In case March turns out very wet, and your sowing is driven into April, it is twenty to one that the crop will suffer.

The preceding is the method pursued by some persons, and with success ; but I must observe, that the Suffolk system is quite different ; and as the crop in that county, abounds far more than in any other of the kingdom, there being perhaps more carrots in it than in all the rest of England together, much attention is necessarily to be paid to their ideas and practice ; and that is, universally, to sow nearly about the 25th of March, and not to plough till then.

CABBAGES.

The fields designed for cabbages in April or May,· and ploughed in October on to the ridge, should this month, if the weather will admit, receive an earth, reversing the ridges, but not stirring flat. This will have good effects in pulverizing the soil, which it may be supposed to want, as it consists only of stubbles turned up in autumn. This is a point that should be attended to ; for cabbages are always to be considered as a fallow, in which light their importance must appear sufficiently great. As this tillage is the first that marks the land for the crop (all stubbles being ploughed in autumn, for whatever crops designed), it will be proper here to speak more particularly of the preparation and design of the culture.

Cabbages flourish to very great profit on all good soils,

soils, and have the particular property of enabling
the farmers of clays and wet loams, to winter more
cattle than those of lighter lands can effect, by
means of that excellent root, the turnip. The
great evil of clay farms used to be, the want of
green winter food, which confined their stocks
to hay alone, and consequently prevented their
reaping those extended articles of profit, that arise
from numerous heads of cattle : and besides the
immediate benefit from the cattle, they lost also
the opportunity of raising large quantities of dung,
which never can be effected so well as by keeping
cattle. But all these evils are by the cabbage culture remedied, and the clay farmers put in possession, in many respects, of an equality with the
turnip ones. If the difference between a summer-
fallow year on clay, and a turnip-fallow on light
land, be considered, the importance of this discovery will appear sufficiently clear. Thirty shillings an acre expence, of the first, are not an ex-
aggerated calculation ; but all is saved on the turnip land, perhaps with profit ; and the barley, that
follows the turnips, is probably nearly as good as that
which succeeds the summer-fallow clay. Supposing
the following clover and wheat equal in both, according to soil, still there remains a superiority in the
article manure ; for all that is raised by the consumption of the turnip crop is so much superiority to
the clay soil. But reverse the medal. Suppose cabbages to be introduced on the clay, and the scene is
changed. That crop will exceed the turnips, yield

G 3   more

more profit, and enable the farmer to make more manure. For these reasons, the recommendation of cabbages appears to be extremely well-founded; and consequently, those farmers who possess the proper soils, cannot determine too soon to enter on the cultivation of them.

But there is another circumstance attending some sorts of cabbages, which make them highly eligible on all farms, which is their lasting for sheep feed longer in the spring. Ruta baga, turnip cabbage, cabbage turnip, and green boorcole, are in perfection in April, and last even to May, the most pinching period in the year. Turnips will do no such thing; consequently those farmers who possess turnip soils, should, on no account, slight the culture of cabbages for this purpose.

WATER-FURROWING.

Care must be taken to cut water-furrows through all new ploughed lands, as soon as the fields are finished. Saving a trifle of money in the omission of such a necessary work, often hazards a crop, and is sure greatly to damage it. In making them, the descents and variations of the surface are to be kept in view, so that no water can lodge in any part, however wet the weather. The old water-furrows in the wheat-fields are also to be examined, as well as those in the fallows ploughed in autumn. If they have been filled at any place, by the crumbling in of the moulds after frosts, or by the passage of moles, or other accidents, they must be
cleaned

cleaned out. Too much attention cannot be given to keep the lands quite free from stagnant water.

MANURE GRASS-LANDS.

This is the proper season for laying on several sorts of manure, such as soot, coal-ashes, wood-ashes, lime, malt-dust, &c. and in general those that are spread in too small quantities to require a whole winter's rains to wash them in. The use of these manures, and other light dressings in February, is very beneficial; but, throughout the management of purchased manures, experiments should be formed for a year or two, before the practice is extended, to see which, at a given price, will suit the land best. Without this precaution, a farmer may probably expend large sums of money to little purpose. Nor would I advise him to trust to the mere appearance of the effect soon after the manuring; for some of them, particularly soot and malt-dust, will shew themselves after the first heavy showers, in a finer green than the rest of the field; but the proof of the effect does not arise from fine greens, but from weight of hay; for I have myself found from experience, that the latter is not always an attendant on the former. Contiguous half-acres, or roods, should be marked out, the prices of the manures calculated, and on each piece a separate one spread, all to the amount of 20s. an acre, for instance. At hay-time, the crops should be weighed. It will then be known which manure, at the given prices, suits the soil best. This know-

G 4   ledge

ledge will prove true experience, and a very different guide from general ideas.

MANURE GREEN WHEATS.

This is likewise the season for spreading superficial dressings on the green wheats, such as soot, ashes, malt-dust, pigeons' dung, poultry dung, rabbits' dung, &c. and many other sorts in the neighbourhood of great cities. It is very good husbandry ; but the profit depends on the expences. I shall venture to recommend trying them in small, (a rood, for instance, to each) before extending the practice to whole fields, especially those which are not dungs. As to the latter, provided the prices be not extravagant, there can be no doubt of their answering on all soils. Whenever a farmer has the choice of manures, never let him hesitate about which to take. He may lay it down as a maxim, that dungs of all sorts are excellent. Other manures may be the same, but they are not so universally beneficial to all soils.

FARM-YARD.

Throughout this month, great attention must be paid to the farm-yard, and all the buildings where cattle are, to see that every place be kept constantly littered, so that the beasts may be clean ; and, if the stock of litter laid in in autumn will not last, it is time now, to agree with some neighbours, for a weekly supply of refuse straw or stubble. At all events, there ought never to be a want of litter, either in the stalls or the yard ; for the only way of
raising

raising large quantities of dung, at a cheap rate, is
to make use of plenty of litter.

PLANT WILLOWS.

I do not, in this Calendar, mean to treat of the
planting of trees, as that is the business rather of
landlords and gentlemen, than of farmers; but,
with the quick-growing aquatics, the case is different.   If any part of the fences of the farm are situated in low, wet, or boggy places, it is a chance if
thorns prosper well.   The best method of repairing
them is to plant trunchions of willow, sallow,
alder, &c. for hedge-stakes, and along the bank,
for plashing down afterwards, which will ensure
the tenant a great plenty of firing; and in such
situations, and in waste spots that cannot easily be
better improved, it will answer extremely well to
set longer trunchions for pollard trees.   They repay the expence with great profit.

PLANT OSIERS.

It is now a proper time to plant osiers and other
sorts of willows.   No part of the farmer's business
pays better than such plantations, and especially
if he has any low, spungy, boggy bottoms near a
stream.   The land should be formed by spade-
work into beds six, eight, or ten feet broad, by
narrow ditches ; and if there is a power of keeping water in these cuts at pleasure by a sluice, it is
in some seasons very advantageous to do so.   The
late Mr. Forby, of Norfolk, knew the value of
these plantations well, for various purposes.   Osiers
planted in small spots, and along some of his
hedges,

hedges, supplied him with hurdle-stuff enough to
make many dozens every year, so that he supplied
himself entirely with that article, as well as with a
profusion of all sorts of baskets, especially one kind
that he used for moving cabbage-plants, for which
purpose they were much better than tumbling the
plants loose in a cart. The common osier he cut
for this purpose at three years, and that with yellow bark at four.

TARES.

This is a proper season for sowing spring tares,
called in some places, vetches, fetches, thetches, &c.
The land I suppose stirred in autumn. The first
season in this month that is dry enough, should be
chosen for harrowing in three bushels an acre of
seed. I suppose them designed for making hay,
or feeding green ; but, if they are for a crop of
seed, two bushels will be sufficient. Tares for hay
make a most excellent fallow year. They are
mown before they draw or exhaust the land at all,
and their extreme luxuriancy and thick shade so
mellow and loosen the soil, and kill all weeds, that,
if the crop is good, and the seed sown not later than
February, there will be a very good chance for a
crop of turnips after them, on one earth ; but,
without such luck, this husbandry is far preferable
to sowing two crops of corn running. If a farmer
thinks of sowing barley after wheat, barley, or oats
or oats after either, let him throw a crop of tares
for hay between two of corn, and he will be sure to
reap the benefit of it. They will give him, on
middling

middling land, from a ton and a half to two tons
and a half of hay per acre, which, with their cleaning and ameliorating nature, will be found to far
exceed any second crop of corn on the same land.

WATERED MEADOWS.

Much attention is now required in the floater.
Mr. Wright remarks, that if the water be suffered
to flow over the meadow, for the space of many
days without intermission, a white scum is generated, which is found very destructive to the grass ;
and if the water be then taken off, and the land
exposed in its wet state to a severe frosty night,
a great part of the tender grass will be cut off.   In
Gloucestershire, two methods of avoiding these
injuries are practised.   One is, to take the water
off by day, to prevent the scum, and to turn it on
again at night, to guard against the frost; the
other method is, to take the water off early in the
morning, and if that day be dry, to suffer it to remain off for a few days and nights; for if the land
experiences only one drying day, the frost at night
will do little injury.   The former of these practices,
where it is found not too troublesome, is preferable
to the latter.   About the middle of this month,
the floater begins to use the water rather more
sparingly than in autumn or winter, for his chief
object now is, to encourage or force vegetation.

In the last week of this month, if the preceding
management has been good, there will be a good
bite for ewes and lambs.

Mr. Boswell prescribes rolling after Candlemas.

POTA-

POTATOES.

This root is one of the most profitable crops the farmer can cultivate; nor does the advantage of it depend on markets for selling them; for they will pay a beneficial price, if given to cattle of various sorts, or hogs. In Ireland they feed their cows on them with profit. The land designed for potatoes, I suppose to have been ploughed in autumn. They are to be planted the beginning of next month; and, as they affect a good tilth, it will be advisable to plough the land this month, preparatory to the planting earth, provided the weather be dry enough : but in the preparation for this, as well as for all other crops, no ploughing should go on while the soil is at all wet.

As this tillage marks the land designed for this crop, it is proper to caution those farmers who are unacquainted with the culture of it, against applying too much land to it. If they have a great plenty of dung at command, they may enter largely into this husbandry; but they should determine to plant no more land than can be manured at the rate of 25 or 30 large loads per acre; for one acre well cultivated, will pay better than five, or even ten, indifferently managed.

PARSNIPS.

Of all the roots which a farmer can cultivate, this is the most valuable; but it demands a better soil than any other crop he can put into the ground. If he has not land of an extraordinary quality, he had better not venture on the attempt. They love

a very

a very deep, rich, dry, sound, friable, sandy loam, ploughed as deeply as possible, towards the end of autumn, and left for the frost to pulverize and sweeten. About the 12th of this month, if the weather be favourable, it will be proper to sow and harrow in five pounds of seed per acre, which will come up in about six weeks. In order that the young farmer may see what inducement there is to apply so good a soil to this use, I shall here lay before him a short detail of advantages, given by a considerable farmer in Surrey, which was communicated to the Society of Arts.

" I will now proceed to relate the use I made of this root. In the first place, I put up 16 hogs a fattening upon them. The method I took in giving them to the hogs, was throwing the parsnips on the ground whole. This I continued for about a month, when finding my hogs grow heavy, I observed they did not go on so well with them as at first. Upon this I boiled the parsnips, and made wash of them : thickening the wash with half a bushel of barley-meal every day. I gave it them in a trough, and continued this method for two months, when I killed them, and found them to be very good meat ; weighing from 28 to 33 stone per hog. One of them, being very large, weighed 58 stone. The neat value of my hogs, when killed, amounted to 52l. 17s. 4d. The whole expence of my barley meal with which I thickened the wash, amounted to 3l. 18s. 9d. ; of the firing to boil them, at 6d. per day, 1l. 10s. ; of a boy to look after them for
three

three months, at 6d. per day, 2l. 5s. which sums,
added to the expences attending the parsnips, prime
cost of the hogs, &c. amounts in the whole to
35l. so that my profit upon this article only, is 17l.
16s. 8d. which remains to be carried to the account of the parsnips. After my hogs were killed
I kept four dairy cows upon the remainder of the
parsnips for three months, which, at 1s. 6d. per
week, amounted to 3l. 12s.; and this sum, added
to the 17l. 16s. 8d. before mentioned, makes the
neat profit on the one acre of parsnips to be
21l. 8s. 7d.

" I must observe, that giving my dairy cows the
parsnips, answered my purpose greatly, by increasing their milk, and making the butter much richer
and finer than turnips or carrots, which I had given
them long before. The manner in which I gave
the parsnips was, cutting them in pieces.

" Finding the parsnips agree with my hogs and
cows so well last year, I now determined to give
them to my horses; and having five that were
making up for sale, I begun with them by giving
them a very few the first week. I observed then
that they agreed with them extremely well, and I
therefore gave them a larger quantity, which made
them thrive very fast, and determined me to continue giving them the parsnips, which saved me a
great deal of hay, as I found they had occasion for
very little of it. I kept them in this manner for
ten weeks, when I sent them to Mr. Bever's repository, where they were sold for 40 guineas each
horse.

horse. The manner in which I gave them the parsnips was, cutting them in small pieces and throwing them into the manger. I calculate the expence half a guinea per week for the parsnips for each horse, which amounts to 26l. 5s. to be carried to the account of the parsnips for this year.

" At the same time I began fattening an ox, which cost me 4l. 10s. from the plough. He was 13 weeks in fattening, and ate nothing but parsnips the whole time. I then sold him to Job Spratley, a butcher at Guildford, for 2s. 8d. per stone, weighing 102 stone 6lb. which amounted to 13l. 14s. 4d. Exclusive of the above, he had within him 22 stone 6lb. of loose fat, which was more than ever was known to be taken out of a bullock of that weight in the town of Guildford, and it was remarked by many, that finer beef never was eaten. I mention these particulars, in order to shew the great use of parsnips, as I am convinced by experience, they are preferable to carrots or turnips. But to proceed on in my account : the profit upon this bullock amounted to 9l. 4s. 4d. which I also carry to the account of the parsnips.

" The remainder of my parsnips I gave to seven dairy cows eleven weeks, at 1s. 6d. per week each, which amounts to 5l. 15s. 6d. so that the neat profit (after deducting 6l. 12s. per acre for the necessary expences attending the parsnips, as per the calculation for last year) amounts to 28l. 10d. besides a great many of the parsnips that I gave occasionally

casionally to my store-hogs and cattle, which I valued at 3l. 10s."

CROP TREES.

Cropping pollard-trees should be finished this
month. Top-wood is usually for the tenant's firing,
and it demands a landlord's attention to see that
they do not, when ditching, convert young timber-
trees into pollards, which certainly was the origin
of all we see.

LAMBING.

The flock will probably begin to lamb in this
month, and there is no business on a farm that
demands more care, attention and assiduity. As
soon as the farmer looks for the ewes beginning to
lamb, they ought every night to be folded in the
standing littered fold, on one side of which there
should be a small cottage-hut, built to be warm,
with a chimney, and stove for heating milk, and a
bed for the shepherd to lie down upon. Here he
is is to sleep through the lambing season, that he
may be ready to watch, assist, and tend, any ewes
that he sees very near lambing, and if necessary, to
give the lamb some warm cows milk. Some of the
considerable Norfolk farmers have these huts on four
wheels, to draw about with the flock, wherever
they may be; but to have one littered and well-
sheltered standing-fold, on a moderate farm, and
two or three conveniently placed on a large one, to
take the flock to, without any distant driving, is far
preferable to that method.—*See the Calendar for
October.*

SHEEP

SHEEP IN ROUEN.

Upon inclosed farms, where the reserve of rouen may be supposed to be much greater than is generally possible on flock-farms, the sheep, as they drop their lambs, should be drawn from the flock of ewes, and put to this food, upon which an entire reliance may be had : and let it be remembered, that all turnips should be consumed this month, which circumstance will prove the vast importance of reserved grass as a succedaneum.

COMPOSTS.

The following observations are worth the attention of a good husbandman. The farmer may have great advantage from composts ; which, when they consist of proper materials, and are skilfully mixed, he may safely depend upon. Where a variety of materials can be had, they may be laid as follows : first, clay or strong earth, next soap-ashes, dung, loamy-earth, lime, tanners-bark, green vegetables before they run to seed, earth, soap-ashes, dung, tanners-bark, earth, or as many of these as can be got : also fat-chalk, sea-weeds, sea-sand, and several others ; which may be so mixed, as not only to raise a general fermentation throughout the whole compost ; but likewise to suit the nature of the land on which it is intended to be laid. The common way is to lay the several materials in layers, one over the other, till a large heap is raised ; and it is advised by some authors, and the practice of many farmers is, to make these layers from six inches to a foot in thickness ; but

H   this

this I have found by experience is wrong. For the fermentation raised in the compost is not strong enough to penetrate these thick layers, especially those of clay, or strong earth ; for after the rest have sufficiently fermented, and the compost is turned, these layers rise almost as whole as when first laid, and must be broken by hand, to mix them with the rest of the compost ; whence arise two inconveniences ; one, an extraordinary expence of labour ; and the other, that twice or thrice turning is sometimes necessary to dissolve these large pieces ; and as a new fermentation is excited every time the compost is turned, the strength of the manure is greatly wasted before it is laid upon the land, where it is then incapable of raising any considerable fermentation, which is one of the principal uses of manure.

" The best way, therefore, of making compost, is not in thick layers ; but after the ground is marked out for the compost, to lay the several materials, after being well broken, in heaps round the space marked out for the compost-heap ; and to place a man between each two heaps, to throw the manure spreading upon that space. In this manner the compost-heap will soon be raised to the intended height, and the several sorts of manure being thus well mixed, the whole will soon begin to ferment, and will incorporate as fully in two months, as the same manures, placed in layers in the usual way, will in four or five. The owner, therefore, in making such compost, should not prepare them

them too long before they are laid upon the land ; otherwise they will be much wasted, and their best parts evaporated.

" Composts prepared in this manner need not be turned, or at most not above once. If the fermentation is observed to abate too soon, make holes with a pole, from the top almost to the bottom of the heap, upon which throw urine, or the running of a dunghill, which will fill the holes, force through the whole substance of the compost, and soon complete the fermentation.

" Such a compost, by duly proportioning the ingredients, may be made to suit any sort of land, and is excellent for meadow or pasture grounds. A way to improve these, is to cut them five or six inches deep with the five-coultered cutting-plough, or scarificator, which cuts the surface in slips four or five inches asunder, but does not raise or turn them. This cutting of the roots of the grass, and the manure laid on at the same time, sinking into these incisions made by the coulters, causes an improvement in the quality of the herbage, and also makes such grass-grounds produce much more than they did before. But here it is to be noted, that cutting the ground first, and then laying on the manure, makes a greater improvement than manuring first and then cutting ; and both are superior to manuring and not cutting; all which have been proved by experiments. The cutting-

H 2　　　　　　　　plough

plough is used with success upon clay-grounds, loams, and gravels; but in very strong grounds, the coulters are apt to be thrown out of their work by stones; and therefore it is not proper to use the cutting-plough where stones abound.

" In such composts, where it is intended to use a large proportion of earth, that lies at a considerable distance from the homestead, to save the double carriage of it to and from the compost heap, the dung and other materials may be carried to a headland of the field to be manured, and there mixed into a compost.

" The best situation for a compost, is upon level ground; or if made upon a descent, a trench should be cut on the lower side to receive the running of the heap, which is some of the best part of it, and should from time to time be thrown up again, which will quicken the fermentation.

" The richest composts may be made in the farm-yard, which should be made deepening all round, from the sides to the middle, in form of a hollow ditch or bason. When the yard is made in this form, little of the urine or liquid part of the manure can run off or be wasted. When the dung is carried from the stables, cow-houses, &c. into the farm-yard, it should not be thrown carelessly in heaps, each sort by itself, but carried in carts or wheel-barrows, and laid regularly, and spread all over the yard. Upon this should be spread a thin layer of earth, mud, the scowerings of ditches and ponds, green vegetables before they
run

run to seed, and other such materials as are most
suitable to the nature of the land, to be manured
with them. The racks and cribs, out of which the
cattle are foddered, should be frequently moved
over the yard, that the offal, straw, and hay may
be equally dispersed, and trod in by the cattle.
This method of spreading the dung and other
materials being continued, the whole will be incorporated with the urine of the cattle, and make
an extraordinary rich compost.

" The only inconveniency of this kind of compost, is its being filled with the seeds of weeds,
from the earth, mixed with it, the hay, straw, and
dung of the cattle. It is therefore a manure best
suited to grass-grounds, and to such arable lands as
are to be hoed, as turnips, cabbages, carrots, potatoes, beans, &c. as these weeds will in great measure be destroyed by good hoeing."

LIME.

The lime-kiln may be kept burning through all
this month, and lime carted and spread whenever
the carts can move without damage to wet soils.
This may be done on dry land at all times.

MARLING

May go on profitably through all this month.
In January, I gave an account of the methods of
one who had marled more than most men ; and
here I shall note some opinions of another excellent farmer, who occupied 1200 acres, and marled
much of it.

" From

" From different trials of my own, at a very great expence, and the observations I have made on my neighbours' and the Norfolk farmers' manner of improving light sandy lands, by clay and marle, I am clearly convinced, that about seventy square yards* is the properest quantity to be laid upon an acre of land, pole measure. If more be laid on, the longer it will be before it incorporate with the soil, and of course, the longer before any benefit can be received from it. I once saw an instance, where a farmer laid on 120 loads, or square yards per acre, and gave this reason for it, that the land was so poor, he was sure he could not hurt it. But the consequence of it was, that after an expence that would have purchased the fee-simple of the land, I could not see, for many years, that he had done it any good, as it produced no better (if so good) crops, as lands by the side of it that had not been clayed at all, but otherwise farmed the same. It has now, however, evidently, the advantage of the other lands, having been done above twenty years.

" This trial was in the middle of a shiftable field, where, by the course of husbandry, two crops are taken to one summer tilth ; and, where this is the case, claying, &c. seldom (or never I might say) answers the expence : for claying and marling

---

* A square yard is as much as is generally carried for a load.

being

being only a first, or beginning of improvement *,
by going on directly with a course of ploughing,
which cannot well be avoided in shiftable fields, it
is often buried and lost before it mix properly with
the soil, especially if turned in too deep the first
earth, of which great care should be taken. I
would therefore recommend claying or marling only
upon inclosed lands, unless where large breadths
lie together, that can be farmed in any manner the
occupier pleases; and in that case (as well as in
inclosures,) I would advise that the lands should be
laid down with clover, rye-grass, and trefoil, the
spring 12 month before laying on the clay or marle,
and to remain at least six months after it, that it
may have time to sink and eat itself into the flag
before it is ploughed up, and then there will be
little or no danger in losing it, as it will already be
in some measure incorporated with the soil.

" No pains should be spared to break all the
lumps, and get it fine by repeated harrowings and
rollings, and having all the stones picked and carried away, that the grass may get through as soon as
possible, for stock to be grazing upon it, which is
the great and finishing improvement ; for, as I observed above, claying or marling seldom or never
answers where you go on immediately with a course
of ploughing in the John-Trott way.

" In my opinion, as much, or more, depends on

---

* An excellent observation. The whole paper is full of truly
practical knowledge. *A. Y.*

H 4   the

the management of lands after claying or marling, than in the mere laying it on, which, however, is very expensive, and therefore a very persuasive argument in favour of that sort of management that will be the most likely to make it lasting.

" Little need be said about the different quality of clay or marle, as every one must be content to use such as is found on his own premises, for I never heard of any in the counties of Suffolk or Norfolk that would answer long carriage * : clay that is freest from sand and marle, that is soft and greasy, are certainly the most valuable ; and even blue clay, that is condemned by most farmers, I have found to answer very well on light sands, but they generally lie at too great a distance from each other to be prudently got together.

" Where there are different sorts of manure equally convenient upon the same premises, which is sometimes the case, *viz*. pure clay, white soapy clay marle, clay with much marle in it, loamy clay and cork ; I should certainly prefer the former for light sandy lands ; on sands of a stronger nature, that have a mixture of loam with them, I should chuse the soapy marle, or that mixed with clay marle, whichever was most convenient ; but any of the inferior ones must be used, rather than submit to long carriage, especially on a large scale.

---

* In the county of Kent I have seen a sort of marle that the Essex farmers buy, which, after being sent many miles by water, I am informed they find answers carrying five or six miles by land.

" In

" In point of the expences, the first is the *fill-ing*, which, including spreading, is 25s. a hundred, or 2½d. a load, with an allowance by some farmers of 2s. 6d. by others of 5s. for opening the pit, and 1s. a load for all the large stones they throw out at the time of filling; the farmer to find drifts and stakes, for letting down what they call the *falls*.

" The team must consist of four strong trace horses, and two shaft horses, which, for such strong work, must have very high keeping. I can-not therefore lay their labour at less than 2s. a day each *, and the carter at 1s. 6d. a day, which, sup-posing they carry, one day with another (allowing for wet weather and hindrance by accidents, &c.), 30 loads a day, will be about 5¼d. per load more, making in the whole 7¾d. a load for filling, cart-ing, and spreading.

" As farmers differ in opinion about the quan-tity that should be laid upon an acre, some prefer-ing 80 loads, and others 70, I will take the medium and say,

---

* Two shillings a day for a cart and horse may be thought a high price, but when it is considered that he is, or ought to be, worth 20l. I believe no person in his senses would lend another such a horse, pay keeping, shoeing, and farrier, and run the ha-zard of his being spoiled by being whipped and strained 50 times a day out of a clay pit, for less money.

Seventy-

	£.	s.	d.
Seventy-five loads, which, at 7¾d. a load, is, per acre, - - - - - - -	£.2	8	5¼
Harrowing and rolling several times, to pulverize and spread it equally on the surface, per acre, - - - - - - -	0	1	0
Wear and tear of carts and harness, including accidents, at a farthing per load, per acre, -	0	1	6¼
Loss of seed, as it should always be laid upon a layer, and be some months before it is ploughed in, per acre, - - - -	0	1	0
	£.2	12	5½

SAINFOIN.

" The sowing of sainfoin seed ought never to be deferred longer than the beginning of March, but it is still better to complete this work in February, as there is generally at that time a degree of moisture in the ground sufficient to accelerate the vegetation of the plants ; whereas, if the seed were sown at a later period, and a dry spring should take place, great part of it would never vegetate, and that which did grow, would be liable to be destroyed in its infantine state by the fly."—*Bannister*.

MARCH.

MARCH.

BARLEY.

This is the proper month for getting seed-barley into the ground. Crops later sown may be very beneficial, but, if all circumstances were equal, the March-sown would be superior to any at a later season, which is here the comparative point of consequence. This grain is sown after various preparations. Turnips are the most common, which root will not last for feeding any cattle, with propriety, upon the average of seasons, longer than the beginning of this month : so that the turnip-land barley must be sown on one earth, or the season be absolutely lost; for April and May sowings are inferior. I am not here asserting, that April is a month improper for sowing barley, I know the contrary from experience ; but if soil, ploughings, manuring, water-furrowing, &c. are equal, a March-sowing will exceed an April one, on an average of several years, by four bushels in the crop. Saying, therefore, that barleys in certain places, sown in April and May, yield great crops, is saying little, unless it be added at the same time, what parallel success other crops had sown in March. Neither do I venture to insinuate, that all March-sown crops will be successful. One great point in putting in most crops, but barley particularly, is to have the land dry. March may pass away without a single ploughing season for wet

lands

lands in the whole month. In such a case barley cannot be sown ; but still this is not in reference to a particular practice, but to a general maxim in husbandry. Ploughs ought never to work if the land be wet ; consequently, advice to sow barley in March must always be under the proviso that the land is dry enough for ploughing.

Summer-land barley, on clay, or other heavy soils, should be sown on one earth, in the first dry ploughing season, whether in February or March. In some clay countries, the farmers have a good system of barley culture. They summer-fallow their land, and lay it up on three feet ridges, well water-furrowed for the winter. In a hard frost they carry on their dung, and leave it in heaps till sowing time, when they spread before the ploughs. This is good husbandry. It is conducted upon the same principle, upon beans, pease, tare, potatoe, or carrot land : all which crops are taken up in autumn, and the land ploughed after them, on to the ridge, and well water-furrowed, ready for spring-sowing. The great point is, to have the soil, previous to the crop, in such good order, that no other spring tillage than the seed earth may be necessary.

The most profitable way of cultivating barley, is to throw it into a regular course, preparatory to the clover. For instance : 1. Turnips ; 2. Barley ; 3. Clover ; 4. Wheat. Or, 1. Cabbages ; 2. Barley ; 3. Clover ; 4. Wheat. Another ; 1. Fallow ; 2. Barley ; 3. Clover ; 4. Wheat. What-

Whatever variations may arise in the crops, still barley must always follow either an ameliorating crop or a follow, and in many cases be followed by clover. In several parts of the kingdom, unacquainted with clover, this latter reasoning may appear bad; but that can only arise from false ideas of the use of clover. Let good grass lands be ever so plentiful, they will in no case be found to preclude the use of clover.

Thus far the culture of barley has been treated, for the use of those farmers who adhere to the common management of spring-ploughing. It is necessary always to keep in mind, that the system mentioned last month, of avoiding spring-ploughings, is applicable to many cases.

BARLEY AFTER TURNIPS.

If the turnips were not drawn and carted, or not eaten by sheep, in time enough to allow sowing with barley in February; or if the farmer does not approve of sowing this grain early, by reason of the quality of his soil; in these cases, March may be the principal period of his barley-sowing. As the fields are cleared, much attention should be given to the state and temper of the surface; for turnips are ventured on so many soils that are not entirely fit for them, that difficulties often occur at the time when it is proper to stir for the spring-corn that is to succeed. The season is now so far advanced that it may be unsafe to trust to such smart frosts ensuing, as shall have any effect in pulverizing the soil. Upon all clays, and loams of any degree of tenacity,

tenacity, which have been sheep-fed lately, the surface may be firm and trodden. The degree will depend upon the weather that has taken place, whether wet or dry; but if the farmer has a strong and
heavy hoe in his hand, or a spade, he will easily
perceive whether or not the temper of the surface
will let the scarifier work effectually. In this respect, more attention is necessary now than in February, as the advanced state of the season has
lessened his chance of frosts, which are more effective in giving friability than any other circumstance. If this tool works well, or is likely to work
well by the 20th, its use should preclude the
plough ; but if, from the state of the surface,
compared with that of the soil, at the depth of five
inches, it appears that a ploughing is really necessary, in such case, the prudent farmer will, of
course, give it. His grand object, in this examination, is to avoid turning down a surface which is
in a friable state, and bringing up another, which
will harden, by north-east winds, into *clods of
brick*, as they are sometimes called. Let him only
have the circumstance in contemplation, and he
will then be guarded, on one hand, against being
wedded to customary tillage, and, on the other,
against being too ready to trust to new methods, of
which he may have had little or no experience.
It must, however, at all events, be prudent to
make a trial in every field, as the result will bring
more conviction than any previous reasonings.
Such trials may be made, whether he sows his
                                                barley

barley broad-cast or drilled. He should keep in recollection, that if the last earth for the turnips turned down a manuring, it is better situated for safety against sun and wind, than if brought to the surface by a new ploughing; that it lies where the barley-roots will find it; and that the urine of the sheep sunk in the soil, is less liable to evaporation without than with ploughing.

BARLEY AFTER FALLOW.

If the weather, in February, prevented sowing the fallows with this grain, there can be no question in what manner to execute it now. *Here* ploughing should certainly be rejected. These fallows have had the frosts of the whole winter, and must necessarily be fit for scarifying or scuffling. It is the same with all land ploughed before winter; such as tare, bean, and pea-stubbles; and also with turnip-lands that were cleared and ploughed early. In all such cases the use of these implements may safely be adopted.

DRILLING BARLEY.

Quere. Whether the importance of this practice does not increase as the season advances? Barley put into the ground in February, has the start of many seed-weeds, which might vegetate as quickly as the crop, in the latter part of March. or in April. In the former case, broad-cast crops might be clean in fields, which, if sown in the two latter months, might be much more subject to the depredation of weeds; if so, the drill will, in this respect, be of great use. It is a dreadful spectacle
which

which some districts exhibit, of crops, yellow
from the quantity of charlock. To free drilled
corn from such enemies, is much easier than to
weed broad-cast crops.

SEED-BARLEY.

Increase the quantity of seed-barley as the season
advances. Early-sown crops have more time to
tiller than late-sown ones. If three bushels be the
quantity in February, three and a half should be
sown the end of March.

" The season of sowing has been recommended
by the Northern Botanic School, to be drawn from
the foliation of vegetables; for which idea, the
following table will be of use.

" As I do not know that any thing of this kind
has ever been published in England, I will subjoin
the order of the leafing of some trees and shrubs,
as observed by me (Mr. Stillingfleet) in Norfolk,
anno 1755.

1. Honey-suckle,	- -	January 15
2. Gooseberry,	- - -	March 11
3. Currant,	- - - -	11
4. Elder,	- - - - -	11
5. Birch,	- - - -	April 1
6. Weeping-willow,	- - -	1
7. Rasberry,	- - - -	3
8. Bramble,	- - - -	3
9. Brier,	- - - - -	4
10. Plumb,	- - - - -	6
11. Apricot,	- - - -	6
12. Peach,	- - - - -	6

13. Fil-

13. Filberd,   -   -   -   -   April 7
14. Sallow,   -   -   -   -   -   7
15. Alder,   -   -   -   -   -   7
16. Sycamore,   -   -   -   -   9
17. Elm,   -   -   -   -   -   10
18. Quince,   -   -   -   -   -   10
19. Marsh Elder,   -   -   -   -   11
20. Wych Elm,   -   -   -   -   12
21. Quicken-tree,   -   -   -   13
22. Hornbeam,   -   -   -   -   13
23. Apple-tree,   -   -   -   -   14
24. Abele,   -   -   -   -   -   16
25. Chesnut,   -   -   -   -   16
26. Willow,   -   -   -   17
27. Oak,   -   -   -   -   -   18
28. Lime,   -   -   -   -   -   18
29. Maple,   -   -   -   -   -   19
30. Walnut,   -   -   -   -   2[
31. Plane,   -   -   -   -   -   21
32. Black Poplar,   -   -   -   -   21
33. Beech,   -   -   -   -   -   21
34. Acacia Robinia,   -   -   -   21
35. Ash,   -   -   -   -   -   22
36. Carolina Poplar,   -   -   -   22

" It is wonderful to observe the conformity between vegetation and the arrival of certain birds of passage. I will give one instance, as marked down in a diary kept by me in Norfolk, in the year 1755. April the 16th, young figs appear; the 17th of the same month the cuckoo sings. Now the word κοκκυξ signifies a cuckoo, and likewise the young

I   fig;

fig ; and the reason given for it is, that in Greece they appeared together. I will just add, that the same year I first found the cuckoo-flower to blow the 19th of April.

" To the instance of coincidence of the appearance of the cuckoo, and the fruit of the fig-tree in Greece and England, I will here add some coincidences of the like nature in Sweden and England.

" Linnæus says, that the wood-anemone blows from the arrival of the swallow :. in my diary for the year 1755, I find the swallow appeared April the 6th, and the wood-anemone was in blow the 10th of the same month. He says, that the marsh-marygold blows when the cuckoo sings : according to my diary, the marsh-marygold was in blow April the 7th, and the same day the cuckoo sung *."

OATS.

White oats should be sown now, in preference to any other season ; and, in the general conduct of them, the farmer should, by all means, avoid the common error of sowing them after other corn crops, by which they exhaust the land. They should always receive the same preparation as barley ; nor ought a good husbandman to think of their not paying him as well for such attention as that crop. It is a very mistaken idea, to suppose it more profitable to sow barley on land in good order than oats. I am, from divers experiments, inclined to

---

* Stillingfleet.

think,

think, that oats will equal, and in many cases exceed barley. The superior quantity of the produce will ever be found to more than answer the
inferiority of the price ; which, however, sometimes exceeds that of barley.

What good reasons are to be offered, for sowing
oats on land in such bad order that barley is not
to be ventured in, I know not. The common argument is their hardiness, which will give a middling produce, about sufficient to pay expences,
and leave a trifling profit, when no other crop will
do the like. But this is only proving them to be
assistants of bad husbandry ; nor is such a paltry
profit, granting false premises (for I am well persuaded that common oat crops, among bad farmers,
are but so much loss), an object that ever ought to
influence good husbandmen. Why should a good
farmer be at all solicitous to gain 10s. an acre profit by oats after barley, &c.? Suppose his course
to be, 1. Turnips; 2. Barley; 3. Oats : or, 1. Fallow ; 2. Wheat ; 3. Oats : in either of these courses,
or in any other, where the oats follow another crop
of corn, the profit of them must be small. What
comparison with sowing clover with the barley,
which will pay far more profit, and at the same
time prepare, in the best manner, for that most
beneficial crop, wheat ? What but a fallow, or a
fallow crop, can succeed the oats ? How unprofitable, compared to the clover system !

For these reasons, I cannot but recommend that
oats should be considered in the same light as
I 2 barley,

barley, and never sown unless the land be in proper order for barley, or to sow them after a fallow crop, and clover with them, in the same manner as barley.

OATS AFTER TURNIPS, &c.

The observations which have been made on barley, are equally applicable if oats be sown. And the farmer should, in the distribution of his farm, consider which of these two crops is likely to pay him best. This will very much depend on his soil. Warm forward sands yield as many quarters of barley, perhaps, as of oats; but upon various other soils, the produce of oats, compared with that of barley, will be as 4 to 3, and on some as 5 to 3. He should also take into consideration, the greater steadiness of price which oats have for many years yielded, in comparison of the price of barley; circumstances which may reasonably induce him to sow them in a larger proportion than is common among his neighbours. On the other hand, it is not to be forgotten that they exhaust more.

OATS ON LAYS.

It is very common husbandry to put in oats on one ploughing of old grass, and on layers of shorter duration. The method is to plough the land before the frosts, and to dibble in the spring, as soon as the weather is dry enough; but the soil must, from its nature, or from rolling, be in such temper as to permit the holes *to stand*, and not to *moulder* in, when the dibble is removed. In some cases, the safe way is, to plough, roll, and dibble immediately.

diately. But in very many cases (possibly in all), it is better to put pease in, on light land, beans on stiff soils, and to follow these with oats or wheat, according to circumstances. I have known oats, which had produced inferior crops, followed by oats again the next year, and produce largely, which proved that they wanted *tilth*. Pease or beans will rather improve than exhaust land, when put in thus in layers : whereas, two crops of oats will scourge the land too much.

Let it, however, be well remembered, that these observations are made (so far as they relate to old grass), on the supposition that the farmer will not, or is not allowed, to pare and burn, a method vastly superior, and which ought to be pursued in all cases where it is practicable.

CLOVER.

There are several methods of sowing this seed, which is so profitable upon almost every farm, that it must be had if possible.

1st, In the drill-husbandry, it may be sown and harrowed in, at the time the barley is sown broad-cast ; a pair of light harrows at the same time following the drill-machine, to cover the clover-seed.

2dly, It is sown before the roller, when the barley is four inches high.

3dly, It is hand or horse-hoed in, when the corn receives either of those operations, if the farmer is in the practice of giving them.

These are the methods most commonly used.

Mr. Duoket drilled the seed in the same drills as the barley, but that way is very uncommon.

Another way I have known, has been that of scarifying the barley-stubble in harvest on light soils, and sowing the seed alone then.

Of these methods, the first is the surest for a crop, and the most to be recommended, notwithstanding the admitted evil which sometimes takes place in a wet season, of the clover growing so luxuriantly as to damage the barley. The second succeeds well, if rain follows in due time, and would perhaps generally succeed, if the farmer ventured to harrow it in, which he might safely do. In the third method it often succeeds, but it also often fails, nor is it necessary, in many cases, to hoe the barley.

In regard to the quantity of clover which the farmer sows, he has several considerations to govern his determination. In the first place, it is in many situations, and on many farms, as profitable a crop as any other he commonly reaps. On tolerably good land, he may expect, at two mowings, three tons of hay; on good, three and a half, and even four. Or, if he applies it to soiling his teams, for want of lucerne, the produce in a different way is equally striking. This produce is also gained at a very cheap rate; cheaper than he gets any other crop. Add to this, that it forms an excellent preparation for either beans or wheat. Still, however, the quantity to be sown will depend in some measure on his having lucerne, sainfoin, or a great

plenty

plenty of meadow-land.   If he is deficient in these, it becomes more than useful, it is essential.

The unfortunate circumstance which attends clover, is its being extremely apt to fail, in districts where it has been long a common article of culti- vation.   The land, to use the farmer's term, be- comes *sick* of it.   After harvest he has a fine plant, but by March or April, half, or perhaps more of it is dead.   This makes a new course of crops neces- sary.   Instead of its occurring once in four years, in the common Norfolk course, it becomes neces- sary to sow it only in the second round alter- nately, beans after barley in one course, and then clover in the next.   This has been found to an- swer.   This observation, however, should be made not without observing, that on a farm at Mor- den in Surrey, Mr. Arbuthnot, by means of deeper ploughing than common, and ample ma- nuring, succeeded well with clover every third year in this course :

1. Beans,
2. Wheat,
3. Clover,

on land that was said to be sick of it, though sown before only once in four years.   I viewed his crops in that new course during three rounds, and never saw finer.

Ten to 12lb. an acre is the usual quantity of seed, but 15 is better.

TREFOIL.

Upon light and poor sandy lands, on which clover does not succeed well, it is common husbandry to sow trefoil, with a portion of white clover and ray-grass, with intention of leaving it two years. Six pound of trefoil, four of white clover, and half a bushel of ray, are common quan- tities. These are for sheep-feed.

WHITE CLOVER.

A very profitable article of cultivation, which has of late years been particularly attended to in Suf- folk and in Essex, is that of white clover alone for seed. The first growth (contrary to the case with red clover) is seeded. Some take a spring feeding first. The returns depend, of course, on the price, which varies much, but it has proved a very pro- fitable article, yielding from 7l. to 15l. an acre. Wheat succeeds well after it.

GRASSES FOR ALTERNATE HUSBANDRY.

Upon impoverished worn-out lands, and others ill-treated by bad management, and over-cropping with white corn, there is no better system than that of the alternate husbandry of corn and grass for sheep-feeding. Such lands are much recruited by these means, and will, after a term of years, surprize their occupier by the superior corn-crops, which five or six, or even four years sheep-feeding will enable them to give. They should, however, be got into clean order. The course :

1. Turnips,
2. Barley,
3. Grasses for 4, 5, or 6 years,
4. Beans or pease,
5. Wheat,

will in one round improve them much. The ob- ject

ject for present consideration is the seeds to be sown. The following may be recommended :

Cocksfoot,　Yorkshire white,
Timothy,　White clover,

and a small quantity of trefoil. If sown alone, the quantities are : cocksfoot, 4 bushels ; York-white, 2 bushels ; Timothy, 6lb. ; white clover, 10lb. ; trefoil, 10lb. but a mixture may be better. In a Prize Essay sent to the Board of Agriculture, the following remark deserves attention :

This is not an uncommon system in Rutland-shire, and is a very profitable one. By loams lying in grass, and being well-covered with sheep, they are prepared in the best manner for corn, and it is the same on fen and peat lands. The following is a proper course :

On soils inclinable to moisture, or of good fertility.	On dryer land.
1. Oats,	1. White pease and turnips,
2. Turnips,	
3. Barley,	2. Barley,
4. Clover, or winter tares,	3. Clover,
	4. Wheat,
5. Wheat,	5. Turnips,
6. Turnips,	6. Barley,
7. Barley,	7. Grass for 3 years.
8. Grass for 3, 4, or 5 years.	

Such husbandry must keep land in a constant state of fertility.

This

This alternate husbandry appears to be well understood in Northumberland. " By means of three years grasses depastured with sheep, the land will grow good crops of oats, which they could never get it to do under their old system. Soil, sandy, and dry light loams."—*Report*, p. 64. " Various systems have been tried in Northumberland, particularly the boasted course of, 1. Turnips ; 2. Barley ; 3. Clover ; 4. Wheat ; till the crops have evidently declined, particularly the turnips and clover; and the only means of restoring such lands, have been the system of three years arable, and three years grass *depastured by sheep*. By this mode, Nature has time to prepare a sufficient lea-clod, which, being turned up for the turnip-fallow, will ensure a vigorous crop of turnips, as it is well known that they always flourish upon fresh land, or where they find the remains of a lea-clod to vegetate in. The portion that is kept in grass for three years, breeds and fattens such a number of sheep, as leave a considerable profit, probably equal, if not more than the arable crops, the yearly profit of a sheep being estimated at not less than 20s. to 30s. six or eight of which, an acre of clover will fatten, and an acre of turnips about double the number." This is to the credit of the Northumberland drill system, or the soil must be very extraordinary. " By this system are obtained the principal advantages of folding, without any of its inconveniences ; for if, on an average,

The

The 1st year's clover and grass carry 7 sheep an acre, for 20 weeks,

2d	-	-	-	-	5	-	-	20
3d	-	-	-	-	3	-	-	20
and the turnips		-	-	12	-	-	20

that is -   -   -   -   27 sheep per acre, for 20 weeks, which is after the rate of 540 sheep per acre for one week, once in six years, leaving 25s. a head profit." *Ib.* p. 163.

The great advantage of having land alternately under grass and corn, was known in the last century. *Considerations concerning Common Fields and Inclosures*, 4to. time of the Commonwealth, page 10. But in France much earlier, De Serres 1629, page 6.

SAINFOIN.

This is the principal season for sowing sainfoin, and if the reader has land that will produce it, he can apply it to no crop equally profitable. Loams and sands upon chalk are the favourite soils; also loams and clays in a shallow stratum, on lime-stone. It does, however, on very dry sound gravels, but not if the under stratum be much mixed with clay. I have tried it without success on good dry turnip loams, but on every species of chalk and white marle its success is certain. The profit far exceeds that of any other application of such a soil. On poor sands in Norfolk and Suffolk, worth only 5s an acre, the crop, for several years (after the first), has been from one to two tons and a half per acre, of excellent hay, and mown every year. Whatever the price of hay may be, such a produce on such land is prodigious,

gious, with the additional circumstance of an after-grass extremely valuable for weaning and keeping lambs.   I know not a more lamentable circumstance than to see such poor soils yielding a beggarly product in corn, other grass and turnips, with not one acre of sainfoin where there ought to be an hundred.

Sainfoin, on extremely favourable soils, will get the better of weeds, but it is always right to sow it with barley after turnips ; four bushels an acre broad-cast, but three are enough drilled.   If with drilled barley, the best way is to drill the corn first, and then the sainfoin across the former drills. Three bushels of good seed sown in this way is enough for an acre.

SOW CHICORY.

For several purposes on a farm, this is of such importance, that a farmer cannot, without its assistance, make the largest profit on various soils. Whenever it is the farmer's wish to lay a field to grass for three, four, or six years, by way of resting the land, or for increasing the food of sheep, he cannot hesitate.   There is no plant to rival it. Lucerne demands a rich soil, and will always be kept as long as it is productive, but upon inferior land it is not an equal object.   Upon blowing sands, or upon any soil that is weak and poor, and wants rest, there is no plant that equals this, which, if sown with a portion of cocksfoot grass and burnet, will form a layer for six or seven years, far exceed-

ing

ing those formed with trefoil, white clover, and
ray-grass, which are the plants usually had recourse
to, and will support so many sheep as very materially to improve the soil. This is one capital motive for cultivating chicory. Another is, for the
application of the produce, on better land, to soiling
horses, oxen, cows, and all sorts of cattle, for which
use three years are the proper duration. A third use is,
for feeding in the field, or soiling hogs, for which
purpose it is very advantageous. These are objects so important in themselves, as to plead power-
fully in its favour. Objections to it have been
made, by reason of its rising, and becoming a vi-
vacious weed, in succeeding crops. If the circumstance be not guarded against this will happen, but
not more, or so much as with lucerne. But who
ventures to forbid that culture on account of this
quality, which is really founded on its merit? When
the land is ploughed, only use a broad sharp share,
and harrow in tares, for feeding or soiling, or break
it up for turnips, and there is an end of the objection. Such observations against a plant are truly
futile, and arise only from ill management. Let us,
therefore, suppose a farmer too wise to listen to
them, and that he determines to sow the plant.
He may do this broad-cast, among spring corn,
in the same way that clover is sown ; or he may
drill it at 9 inches on poor land, and at 12 on better soils, in order to give it now and then a scarifying ; the first for feeding, the last for soiling.
From 10 to 12lb. of seed per acre is the proper
quantity

quantity. Once harrowing after sowing is sufficient.

PEASE.

This month is the proper season for sowing all sorts of pease, that were not sown in February; nor is it proper to delay any of them later, if the weather now suits. White pease should be sown last, and on light land; for they do not succeed well on heavy or wet clays. There are scarcely any soils that do not suit some pea or other. Stiff clays do very well for the hardier hog-pease, and all lighter loams, gravels, chalks, and sands, answer well for the tenderer kinds. In common management they are sometimes ploughed, and at other times harrowed in; which variation often makes a difference in the crop; for, if the land is apt to bind with rain, and the pease are ploughed in, they sometimes do not rise at all, not having strength to pierce the plastered surface. But this evil attends the very binding soils only with late sowing. On the contrary, when the seed is only harrowed in, if the field is not very well watched, the pigeons and birds will carry away much of it, and for that some allowance should be made. If land breaks well with the harrow, it is best to harrow in on all but the very lightest lands. But on loose sands, or very light and porous soils, or those that are extremely dry, it must certainly be preferable to plough in, on account of having a greater depth, and of being further from the sun, which is apt, in hot summers, to burn these soils.

Pease

Pease should be sown after corn. They always come in best after wheat, barley, or oats, generally with good husbandry after layers. I can hardly suppose a situation, where this is not the right management : they come very properly into such courses as these : 1. Turnips ; 2. Barley ; 3. Clover.; 4. Pease ; 5. Wheat : or, 1. Cabbages ; 2. Oats ; 3. Clover ; 4. Pease ; 5. Wheat. When wheat succeeds clover, you may throw in a crop of pease after it, if it suits you better than to come again to turnips, cabbages, or beans, the first of the course.

" If wet weather happens whilst the pease lie in wads, it occasions a considerable loss, many of them being shed in the field, and of those that remain a great part will be so considerably injured, as to render the sample of little value. This inability in pease to resist a wet-harvest, together with the great uncertainty throughout their growth, and the frequent inadequate return in proportion to the length of haulm, has discouraged many farmers from sowing so large a season of this pulse as of other grain ; though on light lands which are in tolerable heart, the profit, in a good year, is far from inconsiderable. The straw (as hath been mentioned before,) is a very wholesome food for cattle of every kind ; and there is generally a considerable demand for pease of every denomination in the market, the uses to which they may be applied being so many and so various. The boilers, or yellow pease, always go off briskly ; and the hog-

pease

pease usually sell for 6d. or 1s. per quarter more than beans. For feeding swine the pea is much better adapted than the bean, it having been demonstrated by experience, that hogs do fat more kindly when fed with this grain than on beans; and, what is not easy to be accounted for, the flesh of swine which have been fed on pease, it is said, will swell in boiling, and be well tasted; whilst the flesh of the bean-fed hog will shrink in the pot, the fat will boil out, and the meat be less delicate in flavour. It has, therefore, now become a practice with those farmers who are curious in their pork, to feed their hogs on pease and barley-meal, and if they have no pease of their own growth, they rather chuse to be at the expence of buying them, than suffer their hogs to eat beans. Nay, so far do some of them carry their prejudice in this particular, as to reject the grey pease for this use, as bearing too near an affinity to the bean, and therefore reserve their growths of white pease solely for hog fatting.

" Pease, if the ground is kindly for their growth, and the summer moist, do generally produce a great abundance of haulm, which takes up a large space in the barn, and for this reason the mow ought always to be trodden with hogs or horses, which will close down the haulm into one sixth part of the compass it would otherwise have occupied."—*Bannister.*

DRILLED PEASE.

This pulse in many districts is drilled, which is
a very

a very good method.   The great use of drilling
pease is, the rendering it so much easier to hand-
hoe them.   Good farmers, whatever their soil, are always desirous of getting the hand-hoe into as many
crops as possible, and few pay better for it than pease;
but, when promiscuously sown, it is difficult and
expensive to perform that work well ; whereas, if
they are drilled in equally-distant rows, one foot
asunder, the hoeing is regular work : it will be executed much easier, better, and cheaper, and the
crop be consequently superior : for that operation,
given while the crop is quite young, checks the
weeds so much, at the same time that the crop is
forwarded, that the tendrils join the sooner, and
are much the stronger for it.   Consequently, a
thick, luxuriant crop, is gained in a much greater
degree.   Another advantage in this method is, the
saving of seed ; for a bushel, or a bushel and a
half less seed, will do in this way than in broad-
cast sowing.

PEASE ON LAYERS.

There is scarcely an article of cultivation to be
treated of in a Calendar at present, in which a
double attention is not necessary, and which may
give an appearance of repetitions that are unnecessary.   But when it is considered that there are
clearly two descriptions of farmers who ought to
find the use of such a work, being both inexpe-
rienced, such an idea will be found erroneous.
There are men who design only to pursue the practice of their own district, in which great improve-

K   ments

ments may not be common ; and there are others, more enlightened, who wish in every month to be reminded of what are the works going on at that time in all other districts, that they may try them or not, at their pleasure. The preceding articles, under the head of pease, are for the former description; but the latter were probably at work for this crop in February, for by means of that new and great improvement of autumnal ploughing, for so many crops to be put into the ground in the spring, without any fresh ploughing, the land may have been laid in stitches, exactly suitable to the drill machine, and to the scarifier, scuffler, and harrow, so that the surface may now be worked to the necessary degree, without a horse ever setting his foot any where but in a furrow, and consequently not treading the beds or stitches in the smallest degree. With this management all spring works are much accelerated. However, February certainly may, in some seasons, pass without the farmer being able to put in his crop of pease. In that case the work must be done in March, by the directions given in February, whether for drilling or dibbling.

BEANS.

February is the month in which the farmer should be active in putting in this crop. Some delay it for the more tender sorts, but I have remarked in many trials, that even these have succeeded better when sown in February. Should the weather prevent early work, of necessity it must

be

be postponed, and then, as with pease, the directions must be postponed in execution for one month.

BEANS TO BE EATEN GREEN.

Some experiments were made by a very ingenious gentleman, in sowing beans for stall-feeding bullocks, while podded but yet green. It was not in my power to ascertain how it answered, but attention to this scheme has been since recalled, by another similar trial, for the use of hogs, by Mr. Cross, and which has been published by Dr. Hunter. The circumstances merit attention. He drilled garden beans at three feet, and afterwards turnips in the intervals. When the beans began to lose their flowers, and to shew a disposition to pod, they were drawn by hand and given to 38 pigs, 10 weeks old, well littered with straw. These were bought the 18th of May, and were kept on clover till the beans were ready. The beans being consumed, the pigs were sold the 18th of September for 40l. beyond the prime cost, and they made 40 loads of rich manure. They consumed four acres of beans. To persons who make it a point of using hogs as the means of raising large quantities of manure (and there is no more effective way of doing it), these hints may be very valuable. Beans used for this purpose may be off the land very early, probably much earlier than these were, and in time for putting in another crop immediately, either late turnips or cole-seed, and the land cannot be in the least exhausted. With this

K 2   view,

view, there should be a succession of plantings in February, March and April.

BEANS BETTER THAN OATS ON LAYS.

" To sow oats on a lay newly broken up, especially if such ground has been many years in grass, is at all times very hazardous, and frequently causes a total destruction of the crop, an instance of which I experienced in the spring of the year 1771. The preceding winter had been very severe, with a continuation of unkindly weather till late in the spring, for at the close of the month of April the ponds were covered with ice, and sharp frosty nights intervened till the 10th of May. Early in the month of March I sowed with oats a sainfoin lay that had been ploughed up some months before, and covered in the seed with the large two-horse harrow, and as soon as possible closed the soil with a five-horse roll, so that the ground seemed to lie as close as one could desire ; but the dry frosty weather abovementioned setting in for a month afterwards, rendered the surface very porous, and the soil was become dry as ashes, and by far too light for the purposes of vegetation. Towards the middle of April, the oats, by favour of some kindly showers, began to make their appearance, but before they were all fairly out of the ground, the worm seized on the fibrous roots below the surface. The land being at that time not sufficiently dry to admit the use of a roll, I endeavoured to close the lightened soil by treading it with horses. My primary view was, to have trodden the upper part of the field only,

the

the lower side remaining at that time unhurt by the worm ; but in a few days these insects spread over the whole close, and although I omitted no opportunity of treading and rolling throughout the spring, the crop at harvest was very slender, as well in straw as grain. From hence we may learn how hazardous it is to sow such lay ground, in the first year after breaking up, with oats, or indeed of cropping it with any other grain than either beans or pease ; for though, in a very kindly year, such corn may not be totally destroyed by the worm, as it turned out in the event, with the greatest part of my oats ; yet there is no doubt but (maugre all his care and pains) the husbandman will then find cause to repent of his conduct, and should a dry frosty time succeed, the destruction of the crop is inevitable. Still more hazardous is it to sow this grain on what is termed a lay-breech, as the worm will in such a season be more likely to destroy the crop, than even after the first breaking up of the lay-ground." *Bannister.*

TARES.

If the weather in February did not allow sowing spring tares, or at least all the crop, the work must not be delayed longer than March, otherwise the crop will suffer. The best way is, to plough the land flat, or on broad lands, and harrow in the seed : but the farmer must observe well, that the soil be dry before his ploughs go on it, which is an universal rule, that ought never to be deviated from.

TARES ON A STALE FURROW.

Sowing on a stale furrow is much superior to a
fresh ploughing. If the land was ploughed before
the Christmas frosts, and the weather in February
has been favourable to the vegetation of weeds,
some may have appeared, but the scarifiers or
scuffles will utterly destroy them, and working only
in the dry pulverized surface, will prepare for the
seed much better than ploughing. Let it be remembered, that the necessity of a suceession of tare
crops (which may be continued through all the
spring) depends on the farmer's being not duly
provided with lucerne, chicory, clover, or other
crops applicable to soiling. If he has such crops,
tares are proportionably unnecessary ; but if he has
not, then he should be very careful to have a due
succession of tares. These are also to be applied
to penning sheep, and are for that use of much
importance.

CARROTS.

This is the season for sowing carrots. The land
should be ploughed in the common manner, but
flat, and 5 lb. of seed to an acre sown broad-
cast, and harrowed in. If the weather is unusually
wet, a farmer may be prevented from getting on
the land ; but, if possible, he should delay it no
longer than the 25th. The proper soil should not
be mistaken through common notions, nor confined
to a compass much within the reality. It is a
general idea, that nothing but sands will do for
carrots ; but this is a mistake. The best soil for
them

them is a sandy loam, rather light, but moist, of a
great depth ; in which there is little difficulty in
ploughing to the very beam of the plough, all the
soil brought up being of the same kind, and as fit
for vegetation as the surface. This sandy loam,
with these properties, should in general yield good
crops of all sorts. But, at the same time that I
mention this soil as most desirable, still the crop
will thrive to great profit on heavier loams, but
not on wet ones or clays. On good wheat loams,
of the gravelly kinds, that plough easily, they do
well. At first sight, such soils might perhaps be
thought too stiff; but they will yield large carrots; but the expences will run higher in cleaning, &c.

I cannot leave this article, without recommending to all the possessors of the lighter sort of lands
that have a pretty good depth, to cultivate this excellent root with spirit ; not to confine it to a little
close of an acre or two, but to introduce it, in the
course of the crops on a farm, regularly, like wheat,
barley, turnips, or any other plant. None will pay
better ; and, if managed tolerably, few so well.

The almost uniform practice in Suffolk, where
this root is cultivated more largely than in any other
part of the kingdom, is to delay all tillage till the
time of sowing, the favourite period for that being
about the 25th of this month. The best of all
preparations for this root is a turnip-fallow, the
crop of which was fed on the land by sheep. The

next best a barley stubble, which succeeded turnips so fed. Some farmers put them in on a wheat stubble, when a manuring of yard-dung has been given for the wheat crop ; but in this way they are more apt to be foul. A modern improvement, and which deserves attention, is, that of steeping the seed from twenty to thirty hours, in order to accelerate its coming up.

Though carrots are *consumed* only in the winter and in the spring, and consequently their use to be treated of under the head of winter and spring food, yet as the young farmer must determine in this month of March, what breadth of land he will apply to carrots, it is necessary to mention here some of the inducements which should instigate him to venture, without apprehension, on so very profitable an article of cultivation.

1st, The teams of horses cannot, in any other way, be fed so profitably as on carrots. If they have only chaff and carrots, the allowance is two bushels per horse per diem. If a bushel of oats per horse per week, then one bushel a day of carrots, and no hay. An acre of 400 bushels, lasts one horse 200 days, or two horses 100 days, during which time they are in perfection. Thus fed, it is an acre per horse : or, at half-feeding, half an acre per horse.

2dly, They are excellent for all sorts of stock-hogs, sows, weaned pigs, and all others. They have been found to fatten well, though some who
have

have tried them for that purpose have had ill success. However, for all lean hogs, there is no
question or dispute, but they *thrive* well on them.

3dly, No food is superior for fattening oxen.

4thly, Nor for feeding young cattle and milch
cows.

5thly, They fatten sheep profitably.

They may be estimated to cost 6l. per acre, or
$3\frac{1}{2}$d. per bushel, prime cost. Supposing 4d., it is
evident that the advantage must be very great.
The common selling price, among neighbours, in
Suffolk, is from 6d. to 9d. a bushel, generally 6d.

Nor is it only in the use of them that this crop
is valuable to the farmer : they are also very advantageous to the land. In the opinion of some
farmers in Suffolk, the barley which succeeds them
is equal to that after turnips fed on the land by
sheep ; and all agree, that they prepare perfectly
well for that crop. A circumstance which speaks
for them, perhaps more than any other, is, that
the culture, within the last ten years, has increased greatly in that county, so that there are
now probably twenty acres, where, twenty years
ago, there were not five. It is common now to
see from twenty to thirty acres on a farm. If a
young cultivator, therefore, possesses any dry and
deep soil, he cannot do better than determine upon
this branch of farming, which will be sure to pay
him well.

CARROTS ON GRASS.

This is not common husbandry any where, but
it

it should be in the farmer's recollection, that they do exceedingly well put in on one ploughing of old grass land, that is on a proper soil. Mr. Gainsborough, of Sudbury, on a farm at Braintree, in Essex, ploughed up a grass-field, the turf seven years old, and harrowed in carrot-seed immediately, the soil a good loam, worth 20s. an acre twenty years ago. The crop varied from 600 to 700 bushels an acre. He practised this husbandry three years running, on different portions of grass, and with uniform success *.

PARSNIPS.

Early in this month parsnips are to be sown. They are not to be recommended, except on the deepest and richest soils. The land should be dry but very fertile. The putrid rich deep sands worth 40s. or 50s. an acre, the deep friable sandy loams, that are as good two feet deep as on the surface, are the soils fittest for this root. On these they come to a great size ; and no other crop on such land can pay better. Where the soil is proper, the inducement to cultivate them is very great, for they will fatten bullocks as well as oil-cake, and are excellent in fattening hogs. Of all roots they are the most saccharine. The tillage and management is the same as for carrots, but they demand deeper ploughing : four pounds of seed the proper quantity, sown like carrots, broad-cast, and the first week of this month the right time. If the weather

---

* Annals, vol. v. p. 414.

is

is favourable, they may be sown the last week in February, 4lb. of seed per acre, and harrowed in. Both these roots have been tried by drilling, by very skilful drillers, but they have not answered like broad-cast crops. Nothing prepares better for corn, if due attention be paid to keeping them entirely clean.

POTATOES DIBBLED.

The land I suppose to have received its first tillage in autumn; and, if it was inclinable to be rough, to have had a second ploughing in February. The first dry season in this month, it should be stirred again flat, turning in the manure, of whatever kind it might be. The best is farm-yard dung; and the more the better, unless the soil be very rich. Upon an average of lands, less should not be laid than twenty-five or thirty large loads per acre, which should be spread equally, that it may plough in well. As soon as the surface is harrowed smooth, the planting should be begun, which is performed in this manner: A man holds in his hand a dibble, which has one point, and a place for him to set his foot on it, to strike it into the earth, in order to make a hole to receive the set. A woman, or boy, follows him with the sets, and drops one into each hole. After this, the land is harrowed twice or thrice, and the business is done. They are in this method set promiscuously, at from nine inches to one foot asunder. The work is done quickly, and is not very expensive. In this
promis-

promiscuous way, from fifteen to twenty bushels of potatoes are necessary to set an acre.

POTATOES ON GRASS.

Grass-land is often broken up for a crop of potatoes, and by most people preferred to any other. The methods are, first, to dung it moderately, fifteen or twenty loads per acre; then to dig up the turf, and work in the dung at the same time, and dibble in the sets, in the way before mentioned. The crop scarcely ever fails of being great in this method. Another is called the lazy-bed way. The grass is dunged as before, and marked into beds five feet wide, with narrow slips between them, two feet wide. The beds are then dunged, about fifteen loads per acre : on the dung are laid the potatoe-slices, after which the turf is dug thinly up in the two-feet intervals, and laid on the sets, which, with another spit, and the loose mould, completes the covering. This is not equal to digging all the ground, on account of its being left whole for the succeeding crop, but the crop of potatoes is generally good ; for, besides the dung, they have the turf below to spread upon, and are partly covered with that from the trenches, so that they lie hollow, and in a rich bed.

POTATOES ON BORDERS.

If you have any rough borders of fields, that were grubbed up to clear away roots and rubbish the preceding winter, it is common to dig them regularly in the beginning of this month, and dibble

ble in potatoe sets, by which means there is a certainty of getting a beneficial crop ; for such places are generally fertile, from the rotting of leaves and wood. They will be left ready in autumn for carrying the earth on the land, in order to level the border, and lay it down to grass. When some rich earth is thrown out of ditches, or mud out of ponds, it is often left long enough for yielding a potatoe crop, which either is capable of giving.

POTATOES FOR HORSE-HOEING.

The new husbandry has been much recommended for the culture of potatoes, and there have been many instances of great crops gained in this manner. The practice of it is various ; but, whatever the manner, the land should be ploughed into ridges for them, according to the rows intended. They have been tried in equally-distant single rows, at two, three, four, and five feet. In double rows, at one foot, on four-feet ridges ; the same, and also three rows, on five-feet ridges. These methods may most of them have succeeded, but the wide distances between the single rows certainly lose too much land. If equally-distant rows were used, three feet is preferable. Double rows on four feet have succeeded. All give the advantage of the horse-hoeing culture, without losing much room. Equally-distant rows, at two feet, with a neat horse-hoe that turns no furrow, but only cuts the surface of the ground, earthing up afterwards, have likewise succeeded. The principle of introducing
the

the horse-hoe is to save some of the expence of
hand-hoeing, and at the same time to make the
crop flourish better. The advocates for this husbandry acknowledge, that there are more plants in
the old method ; but assert, that the tillage of the
plough is so much more effectual than that of the
hand-hoe, and the admission of air among the
plants so much freer, that the loss of number is
more than made up in the gain of size. It has
also been said, that horse-hoeing is so effectual,
that there is no occasion for dung with it ; but let
all good farmers be very suspicious of such assertions. If they give up the benefits of rich ma-
nurings for any purpose so imaginary, they will
certainly repent it. Potatoes may, in certain soils,
be cultivated without dung, with some, but never
with equal profit ; and on most lands it is absolutely requisite. After all, let it be remembered,
that April is a better time for planting this crop.

CABBAGES.

There are two principal seasons for planting
cabbage-crops designed for cattle, viz. the latter
end of April and the beginning of May, and about
Midsummer. The land for both I suppose to have
been ploughed, the first time at Michaelmas. If
February was very favourable for tillage, another
earth should have been given, if the teams had
leisure for it. The fields to be planted in April
and May, must be ploughed again in March.
These stirrings are not to be flat, but the land kept

on

on the ridge, by reversing. If as much time can be spared this month from seed tillage (which are ever the most important part of tillage), as in April, it will be now advisable to plough those lands also that are for the Midsummer crop, by which means there will be a certainty of gaining a fine tilth, late in spring, which is the best method of destroying seed weeds.

In the beginning of this month seed should be sown a second time. The seed, and seed-bed, must be proportioned to the intended quantity of crop. A good rule is, to sow one pound of seed to every three acres of land intended to be cropped. For a seed-bed, a rich piece of land summer-fallowed, and dunged with rotten stuff, is to be chosen. See the Calendar for last month, for paring and burning for a nursery; also for the sorts of cabbage.

REYNOLDS' CABBAGE-TURNIP.

This month, if the weather be favourable, is the proper time for sowing the seed of this plant in a seed-bed, for transplantation in June. It is a remarkable circumstance, that very great and successful exertions were made in the culture of this plant thirty years ago, but that it went out of general use, without any sufficient reason; for its great merit was then well known. Long since that period the ruta baga, or Swedish turnip, was introduced, but in Norfolk the depredations of the fly upon this plant have been so great, that it is also in danger of being given up. As there is no point

of

of greater importance than that of providing green winter and spring food, for cattle, I must recommend a due attention to both these plants. The following particulars, communicated to the Society of Arts, dated in April, well deserves the attention of the farmer, relative to this cabbage-turnip.

" I have sent six more of the roots for the inspection of the Society, indifferently chosen. They weigh 38lb. so that upon an average, at this time (viz. April 29), there is no less than 35 tons per acre. And let it be observed, they are not full grown, the spring being backward, otherwise the product would have been greater. I pitched a fold in the form of an oblong, in two divisions, and placed therein 387 sheep, April 2. The crop has kept them exceeding well, without any fodder, or even any other provision (save only the turning them into a rough pasture a few hours in the middle of the day, for by so doing, they return fresh to the roots in the afternoon) ; and will maintain them in the same manner till the 12th of May, I am confident, which is in all full 40 days. I placed those designed for fattening in the front, and the store sheep in the rear. The plants are drawn up with a three pronged hoe. The fold is removed daily for those in front to have fresh food; and those in the rear eat the remains of what was left the day before. Thus the whole is spent without the least loss, and the land enriched at the same time. This, I say by experience, gained last year; for my barley crop, on land in the same
state,

state, turned out very good, both in quantity and quality inferior to none of my other growth (which was upwards of 80 acres) ; the product full five quarters per acre, sown the 12th of May.

" Eight milch cows have been fed with these roots for this month past ; and are very fond of them ; and I have found great savings in my hay since they began them. Both the milk and butter proved very good, and a considerable increase in both kinds.

" I also find that hogs and pigs like them extremely well. Sows give plenty of milk when fed with these roots.

" Upon the whole, it is very certain that they are found to be of the greatest utility.

" Nothing, that I can find out, is more beneficial to the land-holder, *for spring-food,* especially in hard weather, and times of distress, such as we have of late very severely felt. This consideration *only,* ought to encourage all concerned in raising them, were there no other motive whatsoever. But that is not the case ; for I find that besides this great food, *of giving plenty in time of need,* there is another benefit annexed to it, *viz.* the improving land for the ensuing crop, when these roots are spent with sheep where they grow. These are circumstances of the greatest merit in agriculture, especially to those who have them in possession.

" These roots are proof against frost ; whereas turnips have been but of little service in general this spring, the frost having destroyed many of

L   them

them long ago. But these vegetables are now in a
fine flourishing state, quite sound and good ; well
tasted top and bottom : better food cannot be
desired for horned cattle and sheep.   It seems,
indeed, the very thing long sought for (namely,),
good spring food.

" This is certain, my sheep are now thriving
beyond all expectation ; whilst other flocks, in general, having no such provision, are almost starved
to death for want of sustenance.

" If what is here asserted and proved by experience, will not induce people to raise these roots, I
know not what will."

In a letter from Mr. Reynolds, dated the 15th January, read to the Society, he speaks thus of the produce of the turnip-rooted cabbage. " This is certain,
large crops have been obtained within the two last
years in several counties.   Their product have risen
from 25 to 35 ton per acre; and if my memory serve
me right, there are two accounts from Nottingham
and York, as high as 44 tons.   Kent and Sussex
have obtained near 50 tons ; but one gentleman
in Surrey has outdone all that I have yet heard of.
This plantation, and that no small one, produced
upwards of 56 tons per acre, in 1770.   I have this
well attested ; and that many of his single roots
weighed 14lb. each.   This may seem incredible
to some, but it is not so to me in the least.   For
my shepherd brought me in one single root, on
the 4th of May, 1773, that, when cleansed, weighed
17lb. the most extraordinary plant of this kind ever
                                                        beheld.

beheld. My curiosity led me to see where it grew, and, on viewing the place, I found it stood where a heap of grass-burnt ashes had been ill spread; and this occasioned its extraordinary size." Mr. Reynolds' conclusion, with respect to the great magnitude of this root, is justified by many instances of extreme great crops of the brassica, or cabbage kind, being produced on land, after burn-baiting; and even after burning the turf of heathy land, without any other manure.

This account of the great utility of the turnip-rooted cabbage, is strongly confirmed by several passages of Mr. Reynolds' letter of January 15, 1774, mentioned in the preceding note. He says, " With respect to my turnip-rooted cabbage, I find it is now propagated more and more in this neighbourhood, with all desired success; and begins to spread apace throughout every county in England and Wales, and in some parts of Scotland. It has been cultivated by an eminent North Briton, as he expresses it, *to their very good liking and advantage, and withal, is much admired in bearing frost very well*; which, according to his information, is more intense and severe than with us. I have letters to prove they are in no small esteem in the kingdom of Ireland: and I find they are recommended by their Society to all concerned in breeding and feeding cattle to propagate." Mr. Reynolds supports his assertion in this letter, of the great utility of this plant as a spring-food for cattle, and more particularly for sheep, by the in-

stance

stance of what happened to him in 1773, "when most of the turnips that had been sown the preceding year, had failed; or those few that had escaped run away to seed much earlier than common," and the farmers therefore were in the utmost distress, at the same time that he, having no less than seven acres of this plant, felt no inconvenience.

TURNIP-FALLOW.

The land designed for turnips, I suppose to have been ploughed from a stubble at Michaelmas.   In this month it should receive the second earth, which is very necessary, that it may, by harrowing well, or by another stirring in April, if the land is stubborn, be made so fine, as to ensure a thick crop of weeds in May.   A succeeding ploughing turns them in, and quite destroys them.   This is a method that is very effectual in destroying seed weeds, and particularly suits turnip-fallows, as it is a crop that requires a very fine tilth.

But modern improvements have discovered better ways of effecting these purposes.   The turnip-fallows that received an autumnal ploughing should not be ploughed again till surface operations in March have stirred and loosened the earth, to favour the vegetation of weeds.   This is best done by scarifying or scuffling.   These tools will do their work if they are well constructed, to any depth the farmer may wish, and this tillage keeping the surface that has been pulverized by frosts, unburied, is much more favourable to the growth

of

of weeds, than turning it down by ploughing. The work also is done much more rapidly, which, at so busy a season, is an object of great consequence.

LENTILS.

This crop is not uncommon about Chesterford in Essex. They sow a bushel an acre on one ploughing in the beginning or middle of the month. They make hay of them, or seed them, for cutting into chaff for trough-meat for sheep and horses, and sow them on both heavy and dry soils. The whole country is calcareous. Attention should be paid, not to water horses soon after eating lentils, for they are apt to hove. They are cultivated for the same purpose in Oxfordshire.

LETTUCES FOR HOGS.

I first saw the sowing of lettuces for hogs practised in a pretty regular system, on the farm of a very intelligent cultivator (not at all a whimsical man) in Sussex. He had, every year, an acre or two, which afforded a great quantity of very valuable food for his sows and pigs. It yields milk amply, and all sorts of swine are very fond of it. The economical farmer, who keeps many hogs, should take care to have a succession of crops for these animals, that his carts may not be forever on the road for purchased grains, nor his granary opened for corn oftener than is necessary. For lettuce the land should have been ploughed before the winter frosts, turning in by that earth 20 loads of rich dung per acre, and making the lands of the right breadth, to suit the drill-machine and horse-

L 3   hoes,

hoes, so that in this month nothing more may be
necessary than to scarify the land, and to drill the
seed at one foot equi-distant, at the rate of four
pound of seed per acre. If half an acre be tried,
or even a rood, near the farm-yard, the advantage
will not be inconsiderable.

CHICORY.

This most hardy plant will do well whenever
sown. It is indigenous over the greater part of
the kingdom. There are several views with which
this grass may more particularly be cultivated.
1. On poor barren blowing sands, such as many
districts abound with, especially in Norfolk and
Suffolk, it will yield a greater quantity of sheep
food, than any other grass at present in cultivation. 2. On fen and bog-lands and peat soils, it
thrives to much profit. 3. On all land, whatever
the soil, on which clover, from having been too
often repeated, is apt to fail, chicory may be substituted to great advantage. 4. It does very well
for soiling cattle, both lean and fattening. 5. It
is of excellent use for those who keep large stocks
of swine. 6. It does exceedingly well in an alternate system of grass and tillage, as it will last four,
five, six, and even more years, but it should not
be sown with any view of making hay in this
climate, though it forms a considerable proportion
of many of the best meadows in the South of
France and in Lombardy. Objections have been
raised to it, from its rising again after tillage, but
these will be fully obviated when I mention the
                                                    proper

proper husbandry by which the layers should be broken up.

MUSTARD.

In breaking up the rich common of Marshland Smeeth, in Norfolk, the crop that was supposed to pay better than any other, was mustard. The soil is a rich silt and clay, worth 60s. an acre. They ploughed once, and harrowing twice, sowed one-fourth of a peck of the seed per acre, from Candlemas to the end of March, according to the weather. Few farmers have a soil that answers for this crop, but it is necessary to name it, that if markets are promising they may at least have it in their mind. It may now also be added, that they hand-hoe the plants twice. The crop is reaped in the beginning of September, being tied in sheaves, and left three or four days on the stubble. It is then stacked in the field. Rain damages it. A good crop is six or seven combs an acre : the price from 7s. to 20s. a bushel. They take three or four crops running.

MANGEL WURZEL

Is dibbled in this month, along the tops of ridges two or three feet wide, and which have been previously manured, and reversed to cover the dung. It is very little cultivated at present ; but Sir Mordaunt Martin, of Norfolk, adhering to the cultivation, and finding the root very advantageous for his cows, it is right to name it in a work of this nature.

WOAD,

WOAD.

According to Mr. Cartwright, the middle of this month is the proper time to begin to sow this plant. I no more than name it, as all that a young farmer can, with any propriety, have to do with it, is to let some very rich grass-land to the travelling woad-men, who will give 4l. 5l. and even 6l. per acre, per annum, for liberty to cultivate it for two or three years. If he has such land, he will hear of them. In Somersetshire, they sow it in June.

CHAMOMILE.

This is an article of culture in Derbyshire : they chuse a good loamy soil, well prepared, and plant the roots from an old crop. It remains three years on the ground. It is a troublesome culture, and by no means tempting in profit. To name it once is sufficient.

SCARIFY WHEAT.

This operation, which should not be delayed longer than March, is so material a part of the drill-husbandry, that much attention should be paid to executing it at the right moment, and also to doing it in the best manner.

Opinions are various, and many farmers do not approve of horse-hoeing at all, probably from having done it too late, or too severely. Mr. Cook has invented two implements applicable to this work, his fixt harrow and a scarifier. The former works merely by plain harrow-teeth : it has three rows, and, by varying the position diagonally, one, two,

or

or three teeth may be worked in the space of nine inches, without damage to the rows of wheat. If two, they may be drawn in a breadth of three inches; if three, in that of four inches, and these spaces widened at pleasure, but still so as to keep quite clear of the rows of wheat. By loading the harrow, the teeth are forced to the proper depth. The scarifier has teeth of various breadths, but for working at this season between nine or twelve inch rows, the narrowest are to be preferred. The intention of the operation now performed is, to loosen the surface earth and let in the air. The hoe of the scarifier may cut two inches deep safely. It will do this without raising any such surge of moulds as to bury the plants, and in its motion through the earth, it loosens without removing it. There the air immediately penetrates, and comes in contact with the roots, which, from the soil being moved, can spread with the greater ease. Another motive for hoeing is, to loosen and pulverize the mere surface when bound by rain or other weather. This has relation to the coronal roots of the wheat, which shoot out at this time of the year, when the seminal ones decay; and there is a curious circumstance noted by *Bonnet*, which is, these coronal roots striking not under the surface, but above it, into the air, and entering the earth at a small distance. It is then of consequence that the surface be in a pulverized state, to permit their free penetration. All hoeing before that moment must necessarily be beneficial, because preparing the sur-

face

face to receive these roots ; but, if given *afterwards*, and so near the rows as to disturb them, it must do mischief for the time, as Nature has then her operation to perform again : and though the wisdom of Nature's Author without doubt provides for the case, as in that of the blossoms of wheat being blown off, yet a delay takes place, and a succession of injuries may be felt in the crop. The circumstance should at least induce the farmer to be early in his operations : if he is early enough, he may break the surface of a nine-inch interval, to the breadth of five inches, but afterwards he must recede from the rows, for fear of mischief. These tools of Mr. Cook are very effective, and if used with attention will be safe, while they dispatch a great deal of work in a short time. When, however, it is considered, that at this season, *in common management*, the teams are all in full employment, it must be admitted, that if there is much horse-hoeing to do, the common charge of a few pence per acre, is utterly inadequate for a period when a farmer would be glad to hire horses, could he get them, at 5s. per day, and even at a greater charge ; but if the modern system, of avoiding as far as possible, all spring ploughing, by the use of the same and other tools, be taken into the account, then the drill-husbandry will not demand more horses, possibly not so many. Whatever the operation, never loosen the *bottom*, for that should be left *firm* for wheat.

HAND-HOE WHEAT.

If the farmer does not chuse to practice the preceding operations of horse-hoeing his wheat, he
must at all events hand-hoe the drilled intervals ;
but the same attention to earliness, is as necessary
in this operation as in the former.   If he gives the
first hoeing early in this month, he may do it a
second time the last week, or early in April.

HAND-HOE BROAD CAST WHEAT.

There are several districts in the kingdom, in
which this is generally done by every good farmer,
and in some at a considerable expence.   Some of
the most careful and attentive cultivators, and men
of much intelligence and observation, have however
assured me, that they found mischief from it,
and left it off: and I am much inclined to think,
that if any such operation is intended, the wheat
should by all means be drilled.

HAND-HOE DIBBLED WHEAT.

Two methods of dibbling will be explained under
the proper month.   The one is, planting two
rows on a flag; the other one row, in which
latter way, there is an ample space for effective
hand-hoeing, a method which answers exceedingly
well.   When wheat is thus put in, the farmer
should on no account omit hoeing.

SHEEP.

At this season the stock, whether ewes, lambs,
or fatting sheep, must be kept extremely well.   If
they are pinched now, all the money before ex-
pended will be nearly thrown away.   Turnips can

no

no longer, with any propriety, be depended upon. If the farmer has not a great breadth of ray-grass, or some other dependence, he must sell off his fatting sheep sooner than he desires, and stop the growth of his lambs, at a time when they ought to be half fat.

In fattening wethers, the great object is to keep from selling till the middle of April. To begin then, and keep killing till the end of May, is the most profitable conduct, for the price that a butcher will then give, exceeds what he would have given two or three months earlier. How advantageous is it, therefore, to have food ready to take all the stock by the middle of March, and to have enough to last till May.

SHEEP IN STUBBLE TURNIPS.

One of the resources for sheep feeding, at this pinching season, is stubble turnips. A good manager, who finds a field of warm, forward, rich, land harvested early (whatever may have been the crop), will sow turnips, with a view to sheep-feed in March and April. If they are not sown early they will consist of little more than shoots, but as they run up very rapidly at this time of the year, they afford much food, and are truly useful, at a season when every blade is valuable.

SHEEP IN ROUEN.

As turnips are by this time done, or ought to be done, the farmer will now find the immense importance of that reserve of kept grass, called *rouen*

in

in Suffolk. All his ewes and lambs may now be
in it.

SHEEP IN BURNET.

The autumnal growth of burnet may now be fed
with sheep, to great advantage, and prove of singular importance.

SHEEP IN SWEDISH TURNIP.

This plant should no more be left this month,
where it grew, than common turnips, for as the
shoots for seed spring, they lay fresh hold of the
earth, and exhaust it considerably. Such portion of
the crop, therefore, as was not consumed in February,
should be so early in March, but when the weather is
mild the last week of February, they should then be
removed to a grass field, for feeding. If, however,
the land where they grew be dry, it is of such advantage to eat them on the spot, that another method may be pursued, which is a very good one,
and that is, to draw them with one hand, and by
giving them a chop with a knife by the other, to
strike off the tap-roots, and drop them on the surface of the land. In this way, they will draw the
soil infinitely less, than when left undisturbed, and
women do the work easily. Thus treated, the
young white fibres are broken in drawing, which
checks them, and though fresh shoots will push out
from the part of the root in contact with the earth,
still they will take some time, and be weakly.

COWS, &c.

Throughout this month the cows, lean and
young cattle, should be kept close to the farm-
yards,

yards, and on no account allowed to wander over any of the fields. If they even steal into a grass field, and it be a forward spring, so that they get a mouthful or two of grass, it will be a prejudice to them; for they will not be so well contented with their dry meat afterwards. Besides, they poach the grass, and lose manure. For these reasons, it is very advisable to have all the yards (I suppose water to be in them) locked up, and then it will not depend on the memory of servants. Every place must be well littered with stubble, straw, or fern, and it is to be remembered that omitting this attention will be just so much mischief to every crop on the farm, in the article of manure.

At this season, a farmer who has weaned any of his calves, should observe that they be well and regularly attended. They should have a small yard with sheds to themselves, and have their bellies full of the refuse leaves of the cabbages given to the milch cows, with whole ones, if these are not sufficient. Carrots are also of admirable use. Young cattle should be kept well, otherwise they will come to a very poor size. Their dry meat should be good cut chaff.

THE DAIRY.

This branch of the farmer's business must necessarily depend so much on his wife, if he has one who understands it, or on his dairy-maid, if his wife is ignorant, that his own share can be rarely more than a general superintendance, to see that every thing is clean, and that products and prices

do

do not manifest neglect. The minutiæ of the dairy concerns would fill a volume, and, after all, would not probably be useful to any extent. Here and there a hint may be dropped, to bring certain points to his recollection ; but all will at last depend on the practice and skill of the operator.

FATTING BEASTS.

Much attention should be paid to the stall or yard-fed beasts, in food, water, and litter. The young farmer will do well to weigh them, once a fortnight at least, before feeding in the morning, as this will tell much better than can be guessed by any one; except a most experienced hand, the degree in which they are thriving. If they do not keep advancing. equally to preceding weighings, their food should be changed or varied.

THE TEAMS.

A diligent farmer will now see that his horses and oxen perform a good day's work. In sowing time, he should not let them work less than nine or ten hours ; but this he will not be able to effect, if the ploughmen have to take care of their horses. It is best to have horse-keepers, for the mere attendance of the teams, so that the men who hold the ploughs may have nothing to do but the mere ploughing. Let the horse-keeper have the horses fed and harnessed ready for the ploughman, to be in the field by six o'clock. At eleven they should come home for an hour and a half to dine and bait, during which time the horse-keeper is in attendance again. At half an hour after twelve, they should

should go out again, and work till half an hour after five, when the horse-keeper should again take the teams. By this method a pair of horses, in a well-made plough, will, without any driver, plough an acre and a quarter, or half, very easily ; and no object is more important, than the ploughs doing good days work in the spring of the year. The consequence especially, of making the most of dry weather in March, is extremely great. One acre ploughed and sowed then may be fairly worth two that are left till the beginning of May. From long observation of the value of dry seasons for tillage in this month, must arise the old proverb of—*A peck of March dust is worth a king's ransom.*

HORSES ON CARROTS.

This is a month in which carrots are in full perfection ; they have now evaporated much of their moisture, and easily bend in the hand, being as it were withered. Then every ounce is nourishment, and they are fully as hearty as oats ; insomuch, that horses that have had a month's carrots will refuse oats. To provide this root in ample quantity, for February, March, and April, is an object that ought never to be omitted.

WATER-FURROWING.

In all new sown or ploughed lands, as soon as a field is finished, let the ploughs, before they leave it, strike the water-furrows, and send in men directly with spades to scower them, that is, throw out the molds. In all lands, sown with clover or grasses among the corn, these furrows should be

dug

dug a spit deep, and the molds thrown carefully
out. Many farmers are not attentive enough to
this point. They only scour the furrows; but
they should consider how long the grass is on the
ground, which may be two or three winters, consequently it must be very material to the crop to
lie dry all that time, which scouring alone will
not effect.

POULTRY.

Much attention is necessary to all sorts of poultry this month ; and as it is the first time of the
subject being mentioned, it will be necessary to
offer a few observations on the system which a
young farmer may adopt with relation to this article of live-stock. If, in the common way, he keeps
but a few of each sort, that take their chance at
the barn-door, for the convenience of eggs, and
not to go to market when a fowl is wanted, no particular attention is requisite; but as, in some situations, they may pay well for more food and closer
attention, it will be proper to bring a few circumstances to recollection. The poultry-house should
contain an apartment for the general stock to roost
in, another for setting, a third for fattening, and
a fourth for food. If the scale is large, there should
be a fifth, for plucking and keeping feathers. If
a woman is kept purposely to attend them, she
should have her cottage contiguous, that the smoke
of her chimney may play into the roosting and
setting rooms ; poultry never thriving so well as in
warmth and smoke ; an observation as old as Co-
    M   lumella,

lumella, and strongly confirmed by the quantity
bred in the smoky cabbins of Ireland. For setting
both turkies and hens, nests should be made in
lockers, that have lids with hinges, to confine them,
if necessary, or two or three will, in setting, crowd
into the same nest. All must have access to a
gravelled yard, and to grass for range, and the
building should be near the farm-yard, and have
water near and clear. Great attention should be
paid to cleanliness and white-washing, not for appearance, but to destroy vermin. Boiled potatoes
are the cheapest food ; and of corn, buck-wheat.
Turkies, while young, demand incessant attention,
and must be fed with allum-curd and chopped
onions, for which purpose, store of those roots
should be kept where they will shoot out and produce much food. If there be not much success
in broods, and a certain high price, they will not
answer, for the expences are heavy.

WATERED MEADOWS.

At the beginning of this month the crop of grass
on the old floated meadows will generally be sufficient, Mr. Wright remarks, to afford an abundant
pasturage to any kind of farming stock, and the
water must be taken off for nearly a week, that
the land may become dry and firm before the heavy
cattle are admitted. It is proper, in the first week
of eating off the spring feed, if the season be cold
or rainy, to give the cattle a little hay in the evening to intermix with their moist food. But the
grand application of the young meadow-grass is,
for

for ewes and lambs, and attention should always be paid to hurdling off the grass, and giving stripes across the meadow, exactly in the way that turnips are hurdled for sheep. The caution of Mr. Boswell, never to feed on these meadows any heavier stock, in spring, than sheep and calves, seems to be judicious, but will, however, depend much on soil, for, upon a sound gravel, a practice may be admitted which would be mischievous on a peat meadow.

DESTROY MOLES.

This is one of the principal months for destroying this species of vermin, as they *run* more now and in April, than at any other time. Mole-catchers who do not keep farms clear by an annual contract, but are paid by the head, are too apt to neglect their business when their attention is of most consequence.

MANURE GREEN WHEATS.

February, March, and April, are the months for sowing top-dressings on the young wheats. There are a variety of articles which answer for this purpose, of which Mr. Farey gives a detail, as used in the vicinity of Dunstable, &c.

1. *Soot*, from coals, is bought in London at 7d. to 9d. per bushel, struck. The measure of London soot is very deficient, viz. about four-fifths of Winchester, which makes the price 9d. to 11d. per Winchester bushel, struck. It is brought from London to the lands, and there deposited in a heap (which is their practice also with the other light

M 2

dress-

dressings), at 3d. per bushel. From these heaps a
common seed-scuttle is filled, and a man walking
the length of the lands, sows the soot in the same
manner as corn is sown. The expence of sowing
is a halfpenny per bushel. The quantity used per
statute acre, is from twenty to forty Winchester
bushels. In general thirty bushels are used for a
complete dressing ; *i. e.* when dung, or some other
manure, has not been previously applied to the
same crop, which is very frequently their practice,
and the quantity of top-dressing is then diminished
to about one-half of a complete dressing. Of soot,
a complete dressing as above, costs 30s. to 36s. per
acre. Soot is found to answer best on wheat in
April. It likewise succeeds on pease or clover, in
the same month, and has a good effect sown *with*
barley in the beginning of April, and harrowed in.
A slight dressing of soot is used at any time in the
spring, when grubs or worms appear to injure the
young corn. The worms frequently make great
havock here, by drawing the blades of young corn
after them into their holes :   this, soot prevents
best. Soot thinly distributed on newly-sown turnips, *just before* they come up, prevents the fly or
grub from injuring them, provided no rain falls to
wash it into the soil. Soot answers best on light dry
chalk soils, and in moderately wet seasons. It
does little good on strong or wet land, or in very
dry seasons, unless sown earlier than usual. The
London soot from coals is rarely bought unmixed
with cork-dust, coal-ashes, or sweepings of the
streets ;

streets; yet even in this adulterated state, it is found to answer much better than real country soot from wood.

2. *Coal-ashes* cost in London from 6s. to 14s. per waggon-load (narrow wheels and four horses), the price depending on the business doing in the brick-fields near town, in which considerable quantities of ashes are used. Carriage included, they cost on the land about $5\frac{1}{2}$d. per bushel. Coal-ashes are bought in small quantities in the neighbourhood at 4d. per bushel, and collected to the land at about 1d. per bushel. They are distributed on the land with a shovel, from a cart or wheel-barrow moved along the land. Another, and perhaps the preferable mode, is sowing them by hand. The former way costs 12d. per waggon-load, the latter 18d. Coal-ashes are used from 50 to 60 bushels per statute acre for a complete dressing, which amounts from 23s. to 26s. per acre; they succeed well, sown on clover in March or April, on dry chalk lands; and also do much good to sward, applied during any part of the winter or spring. They are never used on wheat. In very dry seasons coal-ashes do little good: they, as well as most other of these dressings on light land, require rain after being sown, to set them to work *.

3. *Peat-ashes,* brought from the neighbourhood of Flitwick on asses, are delivered on the land

---

* In 1790, Mr. Dann procured from London the finely-sifted coal-ashes, which are sold there, and spread them, 45 bushels per acre, on clover and sainfoin, and the benefit was very trifling;

but

land at $4\frac{1}{2}$d. per bushel, struck; being distributed in the same way, and at the same expence as the last article. Forty bushels per acre is a complete dressing, and costs about 16s. Peat-ashes succeed well, used at the same time, and on the same crops as the last article, except that they apply it on wheat in April, with good effect. Peat-ashes greatly improve dry chalk soils, but will do little good on wet land, or cold sward, or on hot sandy lands. This, like most other of their dressings, is little affected by the season, provided wet falls soon after it is laid on the land.

4. *Peat-dust* costs the same as the ashes, and is sowed in the same manner and quantities. It answers equally well, and in every way the same as the ashes. Peat-dust is esteemed the best possible dressing for an onion-bed in a garden, and is not found to promote weeds more than other dressings. It has great effect on thistles, which it is scattered upon, causing them to wither, as if scorched; but they generally recover, unless the dust be repeated. The occupiers of the chalk-lands in this neighbourhood are under considerable apprehensions for the loss of the valuable peat manures, by the proposed drainage of Prisles Moors, intelligence of which has reached them already.

5. *Folding* is used as a top-dressing, and on

---

but the ashes from Chatham Barracks (not kept under cover), and not finely sifted, have improved those crops to the amount of a load of hay superiority per acre. " The cinder, therefore," says Mr. Dann, " is better than the ash."

these,

these as on most other soils, answers to good purpose : it succeeds best on dry land. Its effect on these light soils is not entirely attributed to the sheep's dung, but in a great degree to the stiffness the land acquires by the treading, which is here found so very beneficial, that they frequently lead the plough-horses a-breast up and down the lands several times after sowing wheat, or other grain, to tread it.

MANURES TURNED IN.

1. *Furriers' clippings,* which are bought in London at 12s. to 13s. per quarter (being a ten bushel-sack crammed full) weighing about $2\frac{1}{2}$ cwt. The carriage to the lands costs 3d. per quarter. They are sown by hand from the seed-scuttle, at about 3d. per quarter, on the land intended to be sown with wheat or barley, and immediately ploughed in, after which the seed is sown and harrowed in, when such pieces of the clippings as are left above ground by the harrow, are pricked or shoved into the ground, by the end of a stick, to prevent their being devoured by dogs or crows, who seize them greedily. From two to three quarters are usually sown per statute acre. Clippings answer well on light dry chalk or gravelly soils, where they are supposed to hold moisture, and help the crop greatly in dry seasons. They have but little effect on wet soils.

2. *Horn-shavings,* which are of two sorts, *small,* or turners, and *large,* which consists of refuse pieces of horn. The small are bought in

London in the same way, and generally at the same prices as the last article. The large shavings cost about 2s. less per quarter. Horn-shavings are used in the same way and quantities as the last article, except that they want no pricking, and the large are generally ploughed into the land three months before sowing wheat or barley. Horn-shavings answer in most soils and seasons, except very dry ones, when they will not work. The small shavings are much the most useful.

3. *Woollen rags* cost in London from 3s. 6d. to 4s. 6d. per cwt. : the carriage home 1s. per cwt. In the country they are bought at 4s. 6d. to 5s. and are collected at about $2\frac{1}{2}$d. per cwt. The rags being generally in large pieces, are housed and chopped at the expence of 5d. or 6d. per cwt. : the extra expence of housing and carting to the land is about 4d. per cwt. They are sown by hand, and ploughed in three months before sowing wheat or barley : the quantity used is six to ten cwt. per statute-acre. Woollen rags, like furriers' clippings, hold moisture, and are adapted for dry, gravelly, and chalk soils, and succeed in dry seasons better than most manures, but they do little good on wet soils. London rags are found much better than those collected in the country ; but the danger of catching the small-pox in chopping and sowing them, deters many farmers from their use.

4. *Sheep's-trotters,* and *Fellmongers' cuttings,* are bought of the neighbouring fellmongers at about 6d. per bushel, heaped loose : carriage to the land

land is about $2\frac{1}{2}$d. per bushel. They are used in
the same way as furriers' clippings, from 20 to 40
bushels per acre, and need pricking in, as dogs and
crows are very fond of them. They do not answer
on wet land, or in very dry seasons : indeed nothing does succeed in excessive dry seasons on these
soils. The trotters contain a considerable quantity
of lime, and are often adulterated with sand, and
sometimes considerable quantities of oak saw-dust
are mixed with them, which has been found not to
injure them.

5. *Malt-dust* costs at the neighbouring malt-
houses 1s. per bushel heaped, and about a halfpenny per bushel carriage to the land. It is sown
by hand, from 24 to 32 bushels per acre, at the
same time with barley, and harrowed in with the
seed. It suits most soils and seasons. Malt-dust
quickly spends itself, and is therefore never sown
with wheat : as a top-dressing to wheat in March,
about 30 bushels per acre, it probably would succeed on these soils. Black malt-dust, or such as
falls through the kiln-plate in drying, is greatly
preferred to the white, on account of the seeds of
charlock *(sinapis arvensis)* with which it abounds
being destroyed by the heat. These chalk-lands,
under their present management, produce such
abundance of charlock, that they are generally
obliged to mow off the heads and flowers about the
middle of June, to prevent its entirely smothering
the corn. Charlock-seed so abounds with oil, that
it will lay for any length of time in the earth without

out vegetating ; which, however, it never fails to do, when brought near enough to the surface by the plough. Pigeons are supposed to pick up considerable quantities of charlock-seed after land has been ploughed. These lands are very subject to be over-run with black grass *(alopecurus agrestis)*, which is said to impoverish it much.

6. *Pigeons'-dung* costs 1s. per bushel heaped, and about a halfpenny per bushel bringing tc the land : it is used as malt-dust, and does good in any soil or season.

7. *Soap-boilers' ashes*, or wood-ashes from which ley has been made, is to be had (in small quantities only) at 6d. per bushel heaped, and costs about 1d. carriage : the effect of these on cold sward is very great.

8. *Hogs-hair*, which is sometimes to be had in London at about 9s. per quarter (the same quantity as furriers' clippings), and carriage 3d. applied in the same manner with clippings, answers well.

Seal-hair, rabbits'-dung, and *lime*, have been tried upon these soils, and found to answer no good purpose.

ALDER.

If a farmer has an alder-car on his farm, or stubs of them by a river, he should be attentive to cut them when the bark will peel, and immediately soak them two months in a pond. This hardens the wood to such a degree as to improve it greatly *

---

* Annals, vol. ix. p. 485.

It

It merits experiment, to ascertain whether the effect would not be the same with other aquatic woods.

PARING AND BURNING.

This is the first month in which a farmer can execute the burning part of this operation, upon any large scale ; but if the north-east winds prevail, which so much excel in the power of evaporation, and which often blow through the greater part of this month, it may go on without interruption. The cases in which this management ought to be embraced are so numerous, that the man who rejects it cannot profit by many situations and circumstances he may be in, without the application of this admirable system.

Before we come to the distinctions of soil, it will be proper to offer some general observations, on the diametrically opposite systems embraced by such numbers of persons, on the general question, whether this practice is excellent or worthless, as two parties have decisively pronounced it.

By one set it is pronounced, contrary to every principle, that it is a wasteful, extravagant operation, which dissipates what should . be retained ; annihilates oils and mucilage ; calcines salts, and reduces fertile organic matter into ashes of very weak efficacy ; that the vegetable particles which are brought into play at once, for the production of a single crop, by less desperate management might be husbanded to the support of many   On the contrary, the advocates for this management assert, that these objections are all founded on vain

reason-

reasoning and philosophical theory; that practice
the most decided, and experience the most extended, pronounce it to be an admirable system;
and that the mischiefs often quoted as flowing
from it, are to be attributed merely to the abuse of
the method, and by no means necessarily connected with it.

I must without the least hesitation declare, that
the latter of these opinions is that to which I must
subscribe.   To trust to reasoning in matters of
agriculture, is a most dangerous reliance.   I shall
leave others to detail their philosophical speculations, and rest what I have to offer solely on the
practice, various and extensive, of numerous agri-
culturists, and on the common husbandry of many
spacious districts.

These agree in declaring, and it is most particularly to be had in remembrance, for the enemies
of the practice admit it, that by paring and burning, you may command two or three good corn
crops in succession.   The fact cannot be denied;
for whether you examine the peat of the Cambridge
fens, or the shallow chalk soils of the downs and
wolds of Hampshire, Gloucester, and the East
Riding, it is known, that bad farmers do act thus
absurdly.   They get great crops, but they too often
take them in succession, to the injury of the soil,
though not to its ruin, unless that can be esteemed
the ruin of land, which enables the tenant to pay
a double rent for it.   Such farmers have been in
the habit of burning for wheat, and then taking
                                                        two

two crops of spring corn ; all good. Now, it might be asked, how is it possible, that that husbandry can have all the philosophical evils detailed above, of annihilation, dispersion, conversion and destruction, which enables a soil naturally poor and weak, to give two or three good crops of corn ? Their argument evidently proves too much. The effect shews, that there is a powerful cause or agent in burning, which they do not understand ; which escapes from the retort of the chemist, and from the *rationale* of the theorist. That operation or manure which will give a good crop of wheat, will give a good crop of turnips or cabbage ; and he, who having made this commencement for the food of sheep on the land, and knows not how to go on, preserving the advantage he has gained, is a tyro in the art of husbandry. The farmers that are railed at, know it as well as their philosophical instructors ; but avarice, united with the baneful effect of short, or no leases, make them practice against their judgments.

Paring and burning will, on all soils, give turnip or cabbage ; these fed on the land by sheep, will secure barley or oats, and seeds ; the seeds fed with sheep, whether for a short or a longer duration, will secure another crop of corn adapted to the soil ; and in this stage of the progress, the soil will have gained much more than it has lost. To instance cases which I have seen, and to quote authorities for these assertions, would be tiresome. I could

produce

produce instances from more than half the counties of the kingdom.

It has been often contended that burning lessens
the soil.   If this happen any where, it must be in
peat; yet, in the fens in Cambridge, this husbandry has been repeated once in eight years, for
a century and a half, and the proofs of a loss of
depth are extremely vague, in every instance I have
met with, and hardly to be distinguished from that
undoubted subsidence which takes place in drained
bogs of every description.   In all other soils the
assertion may be safely and positively denied.   I
have calcined pared turf, not calcareous, after careful separation and weighing, and in a heat far exceeding what is ever given in denshiring heaps, and
re-weighing, found the loss too minute to be attributed to any thing but loss of water intimately
combined, but driven off by heat, and re-exposing
the earth to the atmosphere, free from rain, found
an increase rather than a diminution of weight.
The vegetable particles only are reduced to ashes.
These, in any method of putrefaction, would dissolve, and combining with water, be exhaled by
heat, or absorbed by the vessels of plants.   In
ashes, these are in a *more* fixed state, relatively to
the influences of the atmosphere.   That plants
feed on them, the great crops which succeed offer
abundant proof.

There are men who are timid in acknowledging
truth, who admit the practice to be good, in poor
soils, and in peat, and sedgy bottoms, but fear it
                                         on

on better land. Experiment is against them, for in Yorkshire, land of 20s. and 30s. an acre, has been thus broken up with great success : but in what manner do they reason ? Rich soils are full of vegetable fibres. Then there will be the more ashes. These are in proportion to the organic matter. Peat, which these men admit to be burnt, is the richest soil in the world, and therefore burning the most universally practised on it. The soil itself is not reduced ; if it was rich before burning, assuredly it will be rich after it.

In addition to these circumstances, is the capital one, of destroying insects, grubs, and weeds. These are apt to abound most in the richer soils ; no reason for abstaining from this husbandry on such.

I venture therefore to conclude, that paring and burning, with a proper course of crops, is safe on any * soil ; and essentially necessary on some, as I shall presently shew.

Clay.—The gentlemen who have levelled their theories against paring and burning, have not given many reasons peculiarly appropriate to this soil. The only one that merits the least attention, is the assertion, that it converts what is properly soil, into pieces of infertile brick. The fact is not so, for every one that ever burnt clay for manure,

---

* Very rich ones will do exceedingly well in many cases without it. This distinction, therefore, should be made. There is, in some cases, no *necessity* for it : there *is* in others.

knows,

knows, that though there are many lumps of the substance which they allude to, yet, that the mass of the heaps consists of ashes, properly so called ; but when the tenacity of this soil, which is one of its greatest evils, is considered, it will be found, that bricks are an excellent addition to the soil, to loosen and open its stubborn adhesion.   I have seen and examined carefully heaps of clay-ashes, amounting to many hundreds of loads, that have been burnt and applied to great profit on this soil. By paring and burning, you have therefore on it the common manure found in vegetable ashes, and you have n addition a substance which acts mechanically.   Hitt, who wrote from practice, and whose writings abound with many just observations, remarks : " I recommend burning of the surface as the cheapest manure, and most effectual of any ; for it not only adds salts to the soil, which the burning of grass-roots produces, but it opens part of the stratum of clay next the soil so much, that the roots of vegetables can afterwards feed therein, for when the turf of a piece of land has been burnt in heaps, at four or five yards apart, though all the ashes be taken away, with some of the earth, and spread over the other parts of the land, yet neither corn nor turnips will grow so vigorously there, as on those places that were only opened by heat."

Loam.—This is the soil, especially when good, upon which the practice has been most condemned ; but here we have some experiments to recur to, which, in my estimation, set the matter in so clear a light,

a light, that nothing more is necessary than to recite them very shortly. Mr. Wilkes, of Measham,
in Derbyshire, has for many years been in the
practice of ploughing old rough pastures (the soil
a stiffish loam), eight or nine inches deep, and
burning the whole furrow in heaps of thirty or
forty bushels each, the fires lighted by a few coals,
and coal *slack* ; the effect was very great, and the
improvement immense and durable. Mr. Wilkes
is of opinion, from the experience of many years,
that even this burning, which is twenty times the
depth of common paring, does not waste the soil
in the least, but does no more than break the texture of stiff soils, expelling a great quantity of water ; that by exposition to the atmosphere, the land
re-absorbs its water, and by the great immediate
fertility, fills itself presently with more vegetable
particles than it had before. Thirty years ago, his
father burnt, at Overseal, exactly in the manner described, a field of ten acres, which was not then, and
has not since been treated with any more favour
than the fields adjoining, yet it has ever since retained a superiority.

The writer of this Calendar, in 1790 hollow-
drained an old grass-field of four acres and a half,
of cold, wet, poor loam, on a clay marle bottom ;
the rent 9s. an acre, and not worth more in its
then state, perhaps, than 7s. In 1791 he ploughed four acres of it four inches deep, which was the
whole depth of the *soil*, or surface, of different colour from the stratum beneath, between that sur-

N   face

face and the clay marle, and burnt the whole furrow of the part so ploughed. Having no coal-
slack, and wood being dear, he made but four
heaps in the field; the consequence was, the heat
and degree of calcination were far beyond what is
ever practised in common, and many persons who
knew and approved of paring and burning in the
common way, pronounced the field completely
ruined. The ashes were spread, and ploughed in
with a shallow furrow, and turnip-seed sown, and
very slightly bush-harrowed. The crop was very
fine, worth, to sell for feeding on the land, at least
50s. an acre. The crop on the burnt part double
to that of the half acre.

After feeding them with sheep, the land was
ploughed thrice, and sown with oats and grasses.
The oats produced above 7 quarters an acre, and
the grass has ever since been much better worth
20s. an acre, than it was worth 5s. before. The
oats on the half-acre were not threshed separately,
but judged by those who viewed them, to be much
inferior to the rest. About half the field has been
since dressed with earth and road sullage, and once
dunged slightly. It is remarkable, that in three
years crested dog's tail, an excellent grass, common
in the country, the seed heavy, and which, therefore, could not be carried by the wind, began to
appear, and has been increasing ever since. There
is at present no perceptible difference between the
part burnt, and the other not burnt; if any thing,
the burnt is best. These two experiments prove,
                                                    if

if any thing can, that paring and burning does
not lessen the soil, in its most excessive application, and that it works a very great improvement
on loams.

Sand.—Hitt, a practiser of this husbandry, says,
that it improves sandy soils as much as any other ;
and I have seen some fields thus worked in Suffolk and in Cambridgeshire, and improved by it,
though under a course of crops by no means admissible.   There is not the least reason, from ana-
logy, to doubt the effect on this, or indeed on any
soil.

Chalk.—Here we have a much more ample field
of experience, for it has been, and is the common
method of breaking up downs in every part of England.   On the Cotteswold hills, in Gloucestershire,
it is the common husbandry, and often repeated.
The sheep-walks and warrens on the Wolds of the
East Riding of York, and of Lincoln, have thus
been brought most profitably into culture, though
not with the attention in cropping that ought to
have been given.   In Hampshire and Wilts, the
same husbandry prevails.   In these counties I have
been shewn lands that have been pronounced
ruined by this husbandry.   The cropping was bad,
but still the rent had been doubled by the practice *.   In Kent, Mr. Boys shall speak for him-

---

* In the West Riding, Colonel St. Leger remarks, that if
burning wasted the soil, his lime-stone lands, only four inches
deep, would have been gone long ago, as it had been pared and
burnt for ages.—*Eastern Tour.*

self.

self. In a letter to the editor of a paper, he says, " If any persons who condemn paring and burning, should come into Kent this summer (1795), I can shew them several scores of acres of wheat, barley, oats, and sainfoin, now growing on land which has several times undergone that operation. The crops of sufficient value to buy the land at more than forty years purchase, at a fairly estimated rent before the improvement." I humbly presume, that Messrs. Kent, Claridge, and Pearce, the great enemies of paring and burning, will not pronounce this land *ruined* by that *execrable practice*.

Peat.—This article is dispatched in few words. Whatever variety of sentiments there are on this method, for other soils, here there can be none. The universal practice, from the flat fens of Cambridge to the swelling bogs of Ireland, the mountainous moors of the north of England, the rough sedgy bottoms, in almost every part of the kingdom, when they are broken up by men of real practice and observation, are always done by paring and burning. Registered experiments of doing it by fallowing, are to be met with in various works. The Board's Reports of the North Riding of York, and of Somerset, detail some ; others are to be found in my Tours, and the result is either loss, or a profit so very inferior, that the question ought to be considered as settled and done with. Let it sleep for ever, except for the wrong-headed individuals who will, upon every question, arise in

every

every age, to contradict the common sense of mankind. I could detail cases without end, but really think it would be a loss of time to read them.

PARE AND BURN FEN AND BOG.

In the fens of Cambridgeshire, upon a peat soil, free from large roots and stones, the work of paring is always done with a plough, which they make on purpose for the work, and which executes it in the completest manner that can be imagined. It turns off a furrow, from 12 to 16, and even 18 inches broad, and not more than an inch deep. The use of this admirable tool brings down the whole expence of paring, burning, and spreading the ashes, to 9 s. or 10 s. per acre. But upon those soils, when they have not been in a state of cultivation, such a plough would not work.

PARE AND BURN OLD MEADOWS AND PASTURES.

These are done with the breast-plough, as it is called, which is pushed on by strength of body, the thighs being armed with wooden guards. It is hard work, and now commonly paid for, including burning and spreading, from 25 s. to 40 s. per acre. An inch, or an inch and a half, is the common depth ; but some farmers prefer two inches, for the sake of more ashes. The thinner it is pared, the more certain the burning, should the weather prove unfavourable.

PARE AND BURN HEATHS AND DOWNS.

Considerable tracts of this land, on a weak, thin, loamy sand, with a calcareous bottom, have, within the last five years, been thus broken upon Newmarket

Heath,

Heath, which was done at the expence of 36s. per acre, and immense crops the consequence, but in a very bad course, which will by-and-by raise enemies there to this husbandry, though most unjustly.

PARE AND BURN MOORS, &c.

The moors and mountains of the north of England, Wales, Devonshire, &c. when broken up for cultivation, are often, and ought always to be reduced by this husbandry. It has long been common husbandry in those countries, and is, therefore, done cheaper, from 24s. to 30s. an acre. Draining should precede it.

PARE AND BURN OLD SAINFOIN.

If a farmer has any old sainfoin layers that are worn out, and which he means to break up, let him determine to do it in no other method than this. If done by mere ploughing, the chances are much against success, by reason of the red-worm, which is very apt to abound in these layers, to the destruction of any white corn that can be sown. I have known three successive crops destroyed; but to pare and burn for turnips is the safest husbandry.

General Remarks.—In these several cases, there are some points of management which should be equally attended to in all. The heaps should not be made large, twelve or fifteen bushels of ashes are large enough; twenty may be admitted, but, if much larger, the turfs will be too much burned. This must, however, depend in
some

some measure on the weather, for the worse that is, the larger the heaps must be. It will also depend on the thickness of paring. Thin flags will burn in smaller heaps than thick ones. When the ashes are spread, and the sooner that is done the better, the land is to be ploughed thinly, which is the general practice, lest the ashes be too much buried. Upon high moors, in undertaking new improvements, it may be of singular importance to gain straw. In that case, to sow oats upon the first burnt lands, may be admitted, but in general, it is much more correct to leave the ashes ploughed in for sowing turnips, upon all the land burnt in March, April, or May. This preparation is unexceptionable for potatoes, so that if this root be wanted, it may be planted in April on the land burnt in March.

LIME.

Liming is, in many districts, connected with paring and burning, and it is one of the best methods of applying this manure. From a peck to a bushel, according to its plenty, is added to and mixed with every heap of ashes, and they are then spread together. The effect generally is considerable, but proportioned to the soil. The greatest effect of this manure is upon land that has been long in a state of nature, and particularly upon all peat soils, moors, mountains, and bogs. But upon all on which it is known to have effect, it is well applied in the ashes of paring and burning. March is a proper season for liming in other cases, and he

who has great tracts of land to improve, should be
careful that his kilns continue working throughout
the month.

CART DUNG.

The only crops for which dung from yards and
composts should now be carted, are potatoes and
cabbages.   The first to be planted, and the other
to be drilled, where they are to remain, in April.
The time of doing it, whether now, or in that
month, will depend on the pressure of other business ; but the young farmer must remember that
the work is to be done.   Long fresh dung, from
the yards, will do for either purpose, probably better than any other.   We may suppose that the
land now lies in ridges of that size, whether of
three or of four feet breadth, on which the crop is
to be put in.   The dung is to be laid in the furrows, from 25 to 35 cubical yards per acre.   If
very long and strawy, 40 yards, and the ridges being reversed, it lies ready to receive the seed.

HOPS.

This is the season to plant hop-cuttings, a branch
of agriculture which, if treated much in detail, would
fill such a volume as this work forms.   I shall, in
the seasons of the respective operations, remind the
farmer of the work that should be in his recollection.   No beginner in husbandry would introduce
so operose an article of culture from any book.   He
would fix, for some time, in a hop district, to make
himself acquainted with it, and procure an able,
sensible workman, long habituated to it, to assist
                                                      him

him in the undertaking. But as the article is here
named, for the first time, a few cautions will not
be improper, upon the general question of introducing the culture, where it is nearly or wholly
unknown. Whoever has any thought of such an
introduction, should duly consider several circumstances. He should of all others remember, that
hops demand a greater acreable capital by far, than
any other branch of cultivation. To form a new
plantation, and go through the first year, will cost
from 80l. to 90l. per acre, and the annual expence
afterwards will vary according to circumstances
from 30l. to 40l. per acre. Next he is to bear in
mind, that a small insulated hop-ground does not
usually flourish so well as a large one, probably for
want of shelter. He is also to reflect on the great
and constant demand for manure, which his situation may not enable him to command, without very
much impoverishing the rest of his farm. He
should reside in a populous neighbourhood, for the
plentiful supply of hands at certain seasons of the
year. And lastly, he is not to forget, that after all
his expence, attention, and exertions, it is one of
the most uncertain and precarious crops that any
man can adopt. These circumstances certainly demand his serious attention, however a plenty of
hop-poles may tempt him, in some spots, to the
speculation. I would advise a young farmer to
think many times before he determines to undertake it.

There is one situation in which it may be prudent.

dent.   He who possesses a bog, especially a flat
deep bog in a sheltered spot, and yet not too confined, may very profitably convert it into a hop-
ground.   A solid, weighty peat-bog makes an excellent hop soil, when laid into beds, by transverse
trenches.   Such land is a natural dunghill, and will
demand such manures as may perhaps be easily procured.   Here the chances are favourable.

If such a spot be not chosen, the best preparation of the land for hops, is two successive crops
of turnips or cabbages, both fed on the land by
sheep, and off early enough for ploughing and
planting the land in March.   They may be planted
in rows at eight feet asunder, and six feet from hill
to hill, which will give full space for all the requisite operations.   Three, four, or five fresh cuttings are planted in each hill, or spot which is to
form a hill.

In this month old plantations are dressed, the
hills opened, the roots pruned, and mould or compost returned.   The time of poling depends on the
shooting of the plants.

PICK STONES.

In a dry season, an opportunity should be taken
this month, to stone pick the grass and clover fields
intended for mowing.   In this work no stones
are to be taken but such as would impede the
scythe.   The pickers, who generally like this work,
will over-pick, if they are not attended to, and they
will propose to pick other fields which are not to be
mown ; but on no account is this to be permitted,
if

if the stones be not much wanted. It has been often remarked, and is a known fact, that too much stone picking has done a very sensible mischief, in many cases where picked by authority of parliament for turnpike roads. But Mr. Macro, of Suffolk, ascertained it experimentally. "Having often thought that picking the stones off my turnip lands did more hurt than good, I tried an experiment last spring, by gathering up all the stones of one square rod, after the turnips were folded off, and laying them equally over another square rod by the side of it, then sowed them with barley, and marked them out, and, at harvest time, collected them separately, as likewise another square rod by the side of them, which had only the natural quantity of stones.

	Qts.	Pints.			C.	B.	P
Produce from the rod that had the double quantity of stones,	6	1	or, per acre,		8	0	2
Ditto, from that where the stones were gathered off,	6	0	ditto,		7	2	0
Ditto, from that in its natural state,	6	$\frac{1}{2}$	ditto,		7	3	1*

"From this single experiment the result is in favour of the largest quantity of stones; and I verily believe it is quite wrong, after the sheep have

---

* This result coincides with various observations that have been made in several of our counties, particularly Hertfordshire, and also in France. The lesson it suggests deserves no slight attention.—*A. Y.*

trod

trod out a great quantity of stones, in feeding off turnips, to have them raked up clean, which I have known some farmers do, nor can the rake be used without taking some of the tathe, or dung, with them."

FEED NEW LAYS.

Land laid down last August, or the preceding spring with corn, should not have had a hoof in them through all the last autumn and winter. They will now present to the eye a beautiful fleece of young grass, of much value for sheep, and they are to be well stocked and kept down at present, and through all the following summer, by this stock only. Nothing is more pernicious than mowing a new lay, as directed by certain authors. They may have succeeded *in spite* of such bad management, but never *by* it.

SEED OATS.

Mr. Walker, near Belvoir Castle, Lincolnshire, sows eight bushels of oats per acre, and finds the crop much better, and the sample more equal than with less seed : the oats are less *taily*, no tillers to give different degrees of ripeness, and the crop ready to cut four or five days sooner than with thinner sowing. Mr. Ducket is of the same opinion, and holds no idea cheaper than that of recommending the drill husbandry as saving seed ; he drills five bushels of oats per acre.

APRIL.

APRIL.

BARLEY.

THE barley crops not sown in March, should be in the ground by the middle of this month. The land I suppose to lie as thrown up in the autumn before; so that whenever sown, it is (on the old ploughing system) on the spring earth. This supposition is necessary; because, if there had been previous ploughings in March, or in the end of February, the seed should have been sown then; excepting, however, turnip-land, that broke up at first too rough to be sown, which will sometimes happen. The farmers, in some parts of the kingdom, will put off their sowing till the last week in this month, and the first or second of May, for the sake of gaining time for giving three spring earths; but they lose more by far from late sowing, than they gain by making their land fine. If clover is a principal object, and they had not the land fine enough before, delays must be made; but if so, that can scarcely be owing to any thing but bad husbandry : for such events should be had in view, and the tillage given before winter, on lands not cropped with plants that stand till the spring. The utmost exertions of good husbandry should be made to reconcile jarring circumstances, when they cannot be totally prevented.

But

But in the modern system of avoiding spring
ploughings, with a care proportioned to the heavi-
ness of the soil, the main reliance is on frosts for
pulverization, and the object is to keep the surface
so gained, for the seed to be deposited in it. If
the weather was unfavourable for sowing in March,
or, being favourable, the breadth was too great to
allow the operation to be finished, and if weeds
appeared in the lands laid up for barley, it is to be
supposed that they were of course destroyed by the
scufflers; and this month the sowing must be
finished, whether broad-cast or by drilling.   In the
latter case, the directions relative to the right
breadth of the stitches should have been very attentively executed.   The young farmer must have
it carefully in memory, that as the summer ap-
proaches, with hot suns at intervals, any degree
of poaching, or daubing or trampling, becomes
more and more fatal, for the sun binds whatever
earth was touched in too wet a state.   This caution has little to do with the occupiers of sand,
much of which wants adhesion to be given it by
art ; but here, again, if such land has been amply
clayed, it will sometimes be apt to *set*, to bind with
heavy rains, so that the temper of it should always
be examined before the teams at this season are
permitted to go on it.

PEASE,

Should always be put in before this season, and
therefore directions are omitted here ; but if, from
some peculiar circumstances, the farmer wishes

now

now to sow a field, I need only remark, that they may still be sown, with the expectation of a full crop ; but it may prove too late to have good turnips after them

WHITE OATS.

It is the custom in Hertfordshire to sow barley before white oats. Wherever such maxims prevail, this will be the principal month for sowing oats. All the precautions that have been given with respect to barley, are equally applicable to this crop.

BUCK-WHEAT.

The lands designed for buck-wheat, in May or in June, should be well tilled this month, ploughed and harrowed well at least once. It is not necessary for that grain, but for the grasses which should be sown with it, and for the important object of making all the seed-weeds grow, in order to kill them by the following tillage. This April preparation marks the land for buck-wheat. I shall therefore take this opportunity to advise the farmers in general to try this crop. Nineteen parishes out of twenty, through the kingdom, know it only by name. It has numerous excellencies, perhaps as many to good farmers, as any other grain or pulse in use. It is of an enriching nature, having the quality of preparing for wheat, or any other crop. One bushel sows an acre of land well, which is but a fourth of the expence of seed barley. It should not be sown till the end of May. This is important, for it gives time in the spring to kill
all

all the seed-weeds in the ground, and brings no disagreeable necessity from bad weather in March or April, to sow barley, &c. so late as to hazard the crop.   It is as valuable as barley.   Where it is known, it sells at the same price, and, for fatting hogs and poultry, it equals it.   It is, further, the best of all crops for sowing grass-seeds with, giving them the same shelter as barley or oats, without robbing.

BEANS FOR EATING GREEN.

If Windsor-beans are cultivated with this intention, a portion of land should now be planted with them.

LETTUCES FOR HOGS.

If the stock of swine be large, it is proper to drill half an acre or an acre of lettuce this month. The land should have been *well* manured and ploughed before the Christmas frosts, into stitches of the size that suits the drill-machine.   It should also have been scuffled in February, and again in March, and well harrowed ; and this repeated before drilling.   The rows should be equi-distant, one foot asunder.

The crop which was drilled in March (a succession being essentially necessary) should now be thinned in the rows, by hand, to about nine or ten inches asunder.   If this necessary attention be neglected, the plants draw themselves up weak and poor, and will not recover it.   Women do this business as well as men.   When about six inches high, they should be horse-hoed with a scarifier

or

or scuffler, with the hoe about four inches, or at most five, wide.

SPRING TARES.

If the young farmer depends in any degree upon a succession of tares, he ought to have two sowings, one at the beginning and the other towards the end of April. To have these crops to follow one another in consumption, is a very material object. Two bushels and an half of seed per acre are a proper quantity.

SPRING WHEAT.

I cannot in general recommend the culture of this grain, for barley or oats commonly pay better, but as certain circumstances may render it very profitable, it is proper for every farmer to have the object in his mind. Mr. Marshall has a useful minute on it.

" Spring wheat *(triticum æstivum)* is here cultivated, and with singular success; owing principally to the time of sowing : the wane of April !

" This proves that it is a species widely distinct in its nature from the winter wheats. In the practice of a superior manager (Mr. Paget, of Ibstock), it was discovered, that by sowing early, as the beginning of March, the grain was liable to be shrivelled, and the straw to be blighted, while that sown late, as the middle or latter end of April, or even the beginning of May, produced clean plump corn ; effects directly opposite to those of winter wheat.'

O

STEEPING SPRING CORN SEED.

In case of an unfavourable season, by reason of a drought to an uncommon degree, it should be recollected, that steeping barley and oats, &c. has been tried with success. It is a practice rarely necessary, and mentioned here only as a hint, which a farmer may apply now and then to advantage.

MADDER.

This is the principal season for planting madder. I suppose the deep tillage to have been performed in October, and the land to have been thoroughly water-furrowed. Early in this month it should be ploughed again, and harrowed fine ; and towards the latter end of it another earth should be given, and the land harrowed again. It is then in order for being planted.

Great attention must, however, be paid to this tillage being all given in dry weather. If the soil is moist, or apt to bind, scarifying and scuffling will be superior to ploughing.

The sets are to be slipped from an old plantation. When they are about two inches above the ground is the proper size, and they should be slipped off as much below the surface as possible, because they will then have the better chance of growing ; and as fast as they are taken up, they are to be thrown into tubs of water. Other hands are to be employed in planting ; in which work the cultivator of this crop must be ruled by the method of disposing the beds. That which was practised when (more than 30 years ago) the Society
offered

offered premiums, is equally-distant rows, two feet asunder, the land flat. If, for laying the land dry in winter, ridge-work is preferred, only one row, of three feet, can be set on each. On four-feet ridges, two rows at nine inches or one foot may be planted. The planting should be performed with care. Women or children should drop the sets, and men follow to plant them. In this month there can be no danger of their not growing, especially if the land is in as good tilth as it ought. Watering will scarcely ever be necessary.

Let the young farmer, however, remember, that the culture of these plants, applicable only to the use of manufactures, and which are also largely imported from abroad, is rarely advisable. I was a madder planter once, and lost by every acre I planted. A man may plant in the moment of a high price, and take up his crop, three years after, at a low one. All such speculations are too hazardous; nor was there even a fair open competition among the purchasers.

Those who have cultivated madder with the success boasted by the writers of husbandry, should not hold these observations in contempt. There appears to me almost as much use in mentioning trials that were unsuccessful, as in those that are ever so profitable : for it is certainly of as much consequence to tell one man that his soil *will not do* for madder, as to assure another that his *will do*. Instead of an acre or two, I might possibly have launched (like many others) into 10 or 15 acres, in which case the loss would have been no trifle. And it surely is highly incumbent on every one, to make known to the world such of his experience as will probably be of any use to it. Bad success of several persons in a culture, is too apt to prejudice others *in general* against it. However irrational, still it is so, and it ought to be a caution not to recommend any thing in general, under the extravagant notion, that because an article of culture is profitable on one soil, it must be the same on very different ones. But the grand obstacle to the culture of madder, is the difficulty of sale : for while a man has not a fair market for his unmanufactured madder, none can with any prudence engage in it, unless on so large a scale as to admit the whole apparatus of reducing it to such a state, as to be absolutely a marketable commodity. In answer to this, it may be said, that madder really dry is a *marketable commodity.* But this matters not, if the purchaser has it in his power to be a knave : he has a pretence, a screen always at hand, that will cloak the greatest knavery, and to a degree known in no other branch of agriculture. Among the gentlemen of trade who have a mutual understanding and confidence, such objections appear trivial ; but to the cultivator, at a distance from the market, it is a different affair. He writes to a madder-merchant to know the price. The answer is, *four pounds an hundred weight.* Up he sends his madder, and instead of 4l. receives but 3l. not from a variation in *price,* but in *weight.*

It

It may be said, that the correspondent in London may be in the right. Very true; but will the countryman believe it? He thinks himself right, and has no other proof that he is not so, but the interested assertion of the man who buys it. Is it not evident, that in such a case the cultivator will be disgusted, and throw aside a business in which he knows neither the market weight nor the market price. If encouragement is designed to this culture from any quarter, it should not be exclusive of this circumstance : manufactories should be erected and established, in which the madder could be prepared for any one at so much an hundred weight, and that by persons not the least concerned in purchasing. Then the cultivator would have a commodity in his hands which he could sell in as simple and fair a way as any other. If nothing of this sort can be effected, all encouragement should be for such a number of acres (and no less) as will answer the expence of a private manu facture, which would prevent persons being unguardedly drawn in, by premiums apparently considerable, to cultivate a root which, when raised, is in its sale absolutely at the mercy of the purchaser.

I am informed, that at present (1803) the largest quantity of madder used in our manufactures, is used without being powdered, as formerly, and that it is saleable with common drying, without stovework ; but that *common degree* is open to much uncertainty, so that the preceding remarks are not

done

done away. The price of 4l. per cwt. marks a considerable desiccation.

LIQUORICE.

The liquorice culture is generally carried on more completely than that of madder, which is owing to the nature of the root. Madder spreads its roots horizontally more than perpendicularly; so that good tillage, and plenty of food on the sides of the bed, make amends for depth; but this is very different with liquorice, whose root is a single tap one; so that the whole crop depends upon the *depth* of cultivation. Hence we find, that the planters dig the land four feet deep. This appears vastly expensive, but it is greatly lowered by always planting on the same land, so that one digging does for taking up one crop and planting another, a saving that renders this culture preferable to ploughing. The perpendicular growth of the crop also makes it necessary to plant the sets much nearer than madder ones. For instance: double or treble rows, at nine inches, with two feet spaces for horse-hoeing.

TEASELS,

Or the Fuller's Thistle. They are sown in April, two pecks an acre. The young farmer, should he wish to make trial of this branch of the art, may consult Mr. Billingsley's account of it in the Somersetshire Report.

LUCERNE.

This is the right season for sowing lucerne, which

which must now be considered under the several
heads.   1. Of the utility of the crop, which should
induce a young farmer to enter freely on the culture.   2. The soil to be chosen.   3. The preparation.   4. The mode of sowing.   5. The quantity
of seed.   6. With or without corn.

1. It is an object of such consequence to those
who have a proper soil for it, that it does not admit of doubt or question ; but, if a beginner in
husbandry has apprehensions, let him mount his
horse, and travel near the coast of the Thames and
sea, from Dartford to the limits of Hampshire,
and he will hear of lucerne all the way, and see
much of it.   The principal, and probably the best
use, to which it can be applied, is that of soiling
all the horses on the farm ; next, to soiling cows,
young cattle, and fatting beasts, soiling hogs, and,
lastly, making hay.   These are objects of such consequence, that they ought to be provided for ; the
last, however, in proportion to the meadow-grounds
of a farm, and to the sainfoin which may be on it.
Where these abound, lucerne for hay is the less
necessary.   The importance of a general system of
soiling can never be impressed too frequently.
The repetition and influence of the benefit pervades
every crop on the farm.   Inasmuch as dung is important, this practice is so.   Dung, without it, is
made during half the year ; with it, through the
whole, and he only who knows the immense consequence of raising dung, can duly appreciate the
necessity of soiling.

2. The

2. The soils that suit lucerne, are all those that are at once dry and rich. If they possess these two criteria, there is no fear but they will produce large crops of lucerne. A friable deep sandy loam on a chalk or white dry marley bottom, is excellent for it. Deep putrid sands, warp on a dry basis, good sandy loam on chalk, dry marle or gravel, all do well ; and, in a word, all soils that are good enough for wheat, and dry enough for turnips to be fed on the land, do well for lucerne. If deficient in fertility, they may be made up by manuring, but I never yet met with any land too rich for it.

3. The best preparation for this, as for all other grasses, is two successive crops of turnip or cabbage, both fed on the land, and the last before the sharp frosts are over. This management frees from all weeds better than any other, and at the same time greatly enriches. Upon land previously clean, one of these crops may do well enough ; but let not a farmer ever venture lucerne upon land that by some method, whatever it may be, is not rendered quite clean.

4. In regard to the mode of sowing, the greatest success by far that has been known is, by the broad-cast method, which is nearly universal among the best lucerne farmers, even among men who admire and practice the drill husbandry in many other articles. But as they mostly (not all) depend on severe harrowing, for keeping their crops clean, which is a troublesome and expensive operation,

operation, I shall venture to recommend drilling, but very different drilling from that which has been almost universally practised, viz. at distances of 18 inches or two feet. Objections to these wide intervals are numerous. If kept clean hoed, the lucerne licks up so much dirt, being beaten to the earth by rain, &c. that it is unwholesome, and the plants spread so into these spaces, that it must be *reaped*, which is a great and useless expence. For these reasons, as well as for superiority of crop, I recommend drilling at nine inches, which, in point of produce, mowing, and freedom from dirt, is the same as broad-cast; and another advantage is, that it admits a scarifying once a year, which is much more powerful and effective than any harrowing. These facts are sufficient to weigh so much with any reasonable man, as to induce him to adopt this mode of drilling, as nearer to broad-cast by far than it is to drills at 18 or 24 inches, which open to a quite different system, and a set of very different evils. Nine inch rows might, *practically*, but not literally, be considered as broad cast, but with the power of scarifying.

5. In regard to seed for nine inch drilling, from 12 to 15lb. is to be recommended.

6. The material point, of with or without corn, remains to be spoken of; and here two considerations present themselves. One is, the extreme liability of lucerne to be eaten by the fly, which does great mischief to many crops, when very young, and against which the growing corn is some protection.

tection. The value of the barley or oats is another
object, and not to be forgotten. It is also gained
in the first year's growth of the lucerne, which is
very poorly productive, even if no corn be sown,
so that I must own myself clearly an advocate for
drilling it among corn, either between the rows of
nine inch barley, or across drilled barley, at a foot;
perhaps the latter the best method, as there is less
probability of the crop being laid, to the damage of
the lucerne. The quantity of seed-corn should
also be small, proportioned to the richness of the
land; from one bushel to a bushel and half, according to the fertility of the soil; another security
against the mischief of *lodging*. If these precautions are taken, it would be presumptuous to say
that success must follow, *that* being always, and
in all things, in other hands than ours; seed may
prove bad, the fly may eat, and drought prevent
vegetation, but, barring such circumstances, the
young farmer may rest satisfied that he has done
what can be done; and if he does succeed, the advantage will be unquestionable.

SAINFOIN.

There are parts of this kingdom, in which the
farmers could not pay their rents without the use
of this grass. On dry lime-stones and chalky soils,
or on any land perfectly dry and sound, it will
thrive to extraordinary profit. It may be safely
sown in April. The land should be clean, and free
from weeds and the seeds of weeds; and this is the
principal circumstance to be attended to. It
should

should be sown with barley or oats, the land in fine
tilth, and the seed covered by harrowing when the land
is dry. It may also be drilled as in March. Upon the
soils proper for this grass, no man can sow too much
of it; for no other use of the land will pay nearly so
well. It will, on poor soils, not worth more than
from 2s. 6d. to 5s. per acre, yield a ton and a half,
and even two tons and a half of hay, or a ton at
the least, at one mowing per acre, and afford a
considerable after-grass besides. Now, the use of
hay is so universal, that such products can never
want a market; nor such land, thus improved, fail
of becoming a source of profit to whoever pursues
such a beneficial conduct. The products and profit of such land in tillage, or in a sheep-walk, are
quite inconsiderable, compared to what sainfoin
yields. The proper quantity of seed is four
bushels per acre. It flourishes so well broad-cast,
that there is no *necessity* to attempt it in the drill
method.

I have seen it cultivated, however, with great
success, drilled at nine inches across drilled barley,
on the farms of Mr. Overman and Mr. Coke, in
Norfolk.

BURNET.

This is a proper season for sowing burnet; and
the best method of cultivating it is, to sow about a
bushel per acre, with either barley or oats, and to
cover it at two harrowings. It flourishes extremely
well on most soils; but it yields a produce proportioned to the goodness of the land, though it will
do

do on those which are very indifferent.   The great
use of it is for spring feed for sheep.   If left of a
proper height in the autumn, it will improve
through the winter, notwithstanding frost, and be
ready early in the spring.   This is a great excel-
lency, in which it is rivalled by no other grass.
Burnet does well mixed with ray-grass or cocksfoot : about three pecks of burnet, and one bushel
of ray-grass, or cocksfoot, to the acre.

SOW CHICORY.

This plant (most valuable for many purposes)
may be sown with any sort of spring corn all
through this month.   It may be put in either
broad-cast, sowing 12lb. of seed per acre, or
drilled at a foot distance, with. 9 or 10lb. over
broad-cast, or drilled corn.

LAYING DOWN FOR GRASS.

Laying arable land down to permanent grass, is
a work very rarely thought of by tenants.   I have
known it sometimes done on a piece near a farm-
yard, for convenience, but not often.   As this work
is designed equally for the use of both landlords
and tenants, it is necessary to mention all the practices commonly pursued by either.   I treat of the
preparation for it under this month, because the
spring is, with many, the favourite season for that
operation ; but, as I greatly prefer an August sowing of grass-seeds for this purpose, I shall be brief
at present, reserving my principal observations for
that month.   When sown with spring corn, it
should be with buck-wheat, barley, or oats.   Seeds
take

take better with buck-wheat than with any other crop, but April is not the best time for this plant. Should it prove *blind*, as the farmers call it, that is, blighted and abortive, from frosts, the crop may be mown for hay or for soiling. Seeds succeed, however, very well with barley or oats ; and the chief caution is, to prepare the land in such a manner as to render it perfectly clean. Two successive crops of turnips are the most effective way of securing this degree of cleanness. In the Calendar for August, this point will be further considered, and the proper seeds to be sown specified.

SHEEP.

This is the month that tries the farmer more than any other in the year. In the whole range of husbandry, there is no point that puzzles the farmers more, than providing for their flocks through March, April, and the first week of May. It proves the good husbandman as much as any other article in the most extended farm. The common management is to depend on turnips and hay ; and, when the former are done, to turn them into a piece of rye sown on purpose, or into the crops of wheat, to feed them off. These resources not being proportioned to the want, they let them run over the clover and pastures of the farm ; by which means the crops of hay, and pastures for large cat tle, are greatly damaged. Bad as such a system of management undoubtedly is, yet it is too often to be met with, and the bad consequences are felt so strongly, that the number of sheep on such farms,

is

is governed by the food in April. Few farms are
stocked properly with sheep throughout the whole
year, for want of more food at this season. But
there are some farmers, who have felt these inconveniences so strongly, that they have taken steps
to remedy them. They keep their turnips as long
as possible, so as to make their shoots an object of
sheep-food ; and every year they sow a piece of
clover and ray-grass on land in pretty good heart,
to be ready in the spring to take their flocks from
turnips, and keep them till the general turning to
grasses arrives. This conduct, I must observe, is
an improvement on the other, for it gets rid of
three great evils : depending on rye, which is soon
eaten ; feeding on wheat, which is pernicious to
the crop ; and turning too soon into the general
pastures. But, at the same time that it effects
this advantage, it is open to some objections, which
make further improvement necessary. Keeping
the turnips long in the spring is very bad husbandry. It damages greatly the barley crop, both
in robbing the land, and preventing it from being
sown in proper time : nor is the food of great
consequence ; for many acres of turnip-tops are
requisite, the number of which must be in proportion to the stock of sheep ; and, as to the roots,
they grow so sticky and hard after the tops are at
all advanced, that their value is trifling. With
respect to ray-grass, the clover mixed with it
is seldom above three inches high at this season ;
and a great breadth of ground to a given stock,
                                        must

must be assigned to keep the sheep through April.

The number of acres of that young growth necessary to keep an hundred sheep and lambs is surprizing : so that these farmers, although they manage to spring-feed more sheep than the worst of their brethren, yet effect it at a great expence, and at last not in any degree comparable to what might be done.

A turnip should never be seen on the ground after March. For the month of April, the farmer should have a field of cabbages ready, which, yielding a great produce on a small breadth of ground, reduces the evil of a late spring sowing ; and, if he manages as he ought, totally excludes it. The turnip-cabbage, and ruta baga, will last as long as wanted ; and, though it runs to seed, yet the bulb will not be sticky. The green boorcole may be fed off several times. It is impenetrable to frost, and will make shoots in the winter.

Another crop for feeding sheep in spring, which is of particular merit, is burnet. An acre of it, managed properly, will at this season yield much more food than an acre of clover and ray-grass. It should be four or five inches high in November, and left so through the winter. Burnet has the singular quality of maintaining its green leaves through the winter : so that, under deep snows, you find some luxuriance of vegetation. From November to February the crop will gain two or three inches in growth in the young leaves, and then be ready for sheep. It will be better in
March,

March, and if kept, ready in April not only for sheep, but for horses, cows, or any other stock.

SHEEP IN ROUEN.

But before all the preceding dependencies, may be reckoned kept after-grass on dry meadows and pastures. If a field of this rouen be seen at any distance, it appears most unpromising, being of the colour of very bad hay; but enter it, and turn aside this covering with your hands, and the young green growth is found five or six inches high, nursed up by the shelter and warmth of the autumnal growth. I have often shewn this to persons on my own farm, to their great surprize. The sheep eat both together, and it is found to agree with them remarkably, being, as it were, hay and grass in the same mouthful. I do not conceive that it is possible to keep a full stock of sheep so cheaply in April, by any other method as by this. Tolerable rouen will carry ten ewes an acre, with their lambs, through this whole month. Such rouen may be worth, in autumn, 10s. or 12s. an acre. In April it is worth 30s. or 40s. and if it be a backward season, a farmer that has it would not be tempted to sell it for much more.

SHEEP IN WATER MEADS.

The farmer who has a good breadth of these, may depend absolutely on them. He wants nothing else for sheep and lambs.

SELL FAT SHEEP AND BEASTS.

Markets for beef and mutton are usually as high towards the end of April, as at any other moment

in

in the year. At that period, the supply can come
only from corn or cake-fed beasts, for not one
farmer in a thousand has then any winter green
food remaining. Any beasts really fat, are then
sure to sell well at Smithfield. With sheep, the
case is rather different; for spring food is now
come to him who is well provided, but not in such
plenty, on account of the number of bad managers,
as to lower the markets.

SMITHFIELD.

It is proper for a young farmer to consider well
the various ways by which he turns his fat stock
into money. The first and chief of these is Smithfield market. If he lives in a district divided into
small or middling sized farms, and where the
graziers are all or mostly in a regular system of
employing one or more district drovers, in whom
great confidence is placed, he is as safe as his
neighbours, and may not have reason for any particular caution. This is very much the case in East
Norfolk. If he occupies a very large farm, of
whatever kind, whether an arable-grazing one, as
in West Norfolk, or a grass-grazing one, as in
Lincolnshire, on a scale that enables him to send
many droves pretty regularly to his salesman, he
may safely trust to him. The common confidence
and integrity of trade then take place. But I am
sorry to observe, that I scarcely ever knew a man
send accidentally a lot of beasts or sheep to Smithfield, that got as fair a price for them as his great
neighbour, who was in constant dealing, got the

P   same

same day, or his little one, whose stock took the same chance through the means of a confidential drover. The man who thus drops in a lot, out of the regular course of his business, is rarely satisfied with the treatment he receives. There must be a great deal of truth in this remark, because it has been made to me from so many different quarters, and I have suffered in this way myself.

Let the young grazier, therefore, consider the circumstance well, try the country butchers, and feel his way through the difficulty, if his farm be of that size, and in that situation, which lays him open to its influence. The possession of an engine for weighing beasts alive will be extremely valuable to him ; for, by comparing the live with the dead weight, when the beasts are killed in the country, he will soon be convinced of the truth of the many comparative accounts of live and dead weight, which are published in the Annals of Agriculture, and from which he will be able to ascertain correctly the dead weight of any common sized bullock of which he knows the live weight. He may also compare the result when the live weight is taken from Renton's measurement. As to sheep, calves, and hogs, weighing is done with the utmost ease ; for a cage with a door at each end, and a large pair of steelyards, form the whole apparatus necessary. He should never fatten any animal whatever, without regular weighings, by which means he knows how his stock (whatever it may be) thrives, what changes it is requisite to make in their food,

food, and when to sell, if markets suit. All these are very material points, and he will have much satisfaction in being at any time able to ascertain them. Old and very experienced graziers can do without these helps, but they often suffer for want of them. To young ones they are essential. But let a grazier be as experienced as he may in buying and selling, and judging by the hand and eye, the butcher will beat him, from having been able to bring the live to the test of the dead weight, in such a variety of cases, that his knowledge is perfect. The grazier cannot equal him, but his nearest approximation will be by means of carefully weighing.

COWS.

It is no great object to a good farmer to get his cows out of the farm-yard this month, if he has a provision of ruta baga and chaff, as he ought. He must be very amply provided with grasses, indeed, to do it to good purpose, as his flock of sheep must be the first object for spring food. Besides, the raising of great quantities of manure in the farm-yard, is so important an object, that he should keep it in sight as long as possible. Turning out any cattle, before there is a good bite for them, is unprofitable; for a field so begun will not last proportionably with another of a proper growth. The milch cows should have their bellies full of roots and cut straw throughout this month, and be always kept well littered both in the yard and in the house.

HORSES.

HORSES.

The horses ought to be kept in the stable throughout this month, and to have plenty of litter, that they may continue to raise much dung. This is so busy a time, that a close eye should be had to the work that the teams perform, as one day now is worth two by-and-by. The directions laid down for last month, on that head, are to be followed still.

FEED TEAMS ON CARROTS.

Throughout this month the teams should depend on carrots, which are now in that dry withered state, in which their use is incomparably valuable. They are more hearty and nourishing to horses than any other food. Each horse may have two bushels a day, which will be about the quantity they would eat were there no limitation of allowance.

OXEN.

The ox teams being kept to pretty sharp work at this season, should be well fed with good hay, straw, cut chaff, and a daily allowance of roots. If they are large beasts, they should have fifty pounds of cabbages each, every day. This is a use, among many others, that will be found to shew the great consequence of having plenty of roots.

HOGS.

The fattening swine, sows, pigs, and lean hogs, require good attendance. There being nothing yet for them in the fields, they must be kept close to the farm-yards, where the threshers (who should be be kept at work quite through this month) will partly supply them with food, and the wash cisterns and winter stores of carrots, parsnips, potatoes; &c. will keep them in good heart.

POTATOES.

In the latter end of this month, the land to be planted early with potatoes should all be hand-hoed over the whole surface, to cut up weeds clean, and loosen the earth. This management is known only in the neighbourhood of London, but it should be extended over the whole kingdom, for the excellence of it is indisputable. The expence of hoeing, when there is a clear space to cut, is trifling, and the succeeding cleaning which the potatoes receive after they are up, is performed at a much less expence on account of this operation, and at the same time in a more effectual manner.

But the cheapest and most effective method of performing this necessary operation is by a large shim, which cuts three or four feet of surface. For this purpose, there should be a small broad wheel at each end of the beam, to regulate the depth. The work is confined to the surface, the intention of it being merely to cut up weeds and to loosen the earth, which rain and succeeding sunshine may have encrusted. The operation is of great importance, and will lessen the expence of the following hoeings.

BREADTH-PLANTED.

Before a farmer determines what breadth or number of acres he will plant with potatoes, he

should consider several circumstances ; as the number of acres of carrots he has sown ; for if his soil
be suitable to that crop, they are greatly to be
preferred to this root, being cheaper, not requiring dung, and being applicable to all the uses to
which potatoes are applicable.   They do not at all
impoverish the land, whereas potatoes scourge it,
if the expression be permitted, more than any other
crop the farmer puts in.   These are very material
motives to influence a preference.   But if the soil
will not suit carrots, then it will be necessary to
plant so much the more potatoes.   The same observation may be applied to cabbages, which also,
in a great measure, answer the purposes of potatoes.   If he deals largely in that crop, it lessens
the necessity of having this root ; and ruta baga is
as useful to hogs as the potatoe itself ; but being
far more uncertain, and the difficulty of securing
a crop of it being greater, it cannot be depended
on, like potatoes.   The fly and drought, &c. are
so fatal to it, that many farmers in Norfolk have
sowed in vain for several years together.

PLANT POTATOES.

The end of April is the best season for planting
potatoes, but it ought to be regulated by the
finishing of other work, because this should be the
last of the great spring operations of planting or
sowing.   When all others are done, then is the
time to begin this.   It will, some years, be in
May : and I know several potatoe planters of
great experience, and on an extensive scale, that
                                                    prefer

prefer May to April for this work. That opinion, however, is far from general.

POTATOE CUTTINGS.

The first operation is that of cutting; slicing off the *eyes* of the potatoes, in which a good deal of attention is to be used; first, chusing from the potatoe heap only large and fair roots, rejecting all small ones, which should be thrown by for hogs, &c. There should be but one eye to a slice, but rather than have the slice a very small bit, two may be left in it, for sizeable slices are better, especially if a drought succeed, than small ones, as the plant in this, as well as in many other cases, in its first germination derives its nourishment from the *set*. These attentions are not stated as *essential* points, but as circumstances which will, in certain cases, have a degree of influence, which render them worth some portion of thought; and, in the long run, he who attends closely to every part of the business, and to all minutiæ, will on the average of soils, seasons, and manuring, get the best crops. Some planters, who value themselves much on their skill in this culture, prefer having the cuttings ready some time before planting, as they think a moderate keeping in that state beneficial. This point does not seem to be at all essential.

In the scarcity, *scoops*, for scooping out the eyes in semi-globular cuttings were brought into use, to save the fleshy remains of the potatoes for common consumption. This practice was much condemned by some planters, and equally approved by others. From some experiments carefully made, the result of which I am well acquainted with, it appeared that these contradictory opinions might both be just, when founded on variations in practice. When the soil is sandy, or in a very light pulverized, or highly manured state, and every other requisite for success beneficially secured, these scooped cuttings succeeded just as well as larger sets; but when the soil was more stiff, unfavourable, in worse tilth, or not equally manured, or the sets ploughed in, under circumstances not very favourable, then the larger cuttings had a considerable superiority. The propriety, therefore, on any future occasion, of having recourse to this expedient, will depend on the state of the land, the soil, the manuring, &c. If the cutting be done by the bushel, 2d. is a fair price, where women's labour is 8d. a day.

SORTS OF POTATOES.

They are endless, and fresh sorts coming every day into notice, till they give way to others in succession. It would be easy to name many sorts, but quite useless. The ox-noble was, for some years, the most productive for cattle and hogs, but I have known it to decline of late. It is, however, still preferable for largeness of product. The early Scot gives two crops a year for the table, but deserves no attention from the farmer for livestock.

PREPARA-

PREPARATION FOR POTATOES.

The best of all preparations is that of paring and burning, and then planting in the furrow of the succeeding ploughing, which should not be more than four inches deep. If 10 or 12 loads of long dung be spread over the ashes, and both ploughed in together, with Ducket's skim-coulter, it will add greatly to the crop.

PLANTING.

They should be set in every other furrow, which will make them come up in rows at 18 inches asunder. I plant in the same way when a stubble is dunged for the crop, or previous tillage given. I have had great crops by ridging the land in bout ridges of 26 to 30 inches, dunging the furrows, laying the sets on the dung, and reversing the ridge by a bout of the plough. All these ways will give good crops; and probably the Ilford method of dibbling the sets in may be as good, or better than any other, but it is much more expensive.

SEED.

It takes from 25 to 30 bushels, according to the size of the sets, to plant an acre promiscuously dibbled at ten inches, and from eight to ten to plant every other furrow, at one foot from set to set.

CARROTS.

If the carrot-seed was sown very early (earlier than they ought to have been) the crop will be ready for the first hand-hoeing by the end of this month.

month. The rule is, to give it as soon as the young carrots can be distinguished from the surrounding weeds, and it should never be done in wet weather. The men must use four-inch hoes.

CABBAGES.

April is the season for planting the crop of autumn-sown cabbages. It is a work extremely easy to perform, and not at all expensive; but it is necessary to manage it in a judicious manner, so that it may be done to the best advantage. Just before planting, the land is to be ploughed from the ridges of the last earth. This earth should turn in the manure; then the ridges are to be harrowed, and one row of plants set along each ridge.

Women or boys should lead the way with the plants, and drop them, as nearly as they can, where they are to be planted : then the men follow with dibbles, and set them. The work goes off quickly, and is not expensive. Upon an average, it may be done in single rows, four feet asunder, for 4 s. labour being at 1 s. 6 d. a day an acre. It is proper to keep the men at work as long as they can see. The plants should be packed tight into baskets which are made for fitting into the carts.

DRILL CABBAGE-SEED.

The most certain and profitable culture of cabbages, is that of drilling them in April, where they are to remain. This system precludes the necessity of transplanting, which is at all seasons attended with some uncertainty, and in summer can

only

only be performed in, or immediately after rain, and in case of a drought, must be postponed till a good crop may be unattainable. They should, in the drill system, follow some hoeing or cleansing crop, such as turnip, a previous crop of cabbages, potatoes, tares, beans or pease, &c. I suppose the land, in the case of its having yielded turnip or cabbage, to have been ploughed the moment the produce was consumed, into such ridges as are intended for the cabbage-seed, either three or four feet wide. If any of the other crops preceded, this ploughing should have been given before the Christmas frosts. Into the furrows of these ridges the dung, 30 cubical yards an acre, in no case less than 20, should be laid in March, and the ridges reversed directly, covering up the manure and forming new ridges. They should then be left for 10, 12, or 16 days, to the influence of the atmosphere. In that state they lie sound and safe from rain. When it is intended to drill, harrowing should precede, or it may be omitted if the soil is very friable and in fine order, as the roller to which the drill is attached will level the crowns sufficiently, and they should not be reduced too much. The Northumberland drill is to be hung to a roller eight feet long for four feet ridges, or six feet long for three feet ones. Staples are in the frame of the roller for this purpose, and a chain hooks the drill to them. The roller covers the ridge drilling, and one in advance to be drilled by the next turn. So going on constantly, four pieces of a kitchen jack chain

chain about two feet long, attached to the drill, to
be drawn after it in the centre, will cover the seed
better than any other contrivance. The seed is deposited to the desired depth, by pressing on or
weighting the drill. If it be half an inch deep it
is sufficient. As soon as the plants appear distinctly above ground, if a surge of soot be drilled
upon them to the amount of 10 or 12 bushels an
acre, it is a great security against the fly. One
hopper and one round of Cook's cups, but larger,
fixed to such a frame as that of the Northumberland drill, will effect it simply and cheaply. This
is all that is necessary to be done in the month of
April, and is the perfection of the cabbage husbandry.

WATER-FURROWING.

This is a work that should be well performed
on the new-sown lands, as soon as the tillage is
finished. Very small savings in the omission of
this work will be attended with certain and great
losses.

TURNIP FALLOW.

The fields intended for turnips should be scuffled
in this month, and should remain a short time in
that state, and afterwards have harrowings enough
to make all the seeds of weeds grow, that the tillage in the succeeding month may destroy them.

WOODS.

All work in woods should now be over, or damage
will ensue from carting and from cattle. Good
husbandmen will observe to keep their woods well
fenced

fenced from cattle, the mischief they do is very
great.

HEDGING.

This month must conclude the business of fences.
It is bad husbandry to cut any hedges after April,
nor do the plashes have a good chance afterwards.
They will not be so sure of growing ; and nothing
but a most uncommonly late season should permit
any thing of that kind to be done now. All the
faggot-wood arising from hedges should be brought
home immediately.

CLEAR GRASS-FIELDS.

In the beginning of this month particular attention must be paid to the clearing of the grass-lands
from all rubbish that may affect the young grass,
such as the cores of ant-hills, the sticks and bushes
that are left after hedging, and whatever else may
happen to be found that will obstruct the scythe.
Mole-casts should be spread about with a spade and
bush-harrow, and being composed of nothing but
fine loose mould, they will do good to the grass.
Keeping the meadows and pasture in a neat hus-
band-like manner, requires attention of this sort.

ROLLING.

After the grass is cleared, in the manner mentioned in the last article, it should be rolled
to level it for the scythe. The roller must be of
weight enough to level worm-casts, and crush
mold. Some gentlemen are extremely fond of
using very large and heavy rollers, thinking they
are beneficial in proportion to their weight. This
idea

idea has been unjustly disputed. Another practice, founded on direct contrary principles, has begun to take place ; that of scarifying grass with a plough, consisting only of coulters, or harrow teeth. The advocates for this practice assert, that the burthen of hay (not the beauty of grass as a lawn) is much increased by loosening the surface, for the roots to have the power of a fresh vegetation : that the fault of most pastures is the being quite bound and hard ; that rolling increases this tenacity, and is consequently pernicious. Experiments are mentioned, which prove that grass-lands are infinitely improved by this operation of scarifying ; and further, that its use is extremely great when the operation precedes the manuring of grass-lands ; for that much difficulty is found, to get the manure below the surface, for the roots to feed on ; whereas if it be scarified well, the ground is opened so much, that whatever you spread on it gets at once to the roots ; consequently a small quantity so applied, goes as far as a much larger laid on in the old way.

HOPS.

The chief business of this month, in the hop-ground, is that of poling. In an article of cul ture so extremely operose as that of hops, and which, at the same time, employs the largest capital of any branch of English husbandry, the only object which should be expected in a work of this sort, is slightly to touch on the chief works to be done, not by way of directing how all are to be
performed,

performed, but as a mere aid to the memory to
have the several works in mind, so that if a manager should happen to be careless, the master
may be attentive to what ought to be going on.
Thus, in poling there are several points which demand consideration, such as the quality of the soil,
and the degree in which the last crop weakened
the exuberance of the plants. If overpoled one
year they are weakened, and must be underpoled
the next. The time of picking, whether late or
early, has also an influence. These are points
which must be learnt by practice, and not by occa-
sional observation, and they are named here, merely
to call to them the attention of the young planter.
The number of poles per hill vary from three to
five. Their sort, size, length, and position, when
set, are all of consequence.

FLAX.

Flax may yet be sown. The beginning of this
month will do, though not so well as before.

The soil for flax should be a loam, rendered fine
by tilth, and situated in a valley bordering upon
water; or such a soil as is thrown up by rivers.
If there be water at a small depth from the surface
of the ground, it is thought to be still better;
as is the case in Zealand, which is remarkably
famous for its flax, and where the soil is deep,
with water almost every where at the depth of a
foot and a half, or two feet underneath it. It is
said to be owing to the want of this advantage,
that

that the other provinces of Holland do not succeed
equally well in the culture of this useful plant. Not
but that fine flax is also raised on high lands, if they
have been well tilled and manured, and if the seasons are not very dry.

It is justly remarked in the letters which the Dublin Society have published on the culture of flax,
that moist stiff soils yield much larger quantities of
flax, and far better seed, than can be obtained from
light lands ; nay, that the seed procured from the
former may, with proper care, be rendered full as
good as any that is imported from Riga or Zealand:
but as M. Du Hamel rightly observes *, strong
land can hardly yield so fine flax, as that which
grows on lighter ground. In southern countries,
the husbandmen who raise flax, sow part of their
seed in September and October ; so that the plants
which spring from thence remain of course in the
ground all the winter ; and this is a judicious practice in those places, because plants which have not
covered the earth well before the summer heats
come on, are apt to be parched by the heat and
drought which usually prevail in that season.
They sow linseed again in the spring ; but this last
does not yield an equally abundant crop : however,
the flax which it produces is more esteemed, because it is finer than that of the seed sown in autumn. M. du Hamel seems indeed to think, that
the autumnal sowing yields the best seed ; but

---

* Elements d'Agriculture, liv. x. c. 2.

however

however that may be, in places where the winter is
apt to be severe, and where the flax, which is but
a tender plant, would of course be in danger of
being destroyed during that season, almost all the
flax is sown about the end of March, or in the
beginning of April. This spring flax is that which
will be most particularly spoken of here.

It may be laid down as a general rule, that the
land which is intended for flax should be brought
to exceeding fine tilth by repeated ploughings, and
that it should be enriched by a manure suited to
the quality of the soil. On grass-land some other
crops may be got off the land previous to flax, especially of such plants as do not occupy it long, and
particularly of those which are remarkably bene-
fited by frequent stirring of the earth whilst they
grow ; such as beans, pease, turnips, &c. ; because
these repeated stirrings render the mould fine and
loose, and help to kill the weeds, which would
otherwise do great damage to the flax. The memoirs of the Society of Britanny inform us [*],
that the Livonians, when they clear wood-land,
burn the wood upon it, then plough it, and prefer it in this state to any other kind of soil for
flax.

If the land which is intended for flax be stiff,

---

[*] Corps d'Observation de la Societé d'Agriculture de Commerce et des Artes, etablie per les Etats de Bretagne. An. 1759
et 1760, p. 186.

　　　　　great

great care should be taken not to till it when it is wet, for fear of kneading it.

If the ground on which flax is to be raised has been long in tillage, it should be ploughed deep before winter, and be laid up in ridges, in order that the winter's frosts may the more effectually moulder and loosen it.

In the month of February, if the land be not too wet, some very rotten dung should be laid in the furrows, and immediately covered over. In March, for southern countries, or in the beginning of April where the climate is colder, another ploughing should be given, to lay the land smooth ; the clods should be broken by hand, and the seed should then be sown or harrowed in with a light or bush-harrow, so as not to bury it above an inch deep. If the soil is moist and cold, a little pigeons' dung may be sown with the seed, for it agrees admirably well with the flax : but this must not be done if the ground is very light and too dry. It will also be right to lay wet land out in beds thirty or forty feet wide, separated by deep trenches, to drain off the water, and convey it into the surrounding ditches.

Most of our linseed is brought from the North. Linseed is reckoned good when it is large, oily, heavy, and of a bright brown colour. To know whether it be oily, a few grains of it are thrown into a red-hot fire-shovel, and they in that case crackle almost instantly, and blaze briskly. If it is

suffi-

sufficiently heavy, it will sink to the bottom of water ; and to judge whether it be new, a number of seeds exactly counted should be sown on the end of a hot-bed, and notice taken whether they all grow.

When the goodness of the seed is known, more or less of it is to be sown, according as the husbandman intends either to raise a quantity of linseed for sowing, or to have very fine and soft flax   In this last case, the seed should be sown pretty thick, in order that the plants may rise the closer together, and by that means grow slender and tall, which adds much to the fineness of the fibres of the flax. If the linseed is sown with an intention to let the flax remain for seed, a much less quantity of it should be used, that so the plants may come up thin, and thereby have room to grow to their full vigour and extent.

Some sow, with their linseed, either annual or perennial grass-seeds, when they intend to lay the land down for pasture after the flax is taken off. The plants grow but weakly under the flax, which, however, they do not hurt ; but as soon as the flax has been pulled, they increase apace, to the great benefit of their owner.

Flax is sometimes damaged by insects, when it is about three or four inches high. It is said that they may be destroyed by a slight strewing of soot, ashes, &c.

At all events, it is certain that this dressing will give vigour to the flax, though it should not kill the insects.—*Dublin Society.*

WATERED MEADOWS.

Throughout this month, if there are watered meadows on a farm, the use of them in supporting ewes and lambs is exceedingly great, but Mr. Wright is marked in his caution never to continue it longer, as it will greatly injure the quality of the succeeding crop of hay.   Not, however, that there is any positive rule in feeding, as is evident from a case cited by that writer : " Having heard that the proprietor of an old floated meadow at South Cerney had disposed of the produce of it, in the year 1795, in a way that was well calculated to ascertain its real value, I wrote to a person who resides on the spot, requesting him to send me a particular account of the product, and I received the following statement.   In order to make the most of the spring feed, the proprietor kept the grass untouched *till the 2d day of April,* from which time he let it to the neighbouring farmers *to be eaten off in five weeks (which ran a week into May)* by the undermentioned stock, at the following rates per head : a sheep, 10d. per week ; a cow, 3s. 6d. ; a colt, 4s.   The quantity of the land is eight acres.

107 wether

		£.		
107 wether sheep one week, - - -	£.	4	9	2
8 cows, do, - - - -		1	8	0
4 colts, do. - - -		0	16	0
		6	13	2
				5
For five weeks, - - -		33	5	10
Three colts added for three weeks,		1	16	0
	8)	35	1	10
Per acre,		4	7	8

The hay crop was, as usual, about 15 tons,
and was five weeks in growing.

	£.	35	1	10
15 tons, suppose 50s. - -		37	10	0
After-grass 15s. - -		6	0	0
	8)	78	11	10
Total per acre, £.		9	16	5

" The 4l. 7s. 8d. was made at a time when
other grass-land is in a dormant state, or exhibits
but feeble symptoms of vegetation. But the reader
will perhaps see the advantages of this art in a still
stronger light, when he is told that this meadow,
which is now in the occupation of a miller, was a
few years ago in the hands of a farmer, who, being
at variance with the miller, was entirely deprived of
the use of the water for a whole winter, which unfortunately was succeeded by a very dry spring and
summer ; of course the spring feed was lost ; and
the whole hay crop of eight acres was only three
tons."

POULTRY.

This is a very busy month with the poultry-maid.

The young broods, especially of turkies, demand
such a careful and almost constant attention, that
if they are numerous, the servant to whom they
are entrusted should have little else to do. This
circumstance renders it necessary either to breed a
large number, that the expence may answer, or else
to have no other than the common barn-door system.

BUILDING.

This is an article of rural economy which generally belongs to landlords or their stewards; but
as a young farmer may possess his farm by purchase or inheritance, it is highly necessary that he
should be cautioned in certain points, wherein it is
probable he will have had no experience; and these
may be noted without any encroachment into the
bounds of works properly architectural. If he entered to his farm at the more common season of
Michaelmas, he could not begin any buildings that
require the work of masonry till April, but he
should not longer delay it, for there is no point in
building more necessary to be attended to, than
that of finishing, as early in summer as possible,
that all works in mortar may have much time to
dry before winter. If the house the farmer lives
in be a very bad one, or which wants alterations and
great repairs, it should be exceedingly well considered before they be undertaken, for thousands
have thrown away so much money by beginning too
soon, and without due reflection and foresight, that
I shall suppose him to think steadily of it in some
winter month, and not till he has resided in it a
                                                    year,

year, thoroughly to understand every convenience and inconvenience of the old one, before he thinks of going to work. But the farm-yard and offices, if they must be done, rank with other profitable improvements, that cannot (by those who have money ready) be done too soon. At present, I shall lay down such general observations on each office, and on their general connexion forming the farm-yard, as he may himself easily apply to his own particular case.

1. The threshing-mill. The most important object, perhaps, which is answered by this machine, is that of saving barns, which are so very expensive in forming a new farm. I begin with it, as its position determines that of almost every other building in the farmery. There is not the smallest doubt of the propriety or profit of having one of these machines fixed in the principal farm-yard. If the farm be large, and stacks consequently scattered over various fields of it, then it may be right to have a moveable one also; but so many operations are wanting at home, that one should certainly be fixed. I have, in four plates, in the Annals of Agriculture, vol. xxxiii. p. 488, explained the relative position of the stacks to be built, on standings on wheels moving in a circular iron rail-way, so contrived that a very few horses (four sufficient for any common stack) will draw each stack to the mill. This contrivance is essential, as it saves the whole expence of carting the

corn, as well as the necessity of waiting for fine
days to do it in ; and as the expence is moderate,
I cannot suppose that any person will now go to
the heavy charge of barns and capt-stone standings, when less money will give him much greater
conveniences.   The circular form of the rail-way
on which the stacks are brought to the mill, is
necessary, as being the only one which permits a
choice of any particular stack to thresh, without
waiting for all or many others being done, before it
can be got at ; but a straight line leading to and
past the mill is admissible, except for this circumstance, though inferior in some other points to the
circular form.   But whatever plan may be chosen,
the mill should have the granary above it, to hoist
up the corn as threshed.   It must also have the
chaff-house annexed, as the power of the mill must
cut into chaff all straw which is used in feeding
cattle ; and as hay is used in this operation, mixed
with the straw, this decides the position of at least
some hay-stacks.   Close to and connected to the
mill, must be a shed on posts, roofed to draw one
stack under, before the thatch is stripped, and
from which the corn is delivered at once to the
mill.   It is turned (so much as is wanted for chaff)
into a straw-room, and the rest replaced on the
standing of the stack that was last cleared, and being
stacked on it with some care, is ready to be drawn
away in the circle for litter.   This circumstance
decides the position of the sheds for cattle and
                                        horses,

horses, as they should be so placed as to be very
near this litter. Thus situated, they demand hay
also in their immediate vicinity, and as hay was also
wanted for chaff, all the stacks should be within
the circle. Thus far every thing is connected, and
each building so placed, that it cannot be supposed
in any other place, without a manifest inconvenience following. If milch cows be in the circle,
which they ought to be, this article demands another
combination of the dairy and the piggery, which
must also be connected, but at a due distance from
each other. I have, in the plans above alluded to,
supposed the circle of cattle and team sheds to open
on the outer side, to bring in the cattle, and to
void the dung into a circular repository that surrounds all the sheds. A late writer has disapproved
of this, and proposed to have the sheds to open
within the circle; but this I conceive to be very
erroneous. The beasts must for this purpose be
reversed; their heads to the outside, and the dung
voided within the circle. This completely deranges
the whole design, and converts much convenience
into a most inconvenient arrangement. The chaff,
hay, &c. must be conveyed *without* the circle to
the heads of the beasts, by a long walk, instead of
the nearest line; the dung must be within the area,
cutting off all connexion with it; dirt and litter
will be found where cleanliness should prevail; and
nothing gained in return but a little better shelter,
supposing the sheds to be open; but as the contrary is supposed, this object would not be varied;

so

so that I must adhere to the original proposal, as very much superior in convenience to the alteration thus indicated. Farm-yards, &c. have been exe-cuted by the great at an immense expence, which are not to be compared to this circular system, which might be executed for one-tenth of the cost: and let the young farmer remember, that the com-bination for convenience is as applicable to the roughest and cheapest mode of execution, as it is to works of brick and slate, and ornament. Fir-posts, with a covering of stubble or fern, may be thus disposed, as well as columns of stone and mangers of copper.

The dairy should be situated within a certain reach of that part of the circular shed assigned to the cows, suppose 30 yards, and a slight foot-bridge thrown over the dung-pit, from the centre of the cow-standings. In contriving the dairy, there are a few points which should be attended to. The en-trance into the milk-room should be through the scalding-house, and the copper for heating water, &c. should be in a shed without the scalding-house, that the heat may be as far as possible from the milk. The boiling water should pass by a cock in the bottom of the copper, through a trough or pipe, across the scalding-house (another cock being there in the pipe for washing smaller imple-ments) through the wall into the milk leads, that whenever the dairy is free from milk, &c. or with-out being free in winter, the water may pass at once through the whole system of leads or trays,
                                                              and

and be kept standing at pleasure in any of them, which is the most effective way of scalding, and having made the tour of all, may pass out to a drain. The immediate passage of the water through the wall of the dairy, should be in a trough large enough to receive securely a pail of milk emptied by it, that all from the cows may run at once through a hair sieve in this trough into as many trays as are requisite to receive it. This prevents all ingress to the dairy by dirty men and boys who may bring pails of milk to it. The dairy itself may be circular, and, if expence is not regarded, a fountain of water may play in summer in the centre of it, the water falling in a circular *jet*, surrounded by a clean gutter to convey it away. This, however, is mentioned as a hint for expensive dairies, and not by any means as *necessary*.

The establishment of a piggery demands even more attention than that of a dairy, combining as it does with more objects. This must be in a circle, or it must fail in convenience. In the centre, the boiling or steaming-house, with a granary for corn, meal, bran, &c. a range of cisterns in divisions around it, for receiving immediately from the copper or steam apparatus, and also by tubes from the granary; around these a path, then the fence, wall or paling, in which the troughs with hanging lids, for supplying food directly from the cisterns, on one side, and for the hogs feeding on the other; a range of yards next, and another of low sheds beyond, and last of all, the receptacle for the dung.

The

The potatoe stores (pyes as they are called) should at one end point near to the entrance, and water must be raised to the coppers and cisterns at once by a pump; a trough or other conveyance from the dairy to the cisterns, for milk, whey, &c. Such an arrangement will be very convenient, and the expence need not be considerable. To annex a certain space of grass, or artificial grasses, in divisions, into which the hogs may be let at pleasure, is an addition of admirable use, if the spot permit it. Those who do not possess a convenient pig apparatus, can have little idea of the great use of it, in making manure. This alone becomes an object that would justify any good farmer in going to a certain expence, for attaining so profitable a part of what ought to be his farm-yard system. In nine tenths of the farmeries in the kingdom, it is lamentable to see so many parts of a right piggery scattered and unconnected, in such a manner as to preclude convenience, increase labour, and prevent the making of dung.

In 1765 I built a hoggery, nearly, but not exactly on this idea, the expences of which were,

The boiling-house,	-	-	-	£.18	18	0
Copper,	-	-	-	-	-	13	0	0
Pond,	-	-	-	-	-	4	0	0
Pump,	-	-	-	-	-	1	10	0
Cisterns,	-	-	-	-	-	14	0	0
Shed,	-	-	-	-	-	6	15	0
Paling,	-	-	-	-	-	7	7	0

Carry over　　£.65　10　0

Paving,

Brought over - £.65	65	10	0
Paving, - - - - -	10	0	0
Troughs, - - - - -	3	0	0
Total, besides timber, - -	£.78	10	0

By means of one of these yards, I fattened 88
hogs in spring 1766, with only one man to attend
them ; whereas three would not have been sufficient without such conveniences. They were littered with nine loads of straw and haulm, that
cost 6l. 18s. ; and this made 90 loads of very rich
dung, valued by several farmers on the spot at 5s.
a load.

Value of dung at that rate, - - £.23	23	10	0
Straw, &c. - - - -	6	18	0
Profit in dung, - - - £.15	£.15	12	0

But they had not half the litter they ought ;
they would have made 35l. worth of manure, beyond doubt.

Ninety loads costing 6l. 18s. is 1s. 6d. per load.

These particulars surely must prove the vast importance of such conveniences, for fattening great
numbers of swine, for the mere purpose of raising
manure. Suppose the expence, timber included,
to be 110l. and the interest called 5l. what comparison is there between the expence of 5l. a year,
and the prodigious utility of having it always in
your power to fatten, with scarce any expence
of labour, whatever number of hogs you please ?
With such a convenience, all the pease, beans,
barley,

barley, buck-wheat, potatoes, parsnips, carrots,
&c. that are, or can be raised on a farm, may be
applied to the rearing, feeding, or fattening hogs;
by which means the farmer has the opportunity of
improving his land to the highest degree, and at
the cheapest rate possible.

The total expence at present of such yards
would not be less than 150l. And if made conform-
ably to the more correct idea, would be 200l. or
250l. But the governing idea of position should
be followed in sties of 20l.

The last circumstance of rural management that
I should recommend to a proprietor, on his inhe-
riting a farm, is that of building a house. He
may however be in such a situation, in respect of
habitation, that to build a new house is more
prudent than to submit to the very heavy repairs
of an old one; and in other cases he may come
to a farm without any house, it being let to a contiguous renter. In such cases, it will be useful to
have some general hints for his direction.

It very rarely happens that a man has an opportunity of making the experiment of building a
house twice in his life, and, therefore, he
should reflect well before he begins. It is more
common to see people fail in this essential step,
than almost any other. How many new houses,
in which people have no more elbow-room, in a
number of *pieces*, than if they were in the stocks?
How many in which comfort is sacrificed to show,
warmth to space, the sun's rays, in latitude 55, to
                                        the

the sight of a park or a lake, shelter to a prospect, and the convenience of a lumber-room to the arrangement of an anti-chamber ? and as to cupboards, closets, and stowage of many sorts, the fools in middling life allow their puppy architects to sweep them all away, because my lord, with 40 servants, transfers such things to the offices. Where do we meet with a moderate house well calculated for a small fortune ? Where do you find one planned for a man who keeps the key of his wine-cellar ? who has connected a kitchen and dining-room in such manner, that the smell of the former shall be excluded, without a long walk to the latter ? Who has contrived a moving table, served through the wall, without any servants to wait in the room ? There is not one apartment in a house, from the cellar to the garret, that has not been improved for men of large fortune, but, for small incomes, I believe invention has either gone retrograde, or at least stood still.

Circumstances for convenience, which should be attended to, in planning a house to be inhabited by a family whose income is small.

It is not necessary to define what the income may be, but only to mark it, by the points of the mistress being her own housekeeper, and the master keeping the keys of the wine-cellar; and that a general system of economy pervades the management, the proprietor farming his own estate, whether 400 or 1000 acres.

1st, The kitchen should not be a thoroughfare,

nor

nor any *house-door* open directly into it. The scullery as near it as possible, but without opening into it.

2d, The mistress's store-room should have a square opening into the kitchen (with a sliding door), on a level with the dresser or broad shelf which surrounds the whole store-room, through which she may give out whatever is wanted, without the necessity of her or her maids, &c. passing by a round-about way.

3d, The common *keeping-room* to open on one side into the store-room, and on the other, into a passage that leads directly to the wine and ale cellar, which should be near, in order that the eye may attend to what the hand need not perform.

4th, The window of the keeping-room to look full upon the grand avenue to the yards, barns, cattle, &c. and if possible full (but at a safe distance), into the farm-yard.

5th, The farmer to have a store-room, as well as his wife, for sacks, small tools, nails, &c. &c.

6th, The farmer's bed-chamber, with a large window full to the East, that the sun may shine in early.

HEMP.

This crop is often sown in April, but in the more modern practice of the best cultivators, it has generally been deferred till May, probably from experiencing the evil of late spring frosts, when sown early. I shall therefore postpone the particular directions till the Calendar for that month.

SOWING

SOWING GRASS-SEEDS:

Seedsmen are apt to mix seeds of nearly the same size, in order to have the fewer *casts*. This is a very bad way, and always to be guarded against. Five pounds of any clover, &c. cannot be divided and sown at two casts ; but 10lb. may, and ought, and a larger quantity is better done at three ; but for all small seeds, the Norfolk turnip-trough, which is now adapted to clover and ray-grass, is much the better way of delivering these seeds. Those of grass, which are light, ought never to be sown in a windy day ; for an equal delivery is a point of great consequence. All grass-seeds should be covered at one tining of a very light pair of harrows. Of all other circumstances, sowing in the wet, so as to have the least degree of poaching or stickiness, should be avoided.

SEPARATED GRASS-SEEDS.

I cannot advise a young farmer to pay much attention to this object, unless he proceeds upon very sure grounds, by forming a contract for the sale of the seeds at a good price, before he commences his operations. I have myself gone largely into it, and have found it a cheaper method of procuring the seeds, to have them gathered by women and children, by hand, than to raise them myself, under the determination to have them perfectly clean from all mixture. However, as in certain situations, and under certain circumstances, it may become an object of attention, something should be said of it here.

R   This

This is the month for a spring sowing. There is no object in the whole range of cultivation, which demands land to be so perfectly clean as this, nor is any weed so mischievous, as a different sort of grass from that sown, nor any more likely to come. The seed must be sown in drills by hand, at one foot asunder, and from their first appearance above ground, kept absolutely clean. The year following that of sowing, they yield most seed, and presently decline in quantity. I have known several persons who have made the experiment, and who all gave it up. The sorts to be recommended are the meadow fescue, the poatrivialis, the crested dog's-tail, the meadow fox-tail, and the rough cock's-foot. Timothy is always to be had from America, and Yorkshire-white is in common sale. But for the farmer's own use, it is not so necessary to be so very nice, in which case, broadcast crops may be trusted to, for a mixture is no formidable circumstance. I have had the cock's-foot and the tall oat-grass gathered at 4s. a bushel, in large quantities; and the crested dog's-tail at 1s. a pound, and have thus laid down many scores of acres. At these prices I have found them cheaper than when raised in drills with great attention.

SIBERIAN MELILOT.

" The *Melilotus Sibyrica*, from Mons. Thouin, at the King's Garden, at Paris, makes in the garden of Mons. Faujas de St. Fond, a most superb figure. Nobody can view its prodigious luxuriance, without

out commending the thought of cultivating it for cattle. The *coronilla varia* is a common plant here, and of such luxuriance, that it is hardly to be destroyed. The *hedysarum coronarium* does well here." From this hint (which I extract from my own Travels), I introduced the culture of the me- lilot in my experiment-ground, and found it an object much deserving attention.

WELD.

In Norfolk this plant is sown with barley, in April, from one quarter to half a peck per acre of seed, in the manner of clover, and often with clo- ver at the same time, which is fed or mown, after the weld is pulled in the following year. This being a product sold to manufacturers, the price is not fixed in any manner very satisfactory to the farmer, and therefore I merely name it, that the reader may have it in his mind, for trial, should he be so disposed. In other parts of the kingdom it is sown with turnips, which are fed off in the spring, and the weld left for a crop.

TARES IN CLOVER.

Very early in this month, and in some seasons, in March, the young clovers should be carefully examined, for if the land has had this crop too often repeated, it is very apt to be what the far- mer's call *sich* of it. A full plant in autumn often dies away in winter and spring, so that by this month, the farmer, perhaps, is in doubt whether he shall let it stand, or plough it up. In this case, it is highly advisable to dibble into all the vacant

R 2   spots,

spots, spring tares, which thus take extremely well, and between clover and tares a very ample crop is produced.

SOW FURZE.

Dr. Taylor, in Surrey, had a poor field of six acres, worth 7s. per acre, sown with furze, and thus converted to be the most profitable of the farm : sown, the land being cleared from couch, in April 1782; mown in 1784, to thicken it; and cut for the first crop in 1786, and since regularly every two years, three acres per annum. Last year's cut of three acres produced 7700 faggots, and sold at 3l. 3s. per 1000, on the spot; this 24l. 5s. 6d.; cutting and binding 1s. 6d. per hundred, or 5l. 15s. 6d.; clear, 18l. 10s. Suppose tithe, rates, and fences, to equal 5s., rent 7s., in all 12s.; or, for three acres, 1l. 16s.; that, further charge deducted, net 16l. 14s.; or, per acre, 5l. 11s. 4d.; and, per acre per annum, 2l. 15s. 8d. which is a greater net profit than any man receives from wheat upon such land. Dr. Taylor thinks that the product rather increases than diminishes. For the time of cutting, would recommend dry weather in February, or the beginning of March, when severe frosts are over.

MAY.

MAY.

FARM-YARD.

ABOUT the twelfth of this month the farmer
may calculate that he will have a sufficient bite of
grasses to leave off foddering entirely, and before
that, he should not think of it; for, if cattle are
turned into grass not sufficiently advanced in
growth, they will require such a number of acres,
that his mowing ground will be greatly curtailed.
As soon as the yards are cleared, if he is in the
mixing system, the dung in them must be turned
over, and mixed carefully with the stuff beneath,
whether it be chalk, marle, turf, ditch-earth, or
whatever sort. For this purpose, he must set
many hands to work, so as to get it done as ex-
peditiously as may be; because it should lay a
little after turning before it is carried on to the
land. It thereby undergoes a fresh fermentation,
and becomes more rotten. The method in which
the men should do this work is, to begin and throw
the dung up against a wall, or into some vacant
space, so as to have the command of a trench to
work in: they should always keep this trench three
or four feet wide: then they draw down with dung-
cromes the dung, and, breaking it to pieces with a
fork, throw it up on the part already mixed, in a
spreading manner, so as to cover all the chalk or

R 3   earth.

earth. In this manner they proceed with the dung, to the breadth of about 18 inches, or two feet, till they come to the stuff under it ; all which they pull down with pick-axes or mattocks, and, when it is in the trench, break it further to pieces, so as to have it fine ; that is, no pieces larger than a man's wrist. If water hang in any places in their trench, they should have a water-bowl ready to throw it on to the part they have mixed. If this work is well executed, he will have a large hill of excellent manure, ready to lay on to the cabbage or turnip land, to be turned in by the last earth.

Respecting the quantity—therein lies the proof of his being a good farmer ; perhaps the most important, convincing proof, that a farm can offer. If he has managed well, he will have from 15 to 20 loads for every head of great cattle, and about 10 loads for every hog, not reckoning pigs : not above a third of the whole marle or earth. Every trussed load of straw, trampled into dung, will make six cart ones of dung.

The earth, which has lain under the dung all winter, and received all the urine of the cattle, must by no means be reckoned as inferior to the dung itself. It is become a rich manure without mixing with dung, richer than the best of marles : and I am well persuaded, that this retention of the urine in it is of such consequence, that the whole compost, when well mixed together, will be better than if chalk or earth had not been brought into

the

the yard, at least for most soils : but, that the favourable circumstances of the conduct much exceeds the expence of it, for all soils, cannot be doubted.

A great recommendation of this farm-yard system is the cheapness of thus manuring the land: the farmer will find, that he can, in no other method, manure at nearly so small an expence. Purchased manures come higher ; many of them much more expensive, in proportion to their value.

In some situations there are no manures of any sort to be purchased ; in such, the farmers, if they do not adopt such a plan as I have mentioned, must give their land a poor chance ; for it must be an admirable soil, or course of crops, to render manure unnecessary.

Thus far this article stands, as it did in the former editions of this work ; but more modern experiments and observations have given birth to a different system, which must also be noticed here. The question of using yard dung in a long or a rotten state, was stated in the Calendar for last month. The young farmer will act wisely to try both methods carefully, in order that he may have a degree of conviction which experiment alone can yield : but such a trial demands particular attention, or it may deceive. Supposing two half acres marked out, they should be manured, the one with a certain portion of rotten dung, and the other with that portion of the same sort long, which the degree of freshness would demand in order to pro-

R 4   duce

duce in rotting the quantity so carried in a rotten
state : this cannot be prescribed, for it depends on
the state of both dunghills at the moment.   This
is one way of trying it, but a still more accurate
method is to litter two sties, each of 10 hogs fed
equally, or two sheds of four fat bullocks, with a
certain weight of trussed straw, and to use the
dung of one fresh and of the other turned up and
rotten ; the experiment terminating at a time when
the long dung can be used ; the rotten may wait,
but the long cannot.

Should the farmer determine on the older method, he turns and mixes his yard-dung as before
described.   If on the new way, he has nothing to
do this month, but is to wait till he wants it for his
turnip crop.

FEEDING OR MOWING GRASS-LAND.

As this is the usual season of turning cattle to
grass, and consequently the time of determining
what fields are to be fed and what mown ; our
young farmer has some circumstances to attend to
which may demand consideration.   For instance,
is alternate feeding and mowing better than to keep
the scythe out of pastures and the cattle out of
mowing grounds ?   Mr. Goring here offers a valuable remark :

" I do not even admit that grass-land should be
mown and fed alternately ; it is certainly the way
to go on the longer without manure, and as certainly the way to ruin *(pari passu)* both fields in
the end.   In order to maintain its proper quantity
of

of stock (we used to say) the land must be used to it; the more it keeps, the more it will keep ; four this year, five the next: give it a little manure, more stock will follow, and so on till it has attained its *ne plus ultra,* if that point be to be attained. Land that has been used to the scythe will not *(cæteris paribus)* keep so much stock and so well as an old pasture, though it may have been better manured ; neither will old pasture produce so much hay as the other ; each will grow as it has been accustomed to grow ; but the old pasture has an inherent sweetness in it, as well as virtue, which is hardly to be seen upon the ground, but is to be felt upon the rumps and sides of the ox ; or to be discerned in the number of sheep which it maintains.

FIRST YEAR's MANAGEMENT OF NEW LAYS.

In this point there is a great difference of opinion. Some have contended that the new lay should be pastured by sheep; others by cattle; others mown for hay ; others seeded.

In the North Riding, the best farmers feed their new lays with sheep the two first years.

If ray-grass and white clover be meant to remain some years, a gentleman in Strathern, of superior knowledge, eats them the first year with sheep : by this they are rendered thick, close, and durable.

To let heavy cattle in the first year, does mischief which demands years to recover.

If mown for hay it should be cut early, for nothing

thing is worse for new layers than the grass running to seed.

Mr. Wright, of Ranby, pastures them with beasts the first year, as sheep do harm.

Dr. Wilkinson compared sheep-feeding with mowing experimentally, and the superiority of the former was very great.

The Marquis of Rockingham seeded them the first year.

Colonel St. Leger fed the two first years with great success *.

I have practised all these methods; the last merely for gaining the seed for other lands; and I have not the least doubt upon the question : if the grass be kept unfed in autumn, and it be not turned into too early in the spring, sheep do no harm, but much good : the number should not be so great, nor kept so long, as to allow the plants to be nibbled too close ; but sheep-feeding is certainly the best for the first year. If bents rise, as they will do, let them be swept with a scythe before any of them seed, unless the plants be evidently too thin on the ground ; in that case the seed falling may do more

---

* Though I have little doubt that feeding is the right management, yet it is not to be concluded that, with mowing, the grass will not succeed : Lord Rockingham s new lays, viewed in the autumn of the first year, were by description among the finest that have been seen ; they were manured however the autumn after sowing, which is admirable management, provided the soil is sound, and the season very dry.

good.

good by raising fresh plants, than harm to those which yield the seed.

But it is not only the first year that sheep-feeding is the best management for a new lay; it should be so fed also the second year; and if the third, so much the better: there is no *necessity* of continuing it longer; but I have had some fields which succeeded well in feeding four, five, and even six years: and in general it may be laid down as a rule, that the more the land is sheep-fed, the more it will be improved, and especially if it is ever to be ploughed again for corn. But when sheep-feeding *inclosures* is mentioned, it is understood that the sheep are not folded from such fields; a ruinous, impoverishing, unnecessary system, of which the farmers are too fond, as they are of every way of robbing grass to favour corn.

CATTLE IN GRASS.

When cattle, whether cows, fatting beasts, or young stock, are turned out to grass, it is requisite to consider the best method of feeding. There are two opinions on this point directly contrary to each other: first, it is asserted, by one set of graziers, that, let the grass to be fed consist of ever so many acres, that the cattle should have it all at once: if it is divided into eight or ten fields, the gates of all to be set open, for the stock to feed where they like. Secondly, the other set advance, that large fields, of fifty, eighty, or an hundred acres, should be divided, that the farmer may change his stock from one to the other, and give the grass fresh and

and fresh.   And each of these parties assert, that
they know themselves to be right from experience.
But that is impossible ; one must, undoubtedly, be
wrong.   Let us consider the point from reason : it
is one that will never be decided *fairly* from experiment ; for two pieces of grass, each of eighty or
an hundred acres, contiguous and *perfectly* alike,
are not to be met with in the king's dominions;
and, if they were, two sets of stock, exactly similar, would not be found.   The divisions into fields
by hedges and ditches, for the purposes of draining
and shelter, is not the inquiry, the comparison
not being fair ;   as such divisions may be fed at
once, by setting all the gates open, as well as one
field.   The inquiry is, whether the cattle will spoil
the grass more in one way than in the other ? and
whether the grass will go as far in one as in the
other, by fatting or feeding the beasts as well ?
The argument of giving the grass fresh and fresh,
appears to be rather vague ;   for it supposes that
the cattle will not eat it fresh, if they have the
whole range at once, which may be a mistake :
they will not be seen in the evening where they
were feeding in the morning, but vary their food
in the manner   most agreeable to themselves;
and we may in general observe, that the sagacious
animals, when left to their own conduct, manage
such points better than we can for them.   As to
the treading and spoiling, it is an equal objection
to both methods ; the legs of the beasts are not
tied in small closes, any more than in large ones,
                                                        In

In case all the smaller pieces have not water, the objections to feeding them separate are much greater.

On the other hand, it must be admitted that there are disadvantages attending this way of pasturing : for a time the trampling may be greater, as cattle are disposed to beat a sort of march around their fields on first turning in, and also on some kinds of disturbance : but a greater evil is that of disturbing a large herd instead of a small one : this a dog may effect in one inclosure unseen from others, and consequently the stock in them left without interruption ; and, perhaps, a motive greater than this is, its having been observed that cattle, and sheep also, do better when well proportioned to their pasture, when divided into small lots rather than large ones.

In the stocking grass-lands, the farmer should attend well to the proportion between his stock and the quantity of his feed. Let him remember when he stocks his grounds, that he should be pretty nice in this proportion ; for if he overstocks, his loss will be certain and great ; and, if he does not throw in as many cattle as he ought, then he will suffer in his profit.

There are several divisions in fattening : to buy in beasts in October or November, and put them to straw till the end of February ; then to begin their fatting on turnips, and continue it in March ; thence to the middle of May on other food, and then to turn to grass, and kill in August

or

or September. The other scheme is, to buy in smaller beasts in May lean, and sell them fat from the grass in the October and November following. Where winter food is raised with spirit, and the farmer takes a proper care to provide great plenty of litter to turn into dung, the first method is much the most profitable : but, where either of these requisites are wanting, the latter is preferable.

A third system, to buy at that period which will, according to the size of the cattle, admit their being ready for market in April and May, when meat is generally sure of a good price ; one winter not highly fed ; a summer's grass, and a second winter driven on by the best feeding. This for large oxen; if smaller, to be bought in in spring, and have only a summer's grass and a winter's stalling.

CAKE AND CORN-FED BEASTS.

Our young farmer, if he has any cake or corn-fed beasts, not fat enough to go off the end of April or the beginning of May, is under no necessity of parting with them, as they do very well on good grass though taken from oil-cake : I have known, however, in Lincolnshire, cake in a moderate allowance to be given while the beasts were at grass, and to go on very rapidly while thus fed.

BUCK-WHEAT.

This may be sown towards the end of the month. So late a time has offered the opportunity of good tillage to destroy weeds, and of course, the land is fine, and in good order. It is a most profitable crop, and especially on all (except very heavy soils) land

land that either requires late sowing, or that you are disappointed in the design of sowing soon enough to barley. Late sown crops of the latter grain are seldom good enough to pay expences : in such cases, it is useful to substitute buck-wheat; for I do not think that there are many soils on which a crop of buck-wheat, sown in May, will not exceed in value a crop of barley sown in May : yet, in many tracts of country, it is a common custom to sow barley so late as that season.

LUCERNE.

This plant may yet be sown : being a perennial, and, when well cultivated, yielding an immense profit, too much attention cannot be given to lay the seed in the ground with all possible advantages; that is, the land should be perfectly free from weeds, very rich and fine; these requisites a man may not be able to procure in April. In such case let him not sow in April, but wait till May; and this, whether drilled or sown broad-cast : if the latter, let it by all means be sown with buck-wheat, which is preferable to sowing it alone.

The advantages of cultivating lucerne are so extremely great, that the young agriculturist should determine at all events, to have sufficient at the least for the summer support of all his teams and other horses; and if in addition to this quantity he provides also for thus feeding much other stock in his farm-yard, he will find it a most profitable practice. The proper soil depends principally on two qualities, that it be quite dry and very rich. If near the

stables

stables and yard, the convenience will be much the greater; but to chuse the best land on the farm is upon the whole, the best direction he can have. Those who at present cultivate it on the largest scale in Kent, Sussex, and Hampshire, where are to be found large quantities of it, very generally have it in the broad-cast mode, and as far as positive practice goes, this method must be preferred, but as effective cleaning it, and especially from indigenous grasses, is an object of great consequence, executed when broad-cast by a powerful and heavy harrow, it much deserves attention, whether drilling very straight at nine inches equi-distant would not be a preferable method. Drilling has been tried by many, and abandoned for random sowing; but nineteen twentieths of the drilled lucerne which I have seen, have been at 18 inches, two feet, and some even at three; the consequence of which has been, a heavy expence and trouble in reaping instead of mowing; and if these spaces are kept truly clean, the lucerne being damaged by the pulverized earth adhering to it, and carried to the racks. If drilled at nine inches, it might once a year be most effectively horse-hoed with Cook's scarifiers in the iron beam, which would eradicate grass far better than any harrowing that could be given to a broad-cast crop, without a formidable expence, and some danger of damaging the crop, tough as the roots are. The grand object in the preparation of the ground is, to have it as free from weeds, and especially grass, as skill and perseverance can effect. The crop of
the

the preceding year should have been turnips fed on the land by sheep, before the Christmas frosts ; the field immediately ploughed, and laid by that plough-ing ready for broad-casting or drilling. In March it may have been scuffled on the surface, and at the end of the month, or perhaps better the be-ginning of April in a dry time, sown. This may be done with or without corn ; if drilled, it will be better alone ; if broad-cast with barley or oats, under-seeded : but it may be drilled with corn ; the corn first broad-cast, and then the lu-cerne drilled among it ; or the corn may be drilled in one direction, and the lucerne afterwards across it. Not less than 12lb. an acre should be drilled, and 20lb. sown broad-cast. It is apt to be eaten by the fly, &c. ; if it escape that damage, all is safe, and the farmer may be assured that his care will be well repaid. No manuring at this period is necessary : but to sow soot just as the young lu-cerne is got above ground, may be beneficial against the fly. In regard to proportioning the quantity of land thus occupied to the stock in-tended to be fed on it : a quarter of an acre per head is sufficient for all sorts of large cattle, taken one with another, if the land is very rich and good ; but on more moderate soils, half an acre per head will be a proper allowance. It is much better to have too much rather than too little.

SAINFOIN.

This grass may be sown with buck-wheat with success, if the soil be proper ; for hay, in countries

s where

where natural meadows and pastures are scarce, sainfoin is so valuable, that this culture should be attended to more than it is. It is a common notion, that sainfoin will thrive only on lime-stone lands or chalky soils. This idea excludes very extensive tracts in many parts of the kingdom, where sainfoin would be a valuable acquisition; but it is much to be regretted, that we should not experimentally know the exact soils on which it will and will not thrive. General ideas, the result of ages of practice, are admirable guides to tell us, what *will do*; but they are not satisfactory in informing us what *will not do*.

BURNET.

This grass may be sown in May with buck-wheat, with as great propriety as at any other season.

CARROTS.

If the carrot-crop was not hand-hoed last month, it should be done this: and those crops which were then hoed for the first time, will require an harrowing early in this month, and a second hand-hoeing about the last week. The harrowing will not damage the young carrots, nor pull up one in twenty; but it will displace the weeds set again by rain, and check the growth of those that are got up since. The other hand-hoeing should be performed with nine-inch hoes; and they should set out the plants to the distance of twelve inches from each other. Gardeners do not let them stand further than eight or nine inches asunder; but, when the roots are designed to be of a large size, that is too little:

little : the crop will, in very good land, measure more bushels at a larger distance. These observations are equally applicable to parsnips.

POTATOES.

Some time during May, the early planted potatoe-crop will require a hand-hoeing, which should be done with good attention, that not a weed may be left, and the surface of the land be left well cut, and in fine order. Crops in rows should receive, besides this hand-hoeing, the first horse-hoeing, which should be given with a common swing-plough, drawn by two horses, one before another, and turn a furrow *from* the rows, throwing up a small ridge in the middle of each interval. These operations should be well and attentively performed ; for the weeds grow at a great rate, and, without such an attention, will destroy, or at least greatly damage the crop.

PLANT POTATOES.

This root may be planted throughout May ; indeed many planters consider this as the best time in the year for that business. They are a very tender vegetable, and very apt to have the green shoots out off by late spring frosts. I have more than once had them turned quite black by frosts even in June. Deferring it thus late is also a means of lessening the work of a farm in those very busy months, March and April.

CABBAGES.

The crop of cabbages planted in April will re-
s 2   quire

quire a hand-hoeing this month. It should be given only to the tops of the ridges, about eight or nine inches around the plants : the weeds should be cut up clean, and loose moulds drawn to the stems of the plants. In about ten days after, the first horse hoeing should be given, turning a furrow from the plants, and throwing up a ridge of earth in the middle of each interval. This operation will be of great use : it lets the atmosphere into the ridges on which the plants stand, and consequently sweetens and ameliorates the soil ; and it kills the weeds that grow on the sides of the ridges much cheaper than it can be done by the hand-hoe : it likewise pulverizes the earth taken away, and brings it into fine order for returning to the plants in June, when they will strike into it, and thrive the more.

The land designed to be planted in June, should this month receive an earth to throw it on to whatever sized ridges you intend to plant on. This must not be omitted ; becuuse the beginning of the next month may be taken up in carting on the manure.

DRILL CABBAGE-SEED.

The common culture of this plant, every one knows, is that of sowing the seed on a seed-bed, and transplanting where to remain ; but a much superior method was that practised by the late Mr. Bakewell. of drilling the seed at once the beginning of this month where the crop was to remain. This saves the expence of transplanting ;
but

but that is an object of very small consideration, compared with the much superior point of certainty; and being secure against those droughts in June, which sometimes delay the planting so long, that a full, or nearly a full crop, is unattainable. I have often remarked how exceedingly superior plants have been, that were singled out in nursery-beds, and left without moving, when compared with others drawn out and transplanted. It is true, there is one object in which this system is inferior to the common practice: from February to June is gained for the fallow in the usual culture; but not *equally* in this new method. It is therefore to be recommended, not to practice it on foul land; but, in fields tolerably clean, it may be safely executed: however, perfect safety is attained if the cabbages follow some other fallow crop, as turnips, tares, beans, &c. The seed should be drilled in the Northumberland way, for turnips on ridges, upon good land, four feet broad, and upon inferior soils three feet; the manure being laid in the furrows, and the ridges reversed for covering it. This is a very profitable application of dung, which should be taken long and fresh from the yards, not less than twenty cubical yards per acre, measured before stirring.

When the plants are four inches high, thin them by hand-hoeing, with a view to a second hoeing, that shall leave them at the intended distance at which they are to remain, 18 or 24 inches, according to the soil.

SOW SWEDISH TURNIP.

The ruta baga, or Swedish turnip, has, in a various experience, and through several of our counties, proved one of the most important acquisitions
which the husbandry of this country has made
for many years. I cultivated it upon its first introduction, successfully at Bradfield, in the transplanting method : sowing the seed in a nursery or
seed-bed the end of February, and transplanting
it on the first rains in June ; but my success was
not so great as to induce me to be anxious about
it ; and for many years we heard but little of the
plant. In 1801, I surveyed the county of Hert-
ford for the Board of Agriculture, and was much
pleased to find that this plant was so well established in that county, as to be almost common
husbandry.

In 1802 I surveyed Norfolk, and there found
that many of the principal farmers had made
pretty large experiments on it, and been sufficiently successful *in some fields*, to have an high
opinion of the plant : but many of them com-
plained that the fly made such ravages among their
plants, that they had no dependence on ever being
able to secure a crop. The fact is, that the best
culture of this plant is to sow it where it is to
remain, broad-cast, from the 10th of May to the
end of the month ; and of all others, the best preparation to secure a crop, is that of paring and
burning ; for the fly being the grand enemy, from
its coming so very slowly to the hoe, this operation

tion not only proves by far the best preservative against that enemy, but also pushes the plants on in an accelerated vegetation, and thereby doubly secures the crop. If the seed cannot be thus prepared the next best management is, to sow it after common turnips fed on the land by sheep. If neither method suits, it must be put in on well pulverized soil, very amply manured.

The inducements to the farmer to enter freely on this culture, are many and very important. 1. If he has the right stock of seed, the root yellow in flesh and rough in coat, it lasts through all frosts, and may be depended on for sheep quite through the month of April, though drawn two months before, and spread on a grass field. 2. It is an excellent and nourishing food for sheep, and also for any sort of cattle. 3. It is equal to potatoes, in keeping *stock* swine ; a point of very great consequence. 4. It is, next to carrots, the very best food that can be given to horses. 5. It is sown at a season which leaves ample time in case of a failure, to put in common turnips, or cabbages. All these are powerful inducements to urge a farmer to enter readily on the culture.

Bring the field into just such order as is requisite for a turnip crop, and sow two to three pounds of the seed per acre ; one pound is enough, but if the fly is apprehended, it is much safer not to spare seed.

MADDER.

The crop planted last month will want a hand-

hoeing before the expiration of this month : that
work should be done with eight-inch hoes, and
very carefully ; for the young plants will not bear
rough treatment of any kind, being of a most brittle nature. It will not be advisable to horse-hoe
yet.

LIQUORICE.

The young crop of liquorice must be hand-hoed
in May, and carefully hand-weeded at the same
time. In common management, this is not well
done, owing to the cropping the ground the first
year with onions or carrots, both which, or any
other plant, are but so many weeds, that rob the
principal produce.

WHEAT FALLOW.

If the farmer fallows for wheat, which is, however, but an unprofitable practice, according to
the modern ideas of husbandry, the land should receive an earth this month, to turn in the weeds
that have arisen since the last. The maxim of
making the fallows very fine in April, to destroy
the weeds by a ploughing in May, or the beginning of June, is in general good ; for how are they
to be killed, if they do not vegetate ? If the fallows are left rough in the common manner, the
seeds of weeds are shut up in the clods : they are
broken by the time the wheat seed is sown : must
not the consequence be their growing among the
wheat ? But it has been urged, that on rich clays
this practice would not be proper, on account of
such spring tillage as I have described, cutting in
numerous

numerous pieces the root-weeds, every bit of which grows; consequently you would do as much mischief in one instance as good in another; but, being turned up very rough in large clods, the sun bakes them, and completely kills the roots. It is absurd to reason against experience; therefore, if a farmer tries the spring tillage, and finds, contrary to expectation, that it fills his lands with pernicious weeds, instead of killing them, he certainly should desist. But many persons know, from experience, that such management has destroyed seed-weeds, and proved no impediment to the destruction of root ones, by successive attention through the burning parts of the summer; not by leaving the roots in possession of large clods, but by extracting them. This is the case with the grasses, which are among the worst. Docks, indeed, can be no way destroyed, but by letting them grow, and then digging them up and carrying them clear away from the land.

It should, however, be admitted (and it is a grand objection to summer fallows) that if the summer proves wet, root-weeds will not be destroyed, whatever the system of tillage may be: that of clods will no more kill them than any other method, except picking them up by women, &c. following every machine by which the tillage may be given.

HOE WHITE CORN.

Wheat, barley, and oats, that are drilled wide enough for horse-hoeing, must be well attended to through

through this month, as all should be now finished
and early in it : the rows well hand-hoed and
weeded at the same time, by the men stooping
down to pluck out such with their fingers as they
cannot get away with their hoes, without damage
to the crop.

HOE BEANS.

The rows of beans will demand great attention
throughout this month : the shims must work the
intervals well, and the rows must be hand-hoed
and weeded ; at present the plants are not advanced
enough to offer any difficulties, and all operations
may, consequently, be performed effectually. If it
is a wet season interruptions will happen, for all
hoeing is then very badly done, but no dry time
should be lost ; it may, from succeeding bad weather, be invaluable.

HOE PEASE.

This crop cannot be managed so well and so
easily as beans ; if in double rows, with wider intervals, in order for the tendrils to join and form
one row, leaving a better space for hoeing; no proper season should be lost in cleaning. Crops dibbled close may demand a weeding.

SHEEP.

I suppose the spring food has lasted till the 10th
or 12th of May ; then they are to be turned into
their summer's grass, in which they are to be
managed according to the nature of the stock. If
the flock consists of lean stock sheep, whose profit is lamb and wool, then the business throughout
the

the year, on whatever food, is to keep them in good and healthy order : these flocks are proper for farms on poor soils, and belonging to which are extensive commons, wastes, or sheep-walks : such tracts will only keep the sheep.

Another management in enclosed countries, is to buy ewes in August or September, to turn them on to the fallows, or the poorest grass on a farm, till Christmas, and then to begin to give them some turnips or cabbages, keeping them in good heart through their lambing, and afterwards as well as possible, that the lambs may be drawn fat by the butcher, soon enough to get the ewes fat and gone by September or October. This is a profitable practice.

A third system of conducting sheep is, to buy in two or three-year-old wethers in the beginning of this month ; to keep them rather bare till about three weeks after the hay is cleared, then to give them good keeping by degrees, and from it put them to turnips or cabbages to fatten, &c. contrived so as not to be sold till March, during which season they sell better than at any time in the year. This is a good sheep management, and will pay the farmer well.

Whatever the stock is, this is the time for turning them from spring to summer food. In the distribution of it, you should attend to the distinction between those sorts of cattle that do well on clover, and such as require natural pastures.

Sheep,

Sheep, hogs, young cattle and horses, are fed to more profit on clover than in pastures; but fatting beasts, large working oxen, and cows that are milked, are in more want of natural grass. It is true, butter and cheese are in many places made from clover; but then we do not know whether the prices are not lower. If clover is good, it will carry five, six, seven, or eight sheep an acre, and on some lands even more. Good grass will carry a cow to an acre; but it must be above the common run. However, in proportioning the stock to the grass, take care to be rather under than over; because it is an easy matter to mow a few acres for hay, in case you have too much; but cattle cannot be sold half fat, but to loss.

In an inclosed farm there is one point which should be particularly attended to, and that is the division of the flock into different parcels for all the fields intended to be fed. Bakewell, who was a great enemy to folding, and, in this respect, for many years little attended to, because it was imagined that his opinion was founded on his own breed being ill adapted to the practice, gave as one material reason for his opinion, that it forced a farmer to keep vastly too many sheep in one parcel. He contended, that the waste of food, from this circumstance, was great, and that the sheep would never be kept as healthy and thriving in large as in small parcels. For many years I tried this system of division with all the attention I was master of, and am well persuaded that he was right in his

opinion;

opinion; and that it is impossible to keep as many sheep upon any farm in one flock as in 10 or 20. The farmer knows nearly what number each field will carry, and they ought to be distributed accordingly, with the precaution of having a pen at one corner in order for examining them daily when the fly is abroad. Here if they are left, with no other changing than drawing off a few, or adding, according to extremes of season, they will do well whether fattening, or ewes and lambs; and afterwards I minuted the following note:—" I continue of opinion, that the quantity of stock I am enabled to keep, depends much on the practice of dividing the flock into small parcels, and leaving them quiet in their respective fields, without folding; and that if they were in one parcel, and folded, they would on this breadth of land be starved. I cannot but urge gentlemen on enclosed farms to make some experiments on this great question, in order to ascertain the loss they sustain in the number of sheep they keep, by adhering to the practice of folding; of the benefit of which, for corn, there is no doubt; but the price paid for that benefit ought to be better understood than I find it. Vague ideas have been long the guide of numbers: it is high time that, on such essential points, positive experiment should alone be attended to."

CLOSE FEEDING

In the distribution of sheep there is another point which demands attention, which is the benefit of close feeding. Here I shall insert a minute which was made on a year's feeding with attention.

" The

" The next circumstance I would wish to note, is that of close feeding. In the preceding trials there was not, through the 30 weeks, scarcely a bent to be seen; the pasturage was constantly shorn to the ground; and in that state it was remarkable to see how constantly, and even rapidly, it sprung, during the continuance of a drought that was destructive of all produce in fields on the same farm, suffered to run to bent, for hay or other views. The comparison was the most decisive that can be imagined. I had many fields, better than any here registered, that yielded so contemptible a produce of hay, as to be scarcely worth mowing; and I was amazed to see in some of them how poor the *rouen* or after-grass was, so that both united, or the entire growth of at least 40 weeks, has amounted not to the fourth of the value of the produce of similar soils pared close by sheep. " A Romney-marsh grazier would be ruined if he had so much grass on his land," says Mr. Boys, in his farming tour, speaking of a field understocked * " " Nothing so bad," says another, " in Romney-marsh, as mowing, so that some landlords prohibit it." Pliny knew this—*Est enim in primis inutile, nasci herbas sementaturus†*. Of the fact, however, I have not the least doubt, from various experiments and observations, and there is no man but has remarked it in the case of ray-grass, the produce of

---

* Annals, vol. xix., p. 118. See also Mr. Price's powerful observations, in that Marsh.

† Plin. Hist. Nat. lib. xviii. cap. 28.

which

which is lost, if the bent be allowed to rise. In all plants cultivated for pasturage, there is a great effort the moment the seed-stem runs, to which the whole growth of the plant is directed to form the seed; till then the growth is in the leaves: it is therefore palpable, that the way to have the greatest abundance of leaf, is by feeding so close as to prevent those stems rising at all. And I may further observe, that on this system of feeding, those grasses which yield a very great, but coarse produce, become sweet, fine, and valuable, by thus keeping them close fed. The *avena elatior*, or tall oat-grass, is very coarse, but in a field of that grass $13\frac{1}{2}$ acres, it never was suffered to rise, and consequently was found on examination to appear as fine and pleasing to the eye as any of the more delicate grasses. It is with this view that I am cultivating it largely, and also the *dactylus glomeratus*; and both are remarkably early.

It is an inquiry that deserves attention, whether the superior profit of grazing sheep, on comparison with oxen, does not depend very much on this point of close feeding; for large cattle the herbage must be kept to a good head, to give a full bite, and consequently innumerable seed stems form, which tend to reduce the produce greatly.

FOLDING SHEEP.

This month begins the folding season throughout England; and the practice is (when relied on) of such importance, that it should be steadily pursued. Many farmers give a very slight dressing: one night

night in a place, and the fold three square yards
per sheep; instead of which it should be two
nights, and only two square yards, or but one yard.
In a word, the land should be quite black, if arable; and with a pretty good covering, if grass.
The proper arable lands to fold this month are the
cabbage and turnip fallows : those crops will be
sown and planted in June, consequently will reap
the benefit of the manure directly. All this on
the supposition that the openness or other circumstances of the farm demand folding.

When I first began to entertain doubts of the
propriety of folding sheep on any farms in which
they can be kept to certain fields in the night,
without that practice, I desired earnestly to try
some experiments that should give more insight to
the question than it was possible for reason to do ;
but to effect this comparatively was very different,
as the trial I wished for was such as should carry
some positive conviction with it. I have not been
able to effect this ; but from several trials, I am well
persuaded that it would have been impossible for
me to have kept, on the same land, nearly such a
stock as I have done in one parcel for folding. I
do not conceive that my fields would have carried
three-fourths so managed. Four drivings in a day
make them trample much food, disquiet the sheep,
and transfers the choice of their hours of feeding
and rest from themselves to the shepherd and his
boy : while lambs are young they are injured by
this, and the ewes are liable to be hurried and
heated

heated, all which are objects that should weigh in the question. When sheep are kept in large flocks, it is not only driving to and from fold that affects them, but they are in fact driving about in a sort of march all day long, when the strongest have too great an advantage, and the flock divides into the head and the tail of it, which must trample on one part the food to be eaten by another. All point the very reverse of their remaining perfectly quiet in small parcels.

But the question turns on the benefit to be reaped by the fold; for if this be great enough to compensate for the loss by such circumstances, the practice may not be condemned.

I believe the reason why farmers are such warm advocates for folding, arises from the power it gives them of sacrificing the grass lands of a farm to the arable of it. Their object is corn; by which they can carry off whatever improvement they bring on to a farm. Grass improved, is the landlord's purse in future filled : and tenants are too apt to think, that when this is done, it is at their expence. They do not at all regard impoverishing a grass field, in order to improve a ploughed one; and I need not observe, that every sort of sheep-walk is thus impoverished; so that ancient walks, which have been sheep-pastured, perhaps, for five centuries, are no better at present than they ever were before; whereas, any fields sheep-fed, without folding from them, are in a constant state of amelioration.

T

tion, which leads me to remark the effect I observed in several of my own fields.

I attended, through the course of a summer, many gentlemen over the fields, with a view to examine if the sheep had seemed to rest only on spots, to the too great manuring of such, or, on the contrary, had distributed themselves more equally; and it was a pleasure to find, that they seemed generally to have spread in every part, if not equally, at least nearly so; and the improved countenance of some old lays, fed in the same manner, when examined in autumn, convinced me, as well as my bailiff, that the ground had been unquestionably improved. They had carried a bad appearance for some years, but they were now of a rich verdure, and as full of worm-casts as if they had been dunged. I rolled them heavily in November, but they soon became rough again by worms: they have now a greener and more fertile appearance by far than ever they wore before.

The whole of this circumstance belongs to this method of dividing flocks, to the exclusion of folding: the fold is valuable, but so is this improvement of the grass land, and may, for what I know, nearly equal it: when in addition we include the greater number of sheep kept, and the favour done to them by letting them alone, there remains in my mind no further question of the fact.

It is common for flock farmers in open countries to say, they have not the power to manage so; which

which may be very true upon the major part of their farms; but such have often many inclosures in which this management might be applied without difficulty.

The practice of a very intelligent Kentish farmer here deserves notice.

" The circumstance which perhaps most deserves attention in Mr. Boys's farm, is that of dividing his flock : instead of keeping his sheep in one flock, he keeps four, and is at the constant expence of three or four shepherds, rather than suffer many to be kept together : he is so clear of the profit of this conduct, that he would not for a moment admit that any question could be made of it : the lambs suffer, the food is wasted, and the whole flock the worse in proportion to its size : of this he was clear to the most perfect conviction."

SALT FOR SHEEP.

I shall at another season note the circumstance of giving salt to sheep particularly ; at present it ought to be observed, that the practice is proper for summer as well as for winter. It is remarkable that this custom should be common management in almost every country in the world, England alone excepted. It certainly tends to keep any flock healthy; and is necessary in proportion to the soil, food, &c. being ill adapted to them ; and also to the wetness of the season.

HOGS.

When the farm-yards are cleared of cattle, the

hogs

hogs should be sorted, and all those of a proper age for feeding on clover, &c. should be drawn and turned into it : this is a part of farming that has been much expatiated on ; but is not common husbandry in a third part of the kingdom. It well deserves to be considered, which is easy to do, as we have had pretty clear accounts of it.

In the old management of swine, they were kept at home, about the farm-house, or a close of grass, all summer, with times of regular feeding on wash, grains, or corn ; but the error of such a conduct was making no distinction between sows with pigs, or weaned pigs, and large hogs. In the more modern method, all the wash, &c. is reserved for the former ; consequently a much larger stock can be kept ; and the hogs, half and three-fourths grown, are turned into the clover or chicory about the middle of this month ; and it is directed, that the gates of the fields be locked on them, and kept there till towards Michaelmas : but for this conduct the fences must all be in excellent repair, and a pond in the field for the hogs to drink at. This food agrees well with them : they grow fast, and are taken out of either clover or chicory in good order for fatting. This practice must certainly be attended with beneficial effects : enabling the farmer to keep larger breeding stocks of hogs is alone of much consequence, and cannot fail of improving his profit : swine will pay for their food as well as any other application of it ; and the conse-

consequence of the whole system, in raising large quantities of excellent manure, cannot be too strongly insisted on.

As the dairy will this month afford great plenty of butter-milk and cheese-whey, you should reserve all that is not wanted for the present stock of sows and pigs, in brick cisterns, so contrived, that it may run without loss directly from the dairy into them: this will be worth many pounds per annum in a farm of any size : where such contrivances are not used, the wash must all be used as fast as it is made, and whether wanted or not ; which is a vastly greater loss than many persons, not used to the improved practice, will easily imagine.

SOILING HOGS.

The preceding system is good, but I prefer soiling them in their yards, notwithstanding the expence is much increased, and that some food will be wasted. The inducement to this practice is the immense quantity of valuable manure which may be raised in this manner. Our farmer should not however attempt it, if he be not well provided with litter of some sort or other, including in that term sand and peat. The hog-yards should have gates wide enough to admit carts for bringing these materials and food in, and for moving away the dung. Lucerne, chicory, clover and tares are the proper food for this system.

HORSES.

The beginning of this month, the farmer should leave off dry meat for his horses. He should soil

T 3   them

them in the stable on lucerne ; and, if he has not
lucerne, on tares or clover. This is one of the
most important articles in his business ; he should
therefore consider it well, that he may adhere to
that practice which most reduces the expence of
keeping the team.

Food given in the stable goes much further than
in the field, and also enables the farmer to raise large
quantities of dung throughout the summer. These
are both objects of great consequence ; and if he
appropriates a small field of lucerne, near the stable,
to this use, he will find it by far the cheapest way
of keeping his horses. An acre, perfectly well
managed on rich good land, and amply manured,
will maintain four or five horses, from May to October ; but, if a farmer would manage in the most
judicious manner, he should allot an acre to every
two or three horses; by which means he will be
sure to have plenty to spare for any other use.

This system of conducting the team cannot be
too strongly recommended : those farmers who
provide grass or clover to turn their horses into,
know well the great quantity of land that must be
assigned them, and the high expences in general
of keeping horses : they should determine to em-
brace all methods of lowering such great expences,
and none offers more clearly, and with a greater
certainty, than the cultivation of lucerne for summer food.

OXEN.

Ox-teams are maintained in winter at a much
less

less expence than horses ; but in summer they are
nearer an equality : the same reasoning is therefore
applicable to both.   It is as advisable to soil oxen
on lucerne as horses : they will thrive extremely
well on it, and at a much less expence than pasturing them in the common manner.

COWS.

In this month, the cows should be kept in good
food, that the dairy or the calves may return the
farmer a due product.   Clover, and ray-grass that
has been fed off early with sheep, will suit them
well ; but if the clover should, as it is commonly
imagined to do, give the butter a taste, the variation of price should then be calculated, on comparison with the convenience the farmer finds in
feeding with that grass.   Lucerne does excellently
for cows, and gives the butter no ill taste : it will,
mown and given in racks or cribs, go much further
than food eaten in the field, and at the same time
yield an opportunity of raising much dung : a
point that ought never to be forgotten.   If this
method is pursued, care must be taken that the
feeding-places are kept well littered.   In this manner the dairy or calves will not fail of proving extremely profitable.   It is not at all necessary to
assert, that the cows will yield as large a produce
in this manner, as when turned into natural grass
up to their horns ; that is by no means the inquiry ;
but there cannot be a doubt of their yielding a
much greater profit, which is the only point of
consequence.   In natural grass, they will eat, spoil,

T 4   and

and trample a great breadth; in exceeding good
grass, perhaps, an acre a head at least; but, if
your lucerne is good, one acre will feed three or
four cows amply.   Such a state of the case at once
shews, that the *product* of the cows has little to do in
the inquiry : it is the *clear profit* alone that should
be considered.

In the feeding of horses, oxen, or cows, with
lucerne, let me observe, that it should be regularly mown every day; and the best way of carrying
it to the stable, will be in a small skeleton-cart
drawn by one horse, and made for the purpose.
In the cutting it, the plantation should be marked
into forty or fifty divisions, according to its growth;
one to be mown every day, or every two days, and
the cattle so proportioned, that they may eat it
regularly.   This will save trouble, and make the
proportion between the cattle and their food be
discovered with the greater accuracy : the lucerne,
if well managed, may be cut four times.

THE DAIRY.

Now begins the hurry of the dairy-maid's work :
this is one of the most ticklish parts of the farmer's
business.   Unless he has a very diligent and in-
dustrious wife, who sees minutely to her dairy, or
a most honest, diligent, and careful house-keeper
to do it for him, he will assuredly lose money by his
dairy : trusted to common servants, it will not pay
charges.   The dairy-maid must be up every morning by four o'clock, or she will be backward in her
business.   At five, the cows must be milked, and
there

there must be milkers enough to finish by six. The same rule must be observed in the evening. Cleanliness is the great point in a dairy : the utensils should all be scalded every day : and cold water should be poured down on the floor in hot weather, a cock of water running constantly through it : falling on the floor, and dashing a good deal about, would have effect in cooling the air. The fountain recommended before, better still. There is scarcely any part of a farm that wants contrivance more than a dairy : if the number of cows be great, well contrived conveniences would save much expence of labour, and pay a farmer for erecting them himself.

Mr. Abdy, in his account of the Epping dairies, remarks, that their farmers buy pigs at four or five months old (which, in 1788, cost 18s. each), keep them on skimmed-milk for about a month, and sell them with 6s. profit. The general proportion, one to every three cows in milk ; and as the cows (the long-horned Derby breed) in general stand to the pail for nine months, this will make three pigs fatted from the milk of each. The average quantity of butter made by each cow, per week, is 4lb. of 16 oz. and the whole, therefore, of each cow 156lb.

156lb. at 10d.	£. 6	10	0	at 1s. 3d.	£. 9	15	0
Calf,	-	0 18	0		1	5	0
Pigs,	-	0 18	0		1	10	0
	£. 8	6	0		£. 12	10	0

The

The age of cows when most in use, after the second, third, and fourth calf. When the hay is a large productive crop on the ground, the cows eat more, and the milk is not so rich as when the quantity on the ground is less. The dairy-maids are peculiarly attentive to one circumstance, that there must be a certain proportion of *sour* in the cream, either natural or artificial, or they cannot ensure a good churning of butter; some keep a little of the old cream for that purpose; others use a little rennet; and some a little lemon-juice. When the butter is come, the dairy-woman throws it first into clear water, and then on to a board, and with her hand in each situation, squeezes the butter-milk out, and when on the board, sprinkles a very little salt over the whole mass (for one of the properties of Epping-butter is, to have very little salt in it); the lump is then divided into pounds, and as they are weighed, are thrown into fresh water; when they are all weighed they are again squeezed and rolled on the board with the hand till they are about 14 inches long. This is the whole process.

In Suffolk it has been found that $4\frac{1}{2}$ gallons of milk give one quart of cream; which quart gives one pint of butter, which weighs $1\frac{1}{4}$ lb.

THE METHOD OF MAKING BUTTER IN HOLLAND.

BY MR. R. P. CAREW.

Having milked the cow, the milk is not put into pans till it is quite cold. It is then stirred two or three times a day with a wooden-spoon, to prevent the cream from separating from the milk; and if

it

it can be stirred until the spoon will almost stand in it, it is deemed so much the better. When it is found to be sufficiently thick, it is put into the *churn*, and beat for an hour. When the butter begins to form, a pint or more of cold water, according to the quantity of the milk, is poured in, to separate the butter from the milk.

When the butter is taken out of the churn, it is washed and kneaded till the last water is *perfectly clear* and free from milk. By this method, a greater quantity of butter is made from an equal quantity of milk. The butter is firmer and sweeter. It will keep longer than that which is made in the ordinary mode which is in use in England, and the butter-milk is thought preferable.

N. B. A churn is thought better adapted to the purpose than a barrel.

Butter and Milk in Cheshire.—" There does not appear any thing particularly worthy of notice in the process of making butter, unless it be the common practice of churning the ' whole milk,' instead of setting up the milk for the cream to rise, and churning it alone, as is the custom in most other parts of the kingdom. In Cheshire, the whole milk (viz. cream and all, without being skimmed) is churned together ; and preparatory to that, the meal is immediately, after milking in summer, cooled in quantities proportioned to the heat of the weather, previous to its being put together, which from time to time is done in earthen ' cream mugs,' or jars. In these jars (containing
four

four or six gallons each) it is intended to stand till
it is ' cawed,' (as the term is) or clotted in a proper degree for churning, and this is judged to be
the case sufficient for the intended purpose, as
soon as the whole is coagulated, and has acquired
a small degree of acidity, which will generally take
place in warm weather in the course of a day or
two. In winter the cream mugs are placed near a
fire, to forward the ' cawing ' or clotting of the
milk. If the milk in warm weather has not been
sufficiently cooled before it is put to the former
meal, or if in winter the mugs have been set too
near the fire, it curdles the whole mass, making it
(as the phrase is) ' go all to whig and whey,' and
afterwards heave in the mug. Again, if in summer, or when kept in a warm situation, the milk
is not churned within a day, or a little more, after
it is sufficiently ' cawed,' a kind of fermentation
and heaving also ensues : in both cases the butter
will be rank and ill-tasted ; nor will the milk produce so much butter, as when it has been properly
managed and churned in proper time. We do not
find that any comparative experiments have in this
part of the country been made, so as to ascertain
with any degree of certainty, which of the two
common modes of obtaining butter is the best, in
regard to quantity, flavour, &c. This matter, however well worth attending to, might easily be ascertained by experiments, both simple and unexpen-
sive. In most parts of Cheshire, butter is made
up for sale in lumps, that have the term ' dishes '
applied

applied to them : the weight of a ' dish' is one pound and a half, or 24 oz.

" The churns in common use are of the upright kind, and have, in some instances, a lever applied to them : when that is the case, one end of the lever (which is supported by an upright frame) is connected to the end of the churn-staff; the other end of the lever, by means of a rod, is connected to a crank in a toothed wheel, and this is worked by a pinion fixed upon the axis of a common winch. By means of this contrivance the business of churning is performed by one person with the greatest ease."—*Wedge.*

CHESHIRE CHEESE.

" The general mode of making cheeses is from 50lb. to 60lb. weight each, and which now sell, from good dairies, at from 43 s. to 55 s. per 120lb. and upwards.

" The process of making Cheshire cheese is as follows, viz. on a farm capable of keeping 25 cows, a cheese of about 60lb. weight may be daily made in the months of May, June, and July.

" The evening's milk is kept untouched until next morning, when the cream is taken off, and put to warm in a brass pan, heated with boiling water ; then one-third part of that milk is heated in the same manner, so as to bring it to the heat of new milk from the cow. (*Note.*—This part of the business is done by a person who does not assist in milking the cows during that time.) Let the cows be milked early in the morning, then the morning's

new

new milk, and the night's milk, thus prepared, are
put into a large tub, together with the cream;
then a portion of rennet, that has been put into
water milk-warm the evening before, is put into
the tub, sufficient to coagulate the milk; and at
the same time, if arnotta be used to colour the
cheese, a small quantity, as requisite for colouring
(or a marigold or carrot infusion) is rubbed very
fine and mixed with the milk, by stirring all together, then covering it up warm, it is to stand
about half an hour, or until coagulated; at which
time it is first turned over with a bowl, to separate
the whey from the curds, and broken soon after
with the hand and bowl into very small particles:
the whey being separated by standing some time,
is taken from the curd, which sinks to the bottom;
the curd is then collected into a part of the tub,
which has a slip or loose board to cross the diameter of the bottom of it, for the sole use of separating them, and a board is placed thereon, with
weights from 60lb. to 120lb. to press out the
whey; when it is getting into a more solid consistence, it is cut and turned over in slices for several times, to extract out all the whey, and then
weighed as before; which operations may take up
about an hour and a half. It is then taken from
the tub, as near the side as possible, and broken
very small by hand, and salted, and put into a
cheese vat, enlarged in depth by a tin hoop to hold
the quantity, it being more in bulk than when
finally put into the press; then press the side well
                                                    by

by hand, and with a board at top well weighted, and placing wooden skewers round the cheese to the centre, and drawing them out frequently, the upper part of the cheese will be drained of its whey; then shift it out of the vat, first put a cloth on the top of it, and reverse it on the cloth into another vat, or the same, which vat should be well scalded before the cheese is returned into it; then the top part is broken by hand down to the middle, and salt mixed with it, and skewered as before, then pressed by hand, weighted, and all the whey extracted. This done, reverse the cheese again into another vat, warmed as before, with a cloth under it; then a tin hoop, or binder, is put round the upper edge of the cheese, and within the sides of the vat, the cheese being first inclosed in a cloth, and the edges of it put within the vat.

" N. B. The cloth is of fine hemp, one yard and a half long by one yard wide; it is so laid, that on one side of the vat it shall be level with the side of it, on the other it shall lap over the whole of the cheese, and the edges put within the vat, and the tin fillet to go over the whole. All the above operations will take from seven in the morning till one at noon. Finally, it is put into a press of 15 to 20 cwt. and stuck round the vat, into the cheese, with thin wire skewers, which are shifted occasionally; in four hours more it should be shifted and turned, and in four hours more the same, and the skewering continued. Next morning let it be turned by the woman who attends the milk, and put under
another

another or the same press, and so turned at night and the next morning; at noon, taken out finally to the salting-room, there salt the outside, and put a cloth binder round it. The cheese should, after such salting, be turned twice a day, for six or seven days, then left two or three weeks to dry, turned and cleaned every day, taken to the common cheese-room, laid on straw on a boarded floor, and daily turned until grown hard. The room should be moderately warm, but no wind or draught of air should be permitted, which generally cracks them. Some rub the outsides with butter or oil, to give them a coat."—*Chamberlayne.*

" Cheese made from clover is rather more diffi-cult to make, to even the best of dairy-women, but I have seen very good sound dairies of stout, full fla-voured cheese made from clover, especially when a good deal of time is allowed to bring the cheese, and care is taken not to let it lie too hot after it begins to get dry."—*Twamley.*

" It has generally been reckoned, that the milk required to make one pound of butter will make two pounds of cheese, and a larger quantity where land is poor, the milk being weak, will not afford so much cream."—*Twamley.*

" By a pretty accurate calculation, in the upper vale of Gloucester, I found that a cheese weigh-ing somewhat more than 11lb. (namely, at 10 to the cwt.) took about 15 gallons of milk (ale mea-sure) or one gallon and one-third to 1lb. of *two meal cheese.* From two instances, minuted with tolerable

tolerable accuracy in this district, the proportion appears to be in one instance somewhat more, and in the other somewhat less, than 1lb. of curd to one gallon of new milk.

" The season of making.—" Thin cheese " is made from April to November; but the principal season for making " thick cheeses " is during the months of May, June, and the beginning of July. If made late in the summer, they do not acquire a sufficient degree of firmness to be marketable the ensuing spring.

" The rennet made use of at Frocester was prepared in this manner : to two gallons of water, made salt enough to bear an egg, add one penny-worth of mace, one pennyworth of cloves, a handful of sweet-briar and hawthorn buds, a small quantity of alum (about the bulk of a small walnut), the same quantity of sal prunella, a small quantity of cochineal (a small *pinch*—the bulk of half a hazel nut), and, if to be had, two or three bay-leaves. Pound the alum, sal prunel, &c. and having mixed the several ingredients with the salt and water, add 5 vells, or if small, 6 or 7. In about 10 days the rennet will be fit for use.

" Another recipe which I was favoured with in this vale, is the following : three handfuls of common salt to three quarts of water, a quarter of an ounce of salt-petre, and as much black pepper as will lie upon a shilling, a small quantity of agrimony, a sprig of sweet-scented thyme, a handful of sweet-briar, a handful of the red buds of hawthorn, four

u   heads

heads of sage. Add the ingredients, and boil the water a quarter of an hour. To the liquor, when cold, put one vell. The rennet may be used the next day."—*Marshall, Gloucester.*

PARE AND BURN.

Paring and burning the turf is, in some places, begun so soon as March; it holds all through May. In the burning, many hands should be set at work at once, that a dry time may be caught for it, in case the season in general proves wet. The ashes should be spread before the plough, and turned in immediately: one peculiar circumstance attending the breaking up of grass-lands, whether old turf or sainfoin lays, in this manner, is the bringing them in order for turnips with only one ploughing; and it is a general and very just observation, both in the north and west of England, where this husbandry is most common, that turnips scarcely ever are known to fail on burnt lands: the fly, on such, is nearly unknown. Now, any farmer must be sensible of the vast importance of thus bringing turf-land, by only one ploughing, to a turnip crop: much tillage is thus saved, as well as a great expence; and the turnips are generally a crop, that repays the expences of the operation with profit. In a word, this husbandry deserves the warmest praise.

But of late years, an opinion against it has prevailed much in some counties. Several of the nobility and gentry, of very large estates, have interdicted the practice, not allowing their tenants to

pare

pare and burn under any pretence whatever. The reason assigned for this conduct, is an apprehension that the depth of the soil decreases from it: that you burn *the land*, and reduce half an inch to half a line; a great evil, when the land is perhaps only three or four inches deep on a lime-stone rock. But this reasoning, many very sensible and experienced farmers know to be false. They, on the contrary, urge the universal circumstance of no land ever being pared till it has acquired a turf, which, with natural grasses, will be from 12 to 20 years; and, with sainfoin, the duration of the crop, which is from 10 to 20 years: that it is not the *soil* which is burnt, but the bulbs of the plants, the roots, and net-work of grass roots: the earth, which is intermixed, is not burnt; it is calcined, but not reduced to ashes, all of which arise from bulbs and roots: hence the fact, that the staple of the soil rarely suffers from paring and burning. If this reasoning be not true, whence the known fact, that soils not four inches thick, and which have remained at the same thickness as long as the oldest man can remember, have yet been pared and burnt regularly four or five times in a century; and, as the same husbandry is known by record to have been practised for ages on the same land, the staple must have lost three inches every hundred years; in other words, it must have been totally gone long ago, and nothing but rock remained: all which is evidently false, the soil at this day being as thick as ever. We may hence conclude, what such farmers assert to

be true, that the earth suffers no diminution, those
roots and bulbs only being reduced to ashes, which
in breaking up by the plough alone, would rot
away.

HOPS.

Dig the new-planted hop-garden this month;
earth up the plants, and see that no weeds are left
to infest them.   At this time you should also pole
your old plantations, proportioning the poles to
the age and growth of the hops.   Within a short
time after, the binds are tied to the poles.   These
are nice operations, and not to be detailed in a
work of this sort.

BEES.

Watch well the apiary, for you must now expect
the bees to swarm.   This useful insect is not so
much attended to by many farmers as it ought to
be : not a farm-house should be without bee-hives;
for the trouble they give is very trifling, and by
farmers small profits should not be neglected : the
union of them is not trifling.

HEMP.

As this is the season for sowing hemp, and it
will now probably attract the attention of our
young farmer, it is necessary to consider here
the culture generally as well as particularly :
first, the inducement to enter into this husbandry;
and secondly, if that be determined, the means of
doing it.

As hemp is an article in which the farmer produces a raw commodity for the use of the manufacturer,

facturer, he is liable to suffer if he does not move
with caution in a country where it is not in the common husbandry of the vicinity : where his brethren produce this or any other commodity, the
silent progress of competition, and the customs of
trade, which gradually establish themselves, usually
render it fair between the parties : and an article
in which the profit went too largely to one party,
in prejudice of another, would soon be given up.
This situation is quite different from a young farmer attempting to introduce this or any other article in a district where they are not common ; he
is, in such, extremely liable to suffer in the sale
of his product. Before, therefore, he ventures
on any such articles as hemp, flax, woad, madder,
&c. let him most carefully ascertain his market,
and the variations the commodity is usually liable
to ; and, if he cannot contract for the sale, he
should be cautious of engaging in the cultivation.
It may be twenty to one whether he has any thing
like an open market to carry his product to ; and
if he has not, he may find a crop on his hands,
easily sold, but not perhaps at a fair price. The
profit of cultivating hemp, *in hemp districts*, is not
inconsiderable, amounting usually from 5l. to 7l.
an acre, which, under certain circumstances, is an
object worth attention ; but here again he is to
take into his account some other circumstances
which demand attention.

It requires the very best land to be found on a
farm, or which is made such by manuring : a rich,

deep, putrid, friable, sandy loam ; and it cannot
be too rich. It further demands ample manuring ; that is, from 20 to 30 loads of dung per
acre, equal to 40 cubical yards by measure. It
is true, that it will pay for this dung, but every
one knows that in calculations there is apt to creep
in (in the point of charging dung) some degree of
fallacy. Wherever spread, there, probably, will
be the greatest profit ; and ir hemp, hops, madder, &c. rob the more common crops of dung,
which, but for their culture, would be disposed of
otherwise, it is no easy matter to charge them for
it sufficiently high. If hemp enters largely upon
a farm (which it rarely does, and for this reason),
the cabbages, potatoes, lucerne, &c. must be contracted.

Another very material circumstance is, that
hemp returns nothing to the farm wherewith to
raise dung : corn gives straw, and green crops swell
the dunghill ; from hemp the farmer gets nothing
of this sort.

Hence, the husbandman that looks only to the
profit on the estimate, however fairly it may be
drawn up, will not have the subject before him in
all its bearings : he must reflect well before he begins. If, upon the whole, he thinks the undertaking advisable, he will in the next place attend
to the circumstances of the *culture*.

Soil.—This should be the richest on the farm ;
deep, moist, friable, putrid : if none of that description, any deep, good sandy loam, worth 30s.

or

or 40s. an acre should be applied. Mellow, rich clayey loams do well ; and nothing better than old meadow-land, no matter what the soil, turned down by the skim-coultered plough.

Tillage.—There are many crops for which tillage should be cautiously given, as the weeds that may set a growing will choak and get the better of various plants ; but this is not the case with hemp, which is so predominant in its growth, that it kills all weeds. The land should, from the preceding autumn to the time of sowing, have three or four ploughings : or two, and sufficient scufflings, and be well harrowed to a fine surface.

Manuring.—Dung should be amply applied, in proportion to the deficiency of the soil in fertility ; but it is rarely ventured on any without a good dressing. And when the culture is continued on the same land (the most usual system for it), yet they commonly manure it every year if they can ; from 16 to 50 cubical yards are given, according to the soil.

Seed.—The quantity varies according to the opinions of individuals, and the practice of different districts, from 11 to 14 pecks.

Sowing.—Universally broad-cast ; nor can I see any motive for drilling a plant which utterly destroys weeds, except one, that of burying the seed at an equal depth. Should any person be inclined to drill, the rows should be as near to each other as the shares of the drill can be set.

COTTAGERS HEMP.

It is an extraordinary circumstance, that by far the greatest part of the hemp that is raised in Europe, is by cottagers, or very little farmers : this is the case in the Ukraine, as well as in Suffolk. Here, as well as there, and in many other countries, it is sown every year on the same land by cottagers, who provide dung for it by keeping a cow, or some pigs. Whether it is the most beneficial culture in England for such a person, has been disputed ; but, when the benefit of manufacturing it themselves is taken into the consideration, and the advantageous winter employment it affords to the women and children, I have little doubt of its being the best crop they can attend to ; or, of its yielding them much more neat profit than sufficient to buy any or all other products the same land could yield, if not thus employed.

FLAX.

This is another culture that requires extremely rich land. It answers pretty well with due attention ; but I may remark on this crop what I did on hemp, that the same favourable circumstances of soil, manure, and weeding, would repay the farmer much better in other crops, with this general and great superiority : hemp and flax return no manure, whereas many other crops I propose are undoubtedly beneficial to the soil, and vastly improving to a whole farm, in the quantity of dung they enable the farmer to raise. Flax may be sown the end of April ; but more commonly

monly in May, as it is liable to be damaged by frosts.

It does best on grass-land for the first crop, but perfectly well wrought to a fine surface. Two bushels an acre the quantity of seed. It must be kept perfectly free from weeds.

SPRING TARES.

If the farmer depends on a succession of tares for soiling, or for feeding sheep, he must sow for one crop some time in this month ; and better still twice, in case the April sowing was early in that month.

WATERED-MEADOWS.

The ewes and lambs are to be taken out of these meadows the last day of April, by which time it is supposed they are fed quite bare, and it should be remarked, that the barer they are fed so much the better. I have heard farmers express themselves well pleased at finding that their sheep had eaten almost into the ground, as they said the meadow would be the better, and the crop of hay of the finer quality : the observation is general for all spring feeding. Immediately on clearing, Mr. Boswell directs a week's watering, with careful examination of every trench and drain ; and the water shifted into other meadows in succession, alternately watering and draining ; and lessening the time the water remains on the land, as the weather grows warmer, and in five, six, or seve weeks they will be fit to mow for hay.

FORM

FORM NEW WATERED-MEADS.

In situations that possess plenty of water, with meadows for receiving it, they may be formed at any time
in the year except in severe frosts; but when there is
any doubt of the quantity of water being sufficient,
the safer way (if any large expences are in contemplation) is to begin the work in a season when
the undertaker is not liable to be much deceived :
this *may* be the case in any month, if the preceding
period has been remarkably wet : allowance must,
however, be made for any such accidental circumstance ; and it will be a good precaution not to lay
out any large extent of meadow, till some experience has been had, that the quantity of water is
sufficient for the meditated work.　As this is
(warping alone excepted) the greatest of all improvements, it is deserving of the greatest consideration and study of the water and land before a
beginning is made.　I should recommend, in the
first instance, the employment of a professed irrigator, could the young farmer possess knowledge
enough to ascertain the skill of such a man ; but I
have lately seen such gross blunders made in Norfolk by such an one, on the farms of four or five
persons, and yet highly recommended, and coming from Gloucestershire, that I really think a man
may just as well trust to himself, with the assistance of books, as to put any faith in men who are
reputed skilful only in proportion to the ignorance
of those who employ them.　In the cases to which
I allude, this ignorance was unpardonable ; for as
they

they discovered that he drew out all his works
without the assistance of a spirit-level, they ought
to have dismissed him.   Not that such a fellow
cannot make improvements ; no one can well contrive to bring water on to land without improving
it ; but to pay 4l. or 5l. or perhaps much more, per
acre, for using a small quantity of water to *some*
advantage, when the same might be used elsewhere to *the greatest*, is, comparatively speaking,
throwing money away.   If the following observations are carefully attended to, they will, I trust,
enable any man to operate for himself in most of
the cases that can occur, and with a certain degree
of sagacity, in all.

1st, The great benefit to be derived from power
to take water from a river, or stream, or lake, &c.
will much depend on taking the first level from
the highest spot on the water to which the operator's property or farm extends.   If he is a landlord, and has several farms on a stream, and some
out of lease, others in lease, he must either wait
till the leases are expired, or he must purchase of
his tenants liberty to run his grand carrier through
their farms (the property of the water retained to
himself) wherever the level may point out.

2d, If the stream be any thing considerable, he
may probably find water-mills the greatest impediment to his project, whether his own, if leased, or
belonging to other landlords : he must make himself thoroughly acquainted with this circumstance,
or, after having been at considerable expences, he
may

may meet with a prohibition against taking the
water, when he is just ready to open his sluices and
set it a-flowing.

3d, If the river be the boundary of his property,
and his neighbour on the other side has an equal
right to the water with himself, he may be in the
same predicament, and a deed of sale, or transfer,
or permission, must be obtained before he begins,
not possibly in the power of the person to grant.
All this must be well understood before he commences his operations.

4th, If, in running his first and highest level, he
meet with a field or fields not his property, which
intervenes by elbowing into his estate, and cuts
him off from much other land that belongs to him,
he must buy such fields, or perhaps lose the greatest
benefit which would result from his operations.
If he cannot buy the fee-simple, can he buy permission to cut through it ? He must know this
before he begins.

5th, Having given due attention to all these
circumstances, let him repair with his spirit-level
and attendant to the highest spot where the stream
enters his property or farm, where he has the property on both sides, and where he can erect works
across the stream in order to divert the whole, or
any part of the water into a new channel ; and let
him begin to take the level from the surface of the
water, supposing it pent to the highest by such
works.

6th, He is to level from that spot following the
dead

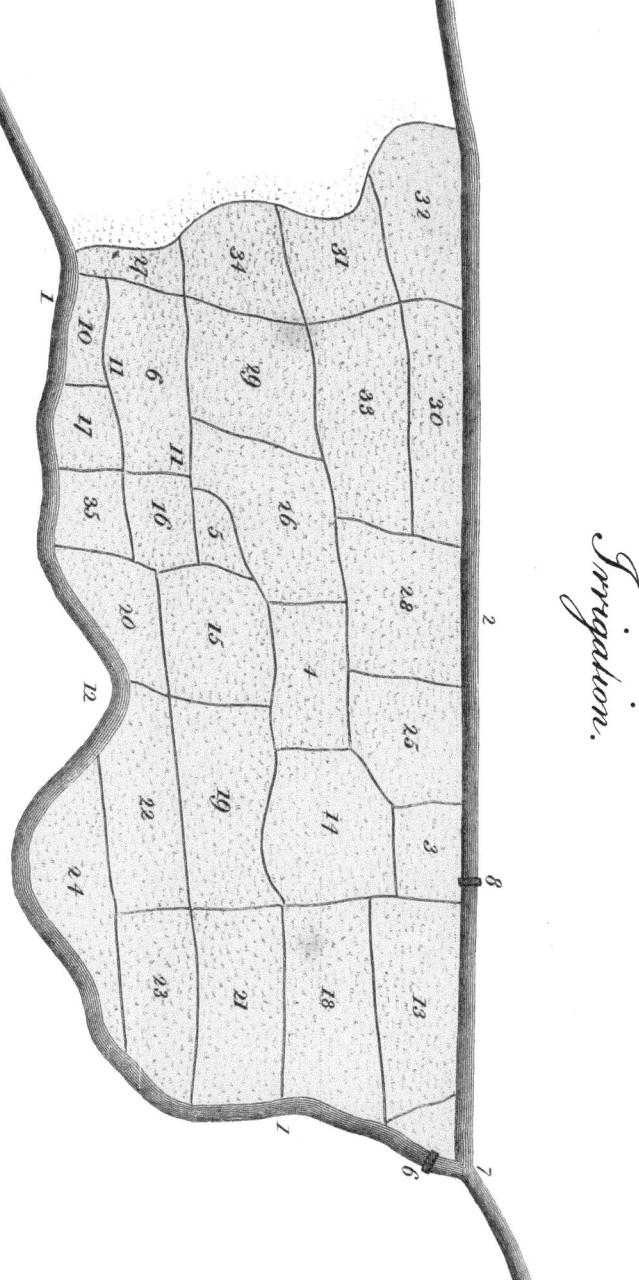

Irrigation.

Slope from 3 to 12 & from 7 to 10

Published Nov. 10. 1804. by Richard Phillips, 71. St Pauls Church Yard.

dead level, and at every three or four hundred yards, staking it out doubly, one stake on the dead level, and another near it, descending from the dead level, so many inches as an allowance to give the water a current; two inches in a mile will move it, but twelve to twenty should be allowed, in order that the current may be sufficient.

7th, Roads need not be any interruption, as they may be passed in a manner hereafter explained; but farm-houses, yards, gardens, cottages, &c. may intervene; and if they do, a much greater descent per mile must be allowed, that such interruptions may not be quite suddenly, but *gradually* provided for; as the former occasions inconveniences.

8th, In this manner let him proceed to the extent of his property, leaving stakes at all his stations, so firm in the ground, that they cannot be removed.

9th, If both sides the stream be his own farm or property, let him go back whence he started at his highest spot, and do exactly the same on the other side of the vale, staking out his level there also.

10th, Having proceeded thus far, let him view, and carefully examine all the land on both sides the river below his lines of levelling, for he has the power of watering all or any part of it, if the stream be sufficient. The breadth will depend on the degree of the declivity between his first and last station; and on the diverging and fall of the higher grounds or hills; but in every case he will find a vastly greater quantity of land than he had

any

any conception of before he took the level. Mr. Bakewell lent his irrigator to a friend, in order to ascertain whether he could water the church-meadow; and on the level being taken, it proved that the water might be carried over the church-steeple, had the land been high enough to receive it. And at Euston, the seat of his Grace the Duke of Grafton, it having been a question in conversation, whether such and such lands could be watered, I took the levels for above five miles from Sappiston Mill, and found that the sand fox covers, on pretty high hills near the Hall, might be converted to water-meadows.

11th, When the levels are taken, in examining all the lands below them, the main point (in which errors are perpetually made) is to determine where to begin. If there be water sufficient, all should certainly be done; but, supposing a choice necessary to make, whether from insufficiency of water to do the whole, or from any other cause, then the operator must exercise his judgment; and in making his estimation, he is to attend to the following circumstances.

1. The expence of digging his grand carriers on the levels first taken, which should be large enough to take *the whole stream* on either side the vale at pleasure; but if the lands on one side are more favourable for watering than on the other, let that carrier be executed first. If the most favourable land to work upon on such side, be at a great distance from the *prise d'eau*, or original spot where the water is first taken, and there be not water for the

the whole below the level on one side of the river,
then he must compare the expence of the carrier
with the superiority of the profit of watering the
most favourable fields, rather than others nearer
not offering in themselves an equal advantage. In
most cases, the benefit of watering at pleasure is
greater than the expence of making the carrier.
And in this point there is also another consideration of great moment; though the river may not
afford plenty of water in summer, or very dry years,
yet it may abound in winter and watering at that
season alone is well worth the expence of all the
necessary works in most cases; and this consideration
tion should influence him rather to extend his carrier,
than, by shortening it, be forced to water fields not
so well adapted as others at a greater distance:
probably the winter watering may go on through
the whole line.

2. When such a level is taken as I have described, there is often found below it a great variety of soils and circumstances; usually a low flat
range of meadows, perhaps wet and boggy, by the
side of the old river, and adjoining, and above
them slopes of pasture, and arable, it may be dry,
gravelly, sandy fields on various angles of declivity;
bogs, ling moors, and in short, every sort of soil
and land; so that the irrigator may chuse what he
pleases to work upon. Here he must be instructed,
that the lands usually chosen for the first operations, are just those that ought to be the last, viz.
the low flat meadows by the river. These are
often improveable to a very high degree by drain-

ing

ing and manuring with sand, gravel, earth, chalk, marle, &c. ; but they are by far the most expensive to irrigate, and when done, unless very well executed indeed, yield the worst hay. They are best watered, and in many cases, only to be watered advantageously, by ploughing them into broad and highly arched ridges ; the delivering trenches to be on their crowns, and the drains in their furrows. But the profit of irrigating dry slopes of sand and gravel, &c. and poor dry ling moors, is immense; the expence is comparatively trifling, and the improvement beyond conception : such lands may be raised from 2s. or 3s. an acre, to 40s. or 50s., while the flat meadows may be 20s. before the undertaking begins, and may not, when ended, be worth more than the others, though effected at ten times the expence. I once found a friend in the full speculation of watering some meadows which were worth 25s. an acre; and just ready to set a man to work, who ought to have known better. I thought by my eye, that the water (the quantity very limited) might be better employed on some dry arable land above the meadows, but further down the vale. I took the levels, and found it as I conjectured : the plan was adopted, and I have since heard, that the undertaking was remarkably profitable. The meadow at Six-mile Bridge, in Hampshire, which lets for above 5l. an acre (a gravel at 10s. before watering), was formed at little other expence than converting a ditch into a carrier ; nor was the conduct of the water, when I saw it, correct by any means.

12th,

12th, In viewing, therefore, the lands below the grand carrier, our operator should chuse, for his first works, those fields, the soil or state, or value of which are the most promising for working a great improvement, and these will be the dry arable slopes, or poor dry pastures. And if he has a choice, let him begin with one which joins his carrier, and mark the lower spot, side or corner of it, where the water may best have its issue, having been worked over its surface; and at that spot taking his station, let him examine what field is so situated as to take the water next. If the declivity is at all steep, any may do; but if gentle, it may be necessary to conduct it diagonally to some distance, before a field is found low enough to receive it; for let our operator have it in his attention, that he is to conduct the water thus first taken till its final exit into the bed of the river, whence it can no more be taken by him, before he meddles with any other work. In some cases it might be more profitable immediately to water other fields nearer the carrier, but as the water used in the first field would in that case run to waste before it arrived at the river (especially if the tract to be watered be of any extent), and as it is beneficial to plan what is to be done with all water taken, in its whole course from the carrier to the river's bed, it is much better to finish with it before a fresh work is opened.

On gentle slopes of country this plan will gene-

rally

rally make it necessary to conduct the water by a
line of fields diagonally across the slope of the
country.   To illustrate this, reference may be had
to the annexed Plate.

Here (1), is the river; (2), the grand carrier;
(3), the field first watered; (4), the field watered
after the first; (5 and 6), ditto, in succession;
(7, 9), the *prise d'eau*; (8), a sluice to throw the
water into the field where first used; (10), final
exit of the water.   But in this respect variations
may be as many as the forms which a tract of
country presents.   (6), in this dragram, may be a
good meadow already; in that case the water may
run to waste in the ditch (11), and so find its way
to the river.   The main object in such a work as
this Calendar, is not to give a treatise on irrigation,
which would demand two such volumes, but only
to bring to the operator's recollection certain points
to which he ought to pay attention.

Having watered one range of fields diagonally
across the space between the grand carrier and the
river's bed; he may then return and work a second
range, and so on till he has acquired a power of
conducting the water to any field at pleasure; and
by means of the ditches which intersect the space
and form the divisions of the fields, he can miss any
fields that are unprofitable on comparison with
others, leaving them for future operations, when
the quantity of water, and the effect of the irrigation shall be better known. But while he thus has

it

it in his power to pick and choose the fields for watering, still the whole is on one regular plan, by which he may at any time be able to execute all the parts, and render the whole, when finished, if it be advisable to finish it, perfect and uniform.

The divisions 13, 14, 15, 16, 17, form a diagonal system.

The others, 18, 19, and 20, another.

21 and 22, another.

23 and 24, another.

25, 26, 27, another.

28 and 29, } others.
30 and 31, }

32 and 33, by themselves, the one from the carrier, the other from the ditch between it and 28.

13th, In regard to the respective operations to be performed in each field, it is necessary in such a work as this, merely to afford such general principles and hints as a man of any sagacity may apply to every distinct piece of ground when he views it. The grand principle which is to govern these works is, to bring water on as plentifully as may be, and let it run off quickly,—*nimbly*, in irrigating language: if it stagnates it does mischief, and if it only creeps sluggishly, much less good than a better course would have enabled it to effect. All spaces that are level, or nearly level, should be ploughed on to lands or ridges eleven yards wide,

x 2   and

and raised, if water be plentiful, three feet higher on the crown than in the furrow, in this form,

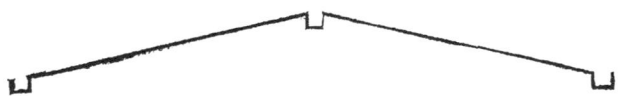

and of course these ridges must be laid out by the spirit level, so that the delivering trenches on their crowns may be able to take water from the larger carriers which lead along one head-land ; and that the furrow-drains may convey the water away to the receiving ditches provided for that purpose. Those trenches are to be so exactly cut, that they will overflow through the whole length equally at the same time, for which purpose they lessen in breadth as they advance. But upon dry slopes, nothing more is requisite than to cut trenches of de-livery, which operate by alternate watering both as deliverers and as drains. This is a point little understood in watering through several districts I have seen, and as it is a very important one, and a branch of that diagonal system I have already ex-plained with relation to the position of the fields, it merits a short explanation.

In the annexed Plate, the slope of the land from A to B, is supposed regular, which, of course, rarely happens in nature, nor is it essential, as any man who has common sense will see that inequali-ties of surface, though they may break the uni-formity of his lines, by causing a necessity of going round

To front page 300

Watered Meadow.

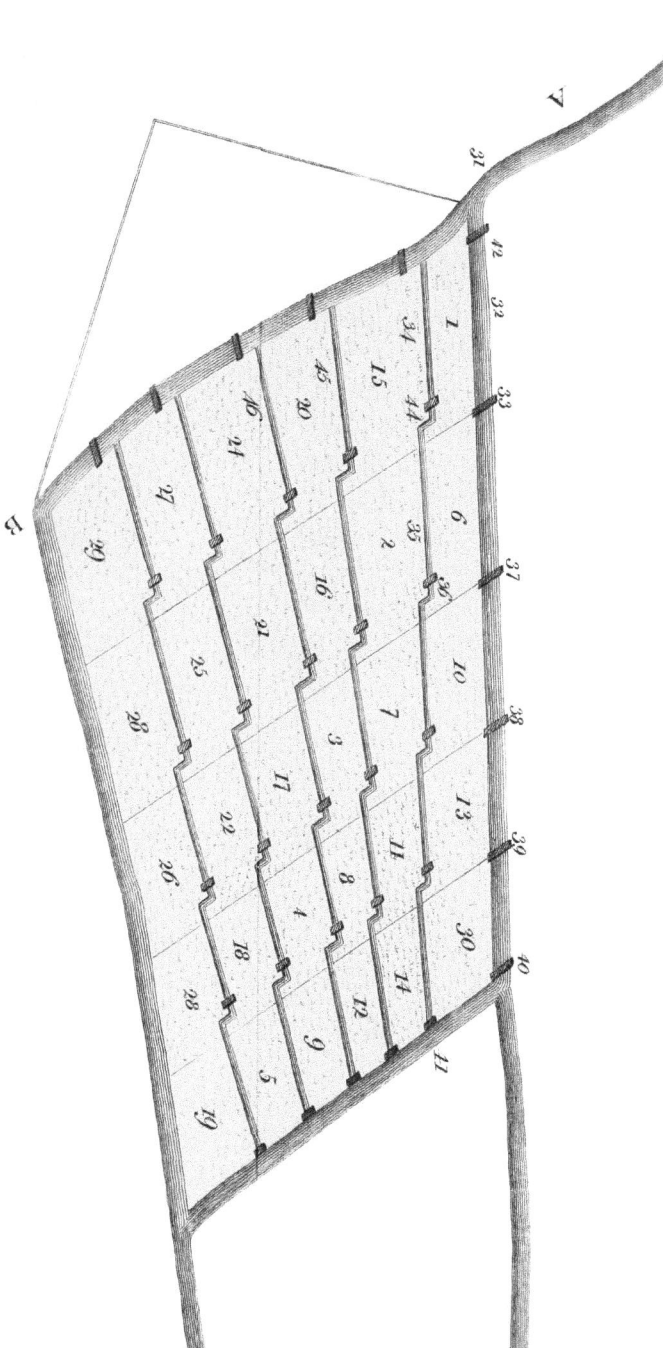

Published Nov. 10. 1804 by Richard Phillips, 71 St. Paul's Church Yard.

round hills or holes, yet will make no breach in the principles which govern the irrigation.

Here, it appears, that if water from the main carrier, river, or ditch, 31, be let into the delivering trench, 32, and the stop, 33, be let down, the water will flow over the division of the meadow (or *pane*, as Mr. Boswell calls it) 1. The delivering trench, 34, then acts as a drain, and conducts the water into the trench, 35, the stop, 36, being let down; thence, of course, it overflows the pane No. 2, and in like manner, successively, No. 3, 4, and 5. If the stop, 33, be drawn up, and the stop, 37, let down, the panes 6, 7, 8 and 9, are watered in the same way; and so on by the stops 38 and 39, which will water the panes 10, 11 and 12; also 13 and 14: and the stop, 40, being let down, and 41 drawn up, the pane, 30, will be watered. Then return to the ditch at the other end of the field, and letting down the stops, 42 and 43, it is evident that the water will flow into the trench 34, and the stop, 44, being down, the pane, 15, is watered, and the trench, 45, becomes a drain, which, successively, conducts the water, as above explained, over the panes 16, 17, 18 and 19. Now it is clear, that when the trench, 46, becomes supplied with water, and the trenches, 34 and 45, are empty, that the panes, 1 and 15, are in a perfect state of drainage; and this may be sufficient to explain the system, and to shew how every trench operates, either for delivering or draining off the water, at the pleasure of the irrigator.

And it should be noted, that this diagonal system enables him to use the smallest quantity of water, as well as the largest, for he can use it only through one system of panes, if necessary, or he can, with great plenty, flow all at the same moment till the meadow has enough, and then stop the whole out, and leave the trenches to operate only as drains, while the water is working in another meadow. It is not uncommon, for want of such a plan of operation, to see trenches of delivery accompanied by drains, which operate only as drains, and which carry away the water without any power of using it even a second time; and in other cases we see the water brought into slopes without any thought of taking it away again, consequently some parts are much watered, some less, and some perhaps not at all.

14th, The application of this system to mountainous moors, is one of the most profitable speculations which agriculture has to offer, and yet there are none so much neglected.

From viewing them I have been greatly surprized at this, because there are scarcely any that do not contain such spontaneous proofs of the advantage, as might have been sufficient for a hint to the stupidest clown. The firm spots by the sides of the torrents, from flooding, acquire a beautiful verdure, that proves a perfect contrast to the dreariness of the waste around ; and where there are little rills on the mountain sides, not considerable enough to cut a regular bed for their waters, but
                                                    which

which spread, they are attended so universally with a verdure, from the grasses getting the better of the heath, owing simply to the water, as shews the advantage in the clearest manner. I am confident that, with a little attention, out of 20 or 30,000 acres on a range of mountains I have viewed in Ireland, water might be thrown over three parts in four. The declivities through which the streams run are considerable, and extensive tracts of land slope off on either side, so that by obstructing those streams, by piling torrent stones across them at various heights, and drawing small channels in the mountain sides, just above such obstructions, to receive the water, this most advantageous work might be done at small expence, and a single experiment of it would presently shew the prodigious advantage of the practice.

In case these papers should come into the hands of any possessors of mountain tracts, willing to try it, but not acquainted with the proper mode of executing the work, I shall here offer a few directions, not by way of going minutely into the whole business, but in order to put every man in such a train as to enable him by practice to instruct himself in the rest, and to carry it further than many books on the subject will teach.

The principle upon which he is to proceed, is to throw as much water as possible over the sides of the mountain, and as equally as possible; and in doing this to guard against two circumstances: first, its remaining in any spots; and, secondly, his works being blown up by sudden floods, from

x 4 heavy

heavy rains, which come in large tracks of mountain with an impetuosity incredible to those who
are used to a flat country.

It would be right to begin, by choosing a place
where the declivity of the mountain is gentle, in
order that the space improved may be more useful
and obvious than it can be when very steep : going
up as high as the water can be conveniently commanded, make a wear of stone across a torrent,
just high enough to form a little bason among the
rocks, if there is none executed to your hand by
nature : in the Galties you find these at every ten
yards.   At the spot where you have made, or found
one of these basons, open a trench from it, a spirit-
level shewing where to conduct it ; taking care to give
it no more fall than necessary to bring the water in a
very gentle current.   The stream is to be made to
overflow out of this carrier-trench all the way it
runs : the trench must be made gradually smaller
to the end, as the body of water it brings lessens
as it advances.

I would advise the proprietor to see the experience of a year or two, watering with no further
expence than I have described (which is evidently
too trifling to be an object).   If he find the effect
great, as in all probability he will, I should then
advise his levelling the spaces over which he throws
the water, to that exactness which is necessary for
mowing ground : this, in many mountains, is the
most expensive part of the business ; for rains
which drive down their sides, in almost universal
torrents, work thousands of little channels round
the

the tufts of heath, that are so deep and sharp, as
every sportsman knows, who has been tired with
walking, or rather tumbling over them: these
must all be levelled, and the water let gently over,
which will soon cover them with grasses, and other
beneficial plants. The heath lives in its own acid
water that stagnates in the moss and peat, as in a
dish, but will die away by being flooded in the
manner I have described. The progress of the
work will naturally arise from success; if the proprietor be attentive, he will find his success so great
and obvious, as to be induced to go into the business with the utmost spirit. He will then level all
inequalities, cut a variety of inclosures, and divide
the declivities into fields by good and sufficient
fences.

15th, Wherever irrigation is applied, it is right,
when arable land is thus intended to be converted
into meadow, or on any other land the surface of
which is much broken by the works, to sow any
sort of grass-seeds that can be procured in the
greatest plenty, before the watering begins. It is
well known that the water will of itself bring
grasses, but it demands some time, and the benefit of sowing them is always found to be considerable.

16th, It is a common practice in Lombardy, to
have a sort of heavy harrow drawn along the bottom of the main carriers, in order to disturb the
mud in autumnal, winter, and early spring irriga-
tions; and it has been practised in England to
throw

throw lime in, the great divisibility of which body
in water is well known : these are means of adding
to the manuring quality of the water very easily to
be practised.

17th, To attempt describing the minutiæ of
erecting wears, sluices, stops, and to note how
trenches or drains are dug, would be unnecessary.
In Mr. Boswell's pamphlet on watered-meadows, a
work of great merit, these particulars are detailed;
and to him I refer for the necessary information.

18th, Wherever roads intervene, the Italian
method is to form a work of masonry to act as a
syphon : the water is made to descend perpendicularly on one side the road in a tunnel of brick
or stone, pass in an arch under the bed of the road,
and rise on the other side in a similar tunnel, and
then pass on in its course. I have seen several of
these between Coni and Turin, and in other parts
of Lombardy.

SOILING.

This month being in general the period for turning out various sorts of live stock to grass or clover,
it is now a question which demands the young farmer's very serious attention, whether he should
comply with the more common custom of feeding
off certain crops, or whether he should determine to pursue the soiling system, of mowing,
and giving them green in the stables, stalls,
yards, &c. Considering the decisive superiority of
the latter mode of consumption, there is not a general

neral fact in husbandry which ought to create so
much surprize as the general custom, all over the
kingdom, of feeding cows, young cattle, oxen,
bullocks, &c. in the fields; and the almost general practice of managing the teams in the same
way. Enlightened farmers have in many districts
adopted this system for horses, but still reject it
for cattle; and it will probably take a century to
render it as universal as it might be, most profitably.
The objections to it are not of any importance, even
if started in the strongest manner: it has been ar-
gued that the expence is an object; and that cattle
will not thrive so well, nor will cows give so much
milk as if fed in the field. That the expence is
something, cannot be denied, but that it amounts
to any thing considerable, is contrary to fact. Mr.
Mure fed 240 fatting oxen in sheds through a whole
summer by the mowing of one scythe: if the attendance upon the beasts be added to this amount,
the whole will evidently come to a sum which,
when divided either per head or per acre, will be so
low as to do entirely away this objection. As to
the question of thriving, the assertion has been
made, as far as it has come to my knowledge, without a trial, and is consequently mere theory. The
beasts mentioned above, were all sold fat at Smithfield, and did as well as similar beasts had done fed
abroad in the most favourable seasons, and better than
in any summer not remarkably favourable. I practised it for several years together very carefully for
fatting cattle, weighing alive periodically, both while
in

in stalls and when at grass, and I found that in
soiling they throve better than when abroad. If
the world will reason upon every question of farming, they should do it without prejudice, and then
their reason would, to my apprehension, agree with
these facts.   Every one knows how tormenting
flies are to cattle when abroad : ride into a field in
summer to look at stock, and where do you find
them ?   Not feeding, but standing or resting under trees, in ponds, in rivers, and, if there is no
better shelter, in ditches under brambles; in a
word, any where but feeding in the open air.
What they graze, is in the morning and evening;
and in many cases they lose in the heat of the day
all they gain at those moments of their comfort.
To this superiority we must add that of the main
object, which is the dunghill : in one case this is
accumulated in a degree even superior to what is
effected in winter ; in the other it is scattered
about the pastures, and nine-tenths of it carried
away by the flies, or dried almost to a *caput mor-
tuum* by the sun.   The warmth of the season in
summer promotes the fermentation in a mass, and
speedily prepares it for use, in whatever state the
farmer wishes to have it.   The prodigious superiority of thus raising a large and very valuable
dunghill in one case, and none at all in the other,
ought to convince any reasonable man, that there is
not a practice in husbandry so decidedly superior as
this of soiling, were there not one other reason
for it than what have already been produced.

                                        Those

Those farmers who have given particular attention to the state of farm-yard manure, as it is made
in winter and in summer, and to the efficacy of both,
can scarcely have failed to remark, that the superiority of the dung arising from any sort of stock,
commonly fed, in summer, is very great to such as
is made in winter from stock no better fed. The
manure yielded by fat hogs, and by fat beasts fed
on oil-cake, is of such a quality, that the season
does not demand attention; but with all other
stock I have great reason to believe, from many
observations, that a farmer should make as large a
reserve of straw-stubble, &c. for littering in summer, as possible.

Cattle, when soiled upon any kind of green food,
as tares, clover, chicory, lucerne, or grass, make
so large a quantity of urine as to demand the
greatest quantity of litter: the degree of this
moisture in which their litter is kept, while the
weather is hot, much assists a rapid fermentation,
and great quantities of carbonic acid and hydrogen
are generated. The winter's cold, with superfluous
water, by rain or snow, has a contrary tendency:
the manure is, comparatively speaking, weak and
poor.

When I view the common spectacle of a large
yard spread with a thin stratum of straw or stubble,
and a parcel of lean straw-fed cows wandering
about it, I think I see the most ingenious way of
annihilating litter, without making dung, that the
wit of man could have invented. Burning such
straw

straw upon the land before sowing turnips, would
be an application far superior.

Cows thus managed, are amongst the most
unprofitable stock that can be kept on a farm.
With the best food and management, their dung is
inferior; but thus kept on a wide expanse of thin
litter, well drenched in rain and snow, running to
ponds and ditches, they destroy much, but give
little.

When a farm is rich enough to summer-graze
oxen, large or small, oil-cake feeding to finish, or
wait for markets, is often profitable, and the litter
sure to be converted into excellent manure; but
when the grass-lands will not permit this system, a
farmer cannot possibly be too sparing of litter in
winter. Hogs form an exception, but I know
not another.

It is a fact, that stock, not in fattening condition, make good dung in summer, but they do it
not in winter.

But there is another equal to this important
one; and that is, the food thus given going so
much farther than it will do when grazed where it
grows. This superiority has in certain experiments
been marked as amounting to double, thrice, four
times, and even five times as far as when eaten in
the field; and when we recollect the old remark,
that a beast feeds (or consumes) with five mouths,
and it might be said to be with seven, we shall
not be surprized at those remarks. However, that
food thus given, goes much farther, cannot be
doubted;

doubted ; thus, a much greater stock may be supported by the same farm, in one system than there can be in the other.

Two circumstances demand attention, which, if neglected, will considerably lessen the benefit to be derived from soiling. The one is, to have a plentiful provision of litter ; and the other, much care in feeding; to give the beasts but little at a time : if much be tumbled carelessly before them, it heats, they pick it over, and the waste may be great ; and if a cart be left in the yard loaded, the contents heat, and then cattle will not eat it. A certain degree of care is necessary in every thing ; and in nothing more than in feeding all sorts of cattle. As to litter, it is an object of such importance, that provision for the system should be gradually made through the winter, if corn enough be not left for summer-threshing to supply the beasts. All wheat-stubbles should be cut and stacked; leaves, in woodland countries, should be collected ; fern procured from commons and warrens, rushes and aquatic weeds stored from fens, &c. ; and if nothing else can be had, heaps of sand formed for this use ; for which peat also is excellent. An enterprizing, vigilant farmer, when he has such an object as this in view, will exert every nerve to be prepared for a system, the profit of which will depend so much on the care previously taken to be well provided with litter of some sort or other.

The first crop that will be ready for soiling is lucerne ; which may be supposed to last all the

stock

stock till the first sown winter tares are ready,
when the lucerne left uncut should be mown for
hay. The second sown winter tares come next;
then clover, to be succeeded by the third sowing of
tares, and by the second growth of lucerne. After
this come spring tares and the second growth of
clover; and the third cutting of lucerne may follow.
If chicory be applied to this use, for which it is
well adapted, it will, on any good land, be mown
thrice, and on very good soils four times. Thus the
whole summer may be provided for, without having recourse to natural grass; but, if wanted, that
also should be used in the same manner. The
quantity and value of the manure thus made will
surprize those who have not witnessed it, whether
the stock be stalled, or kept in well-littered yards,
in divisions, according to sort, size, age, fatness,
value, or any other rule of separation: if they are
fed carefully, have water at command, and are kept
clean, all sorts will thrive to the farmer's satisfaction; and if hay be an object to him, and he has
artificial food sufficient for the whole, he will be
enabled to mow all that is mowable. As to horses,
it is not requisite to say more than this; it is at
present the system with every truly enlightened
farmer in the kingdom.

CART OUT YARD-DUNG.

As this is the first time that it has been necessary
to treat distinctly of this work, it will be proper
to bring to our young farmer's attention, a very
material question, respecting all the dung he
may

may raise on his farm, especially in the yards,
stables, stalls, sties, &c. And this is, whether it
should be removed in a fresh, long, strawy state, or
turned over to ferment and rot; or carted first to
a compost or heap, in order for turning over and
mixing, and for keeping till more rotten still.
There are many variations in management, for
which some better reasons ought to be given than
we commonly meet with. A very common method
is, to leave the dung where made till all winter feeding is over, and then to turn it up where it lies
into heaps, and leave it till wanted to cart on for
turnips. Others, who intend it for wheat, having
turned up a border or hedgreen in the field that is
to be sown with wheat, cart the muck on to the
earth so prepared, and afterwards, sooner or later,
mix them together, and before the wheat-seeding
cart it on to the land. If no border of this sort,
they make a heap of it, which is afterwards turned
over. These are the more general methods. Some
few, thinking it beneficial to have the dung as
rotten as possible, keep it *over year*, as they term
it, and turn a second time. It is evident that these
several methods are founded on certain ideas, that
rotting is beneficial, and that more is gained in
quality than is lost in quantity by that operation.
This is a very important question, and well deserves many careful experiments to ascertain the
real fact; but unfortunately the number of these
hitherto made is so few, that they have not done
much more than excite some attention to the point,

and

and instigate several intelligent and thinking men
to give more consideration to the subject than they
had been accustomed to do.   Mr. Thompson, of
Northamptonshire, observing a spot in a field of
corn better than the contiguous parts, and not being able to account for it, made inquiries among
his people, and found that it was where long straw
dung had been spread, the rest in a rotten state:
he took the hint, and tried the comparison experi
mentally : the result the same.   He repeated it,
and was confirmed in the conclusion he drew, and
from that time changed his practice.   A celebrated
farmer near Lewes, in Sussex, made a similar remark on the comparison between yard-muck turned
up after winter, and some not stirred, and convinced himself, by repeated observations, that the
latter was most advantageous.   In addition to these
cases, it is remarkable, that Mons. Hasenfratz, the
celebrated French chemist, from experiments made
on a different object, and with very different views,
drew collaterally the same conclusion.

" A circumstance in favour of the Picardy
farmers is, the continual transport of their dung
to their fields, rather than leave it to destroy itself
in the farm-yards, by waiting for fixed periods to
move it.   By carrying it still fresh to their fields,
the heat of its first fermentation is employed in
heating the soil ; the little alkali which it contains,
instead of being dissolved in the farm-yard, and
carried away by the rain, remains in the earth and
improves it, if the alkali is useful to vegetation.
The

The straw, yet entire, divides the soil better; its fermentation goes on less rapidly, is less advanced when the seed is sown, and, consequently, the dung is more in a state of furnishing a greater quantity of carbonic acid, which appears, with water, to be the principal aliment of plants."

I have since been informed of a variety of other cases, which seem to give much weight to this new opinion, that long dung is more beneficial in many applications than that which is short and rotten, and particularly upon all heavy and tenacious soils. Its superiority upon grass-land seems equally well established. Should this at last prove the real fact, it affords a curious reflection on the erroneous conduct of such multitudes of practical farmers who have been all their lives putting themselves to considerable expence in carting and re-carting, and turning and mixing, for no other purpose but to do mischief.

MOW TARES.

Very forward winter tares in a mild spring will be ready to mow for soiling this month. Great care should be taken to make the men cut one entire stitch along the field before they begin another; and not in the common slovenly way in squares, or irregularly, so as to preclude the plough or scarifiers till much is done. Whether the work is done this month or in June, it is of great importance that the tillage, whatever it may be, is given immediately; the land is thereby preparing for turnips; whereas if, from irregular mowing, the

teams are kept out, and a drought should succeed,
the tillage will be badly performed, and, perhaps,
not at all.   Immediately after the scythe, whatever
the weather may be, tare land is always in admirable order for tillage : loose, friable, and, if the
crop has been great, putrid.   It is best to plough
it with Ducket's skim coulter, which will completely turn down and bury whatever remnants of a
great crop there may be, which, becoming dry and
fuzzy, are apt to impede the common plough, and
stick out between the furrows : in this situation it
does not rot, is unsightly, and injurious to the
work of any implement that follows, which should
be the scuffler and harrows.   A little attention
thus given, will save much trouble and expence afterwards.   There are many operations on a farm
which can be effectively done only by means of
that coulter : a farmer should at all events possess
it ; on many occasions, he will not have a more
useful implement on his farm.

MOW LUCERNE.

If the lucerne was well manured, there will be
a cutting, as already observed, towards the end of
this month ; however, this will necessarily depend
on the season ; if there are late frosts it will be
much impeded, and this work not take place till
June.   Whenever it is fit to cut, the attentive husbandman should order his men always to mow
longitudinally of the rows if drilled, or of the
field if broad-cast.   Left to themselves, it is twenty
to one but they cut out a square, and enlarge it in
such

such a manner, that however he might wish to give a scarifying, he will be precluded till the young growth is too forward, and will be damaged.

FOGGING.

This is a most peculiar husbandry, no where commonly practised but in South Wales. It consists in keeping the whole growth of grass in upland meadows free from either scythe or stock, and eating it in the following winter. Many years ago I knew a Suffolk clergyman who was in the regular habit of this singular practice, and spoke of it as a most profitable one. I have tried it thrice, and with success : it thickens herbage greatly, and yields far more valuable winter and spring food than any person would expect who never tried it. But it should be practised only on dry, or tolerably dry land.

JUNE.

TURNIPS.

THIS is the great season for sowing turnips :
later sown crops scarcely ever arrive at the size of
those sown in June.   There is a common idea
among the farmers, that the turnip season lasts
just a month, a fortnight before, and a fortnight
after Midsummer.   The land I suppose to have
been ploughed for the last time but one, in May :
the beginning of this month the manure should be
carted on to it, which, in a well-ordered farm,
should come from the farm-yard ; and, if that
does not yield a sufficiency to cover a fifth part of
the arable land, the farmer is negligent.   If he has
a thorough command of litter, and money enough
in his pocket to buy plenty of cattle, it will cover
a third of it ; but, whatever quantity of turnips he
has, let him dung them well.   In this work he
should proceed regularly, beginning on one side of
the field, and laying the heaps in lines from top to
bottom : it should be spread immediately, and the
ploughs follow directly to turn it in.   Upon that
ploughing, the seed should be sown without loss of
time, and covered by two or three harrowings, according to the fineness of the land.   I have sometimes seen the dung carried out long before it is
ploughed in ; but that is bad husbandry : for much

of

of the goodness is carried away by the sun. It
should be taken in full fermentation from the heap,
and turned directly in, so as to ferment under the
moulds, which will ensure a great crop. If the
farm employs many teams, it will be proper to proportion them, so as to let the manuring, ploughing, and harrowing, be constantly going on, without interruption. As to the seed, observe well to
sow the great round Norfolk white turnip, that lies
above ground, and holds to it only by a tap-root.
It grows larger than any other, and has the excellent quality of being used in winter with much
greater ease than those sorts which root quite under
ground, and are consequently not to be got at in
frost. Sow about a quart an acre : less than a pint
is sufficient, perhaps half a pint, if they all grow,
and escape the fly ; but I have seen many thin-
sown pieces eaten up, when the thick-sown ones
have suffered much, and yet enough escaped for a
crop.

In extreme dry seasons, much seed will not vegetate ; but such instances are rare : the most
common misfortune is the fly, which eats them off
before they gain the rough leaf. Many remedies
have been proposed for this evil ; but none that are
effectual. Steeps for the seed are mere quackery.
Strewing soot over the plants, as soon as attacked,
will very often save them, but the remedy is expensive ; because, on numerous soils at this season,
the soot will be of little service as a manure. The
very best dependence is on the richness of the soil :

if it is extremely fertile, or full of dung, the growth of the turnips will be forced; so much accelerated, that they will presently grow out of the power of the fly. I have often remarked in fields partly dunged, that those lands which received no manure, have been totally eaten up, while the dunged parts have escaped. Without manure, the growth is so slow, that the enemy has many opportunities to attack the plant.

When a crop is totally destroyed, the farmers plough or scuffle and sow again, which should never be omitted, if you have time. Probably you may do this, and yet get in your crop in June, which will be a fortunate circumstance attending a first early sowing.

The directions here given for sowing turnips throughout the month of June, are for those put in before the 20th, chiefly applicable for such as are to be used before Christmas; for early sown crops are much more liable to the mildew than such as are sown later; and the young farmer is to remember, that early sown turnips are much more apt to be attacked by that distemper than such as are sown later.

TURNIPS IN THE NORTHUMBERLAND METHOD.

Upon this most interesting subject, the cultivator of 500 acres annually, shall speak to our young husbandman. Mr. Culley says, " The land being made very fine, prepared, &c. as in the broad-cast method, the ploughman (where it is thought most proper to begin) sets up three sticks or poles in a
right

right line ; by having his horses yoked double, and
driven by himself with cords, he sees these poles
between the horses, and by keeping his plough to
bear always upon the poles, he draws his first furrow as straight as possible * : in returning, he keeps
his far-side horse in the new-made furrow, and his
plough at such a distance as to form a one-bout
ridge, like an A ; by proceeding in this manner,
the land, when finished, will appear thus :

the distance of these little ridges is generally from
27 to 30 inches : a less distance does not admit
ploughing between the drills.

" The next operation is spreading the dung, which
is performed as follows : a cart goes down every
third drill, and lays the dung in small heaps ; women and children follow, and with small three-
pronged forks spread the dung evenly in the bottom of three drills, that is, in the one where the
dung is dropped from the cart, and those on each
side of it : when this is done, the ploughman splits
the one-bout ridges, and covers up the dung exactly in the middle of a one-bout ridge ; but before the seed can be sown, they require to be flattened at the top by a small roller, four feet eight
inches long, and 9 inches diameter, which flat-
tens two ridges at once : on the top, and exactly in

* Our Suffolk ploughmen do this in perfection, by a single
white stick, even 40 rods from them at beginning.—*A. Y.*

the

the *middle* of these *flattened ridges,* the seed is
deposited by one or two drill-machines tied to the
roller by a rope six or seven feet long, at which
distance they follow the roller, and each machine
guided by a man : when finished, the work appears
in this form :

where *s* represents the seed, and *d* the dung directly under it. The roller is drawn by one horse,
driven by a boy. Setting up the one-bout ridges,
and covering in the dung, are performed by a
common swing-plough. The drill-machines are of
various constructions : we generally sow about 1lb.
of seed to an acre, as it is better to have an abundance of plants for fear of accidents.

" When the plants have got four leaves we begin
to hoe, and leave the plants at only eight or nine
inches distance in the rows : as they have so much
room sideways, or from row to row, the hoers go
sideways and pull the surplus plants, weeds, &c. into
the hollow or space between drill and drill, and the
turnip-plants are left as regular as if they had been
planted with the greatest care and exactness : the
hoeing is performed by women and children, and
costs about 4s. per acre for two hoeings.

" If the drills be made in the same direction the
ridges lie, at the next ploughing for corn, the surface will be irregular, and the dung unequally distributed. To avoid this inconvenience, if the land
be

be dry and level, the drills are made diagonally across the field, but if the ridges be high, it is best to make the drills directly across the ridges, and draw a plough down the furrows to take off the water.

" The quantity of dung used is from twelve to twenty-two, two horse cart-loads to an acre, according to circumstances.

" It is generally supposed that a weightier crop is produced by the drill than by the broad-cast method ; but, even admitting them equal in this respect, the superiority as a fallow crop must be allowed, because, by the repeated horse-hoeings or ploughings in the intervals, and hand-hoeing in the rows, you have it in your power to extirpate the whole race of annual weeds, and so much surface being exposed through the winter, makes a higher preparation for any succeeding crop. Another advantage is, the facility with which they are hoed, as a boy or girl of nine years old can hoe them with the greatest ease, and indeed generally better than experienced *broad-cast* hoers, because these are more apt to take too many away, and leave them over thin in the rows, while the young ones, from the apprehension of hoeing them too thin, will leave the plants at any distance you fix upon.

" This mode of drilling turnips has fully established itself wherever it has been tried. Very few or no turnips are now sown broad-cast in this country ; the drill system universally prevails, and is now
practised

practised even by its most virulent opponents, to the extent of several thousand acres yearly : the farms are in general very large, and there are many farmers who drill every year from 100 to 200 acres.

" In this neighbourhood, last autumn, several small parcels of turnips, drilled in the manner above described, were sold at 6l. an acre; and upon our own farms we had at least 500 acres drilled, which I have no doubt could have been readily sold for 2,500l. ; or, on an average, at 5l. an acre."

TURNIPS AFTER TARES.

The winter tares that were sown the last week in August, or the beginning of September, will, if the season proved favourable, be mown for soiling early enough to put in turnips within this month. If manure was necessary, it should have been spread for the tares, and by that means the field will be in fine order for this crop. Much depends on management : the tares should be mown stitch by stitch longitudinally, and on no account whatever in the common random way, in a round or square portion irregularly ; for, by that means, the ploughs are kept out so much longer, and if a drought succeeds, the land may not be in a state to plough till too late ; but taken soon after the scythe, no land stirs better than that where tares have been mown. If the crop was large, and beaten at all to the ground, there will be an uncut stubble, which is scarcely ever turned clean in by any common ploughs ; it should be attempted only with the

the skim-coulter, which sweeps it clean to the bottom of the furrow so buried, that the harrows drag out none. The turnip-seed should be sowed immediately, and thus managed, there will be little fear of a crop.

They have on the South Downs an admirable practice in their course of crops, which cannot be too much commended; that of substituting a double crop of tares instead of a fallow for wheat. Let the intelligent reader give his attention to this practice, for it is worth a journey of 500 miles. They sow forward winter tares, which are fed off late in the spring with ewes and lambs : they then plough, and sow summer tares and rape, two bushels and a half of tares, and half a gallon of rape; and this they feed off with their lambs in time to plough once for wheat. A variation is for mowing; that of sowing tares only in succession, even so late as the end of June for soiling. October 6th, I saw a fine crop finishing between Lewes and Brighton, on land that had yielded a full crop of winter-sown ones. The more this husbandry is analyzed, the more excellent it will appear. The land in the fallow year is made to support the utmost possible quantity of sheep which its destination admits : the two ploughings are given at the best seasons; in autumn, for the frosts to mellow the land, and prepare it for a successive growth of weeds, and late in spring to turn them down. Between the times of giving those stirrings, the land is covered by crops. The quantity of live stock supported.

supported, yields amply in manure. The treading the soil receives previous to sowing wheat, gives an adhesion grateful to that plant : in a word, many views are answered, and a new variation from the wretched business of summer-fallowing discovered, which, by a judicious application, would be attended in great tracts of this kingdom with most happy consequences to the farmer's profit.

Another stroke of practice, which Mr. Ellman, of Shoreham, is warm for, and with great reason, is that of breaking up his layers two years (clover, ray, and trefoil) for summer tares and rape. What an immense improvement is this upon the common slovenly custom of Norfolk, of ribbling, or half, or bastard-ploughing such lsyers ! a miserable practice, yet very general amongst the spirited cultivators of that celebrated county.

TURNIPS ON PARED AND BURNT LAND.

The lands that were pared and burned at any time previously to the time of sowing turnips, may now receive the seed. The ashes having been spread, and the field thinly ploughed with an even level furrow, will present to the eye a face of whole furrows. The best operation further to prepare the surface, is to pass over it the Norfolk heavy drill-roller, drawn by four horses, which will cut the furrows in pieces without disturbing either them or the ashes, and has an effect in executing this work, which will be admired by all who view it. The cutting circles which move around the iron axle are only four inches asunder : if the seed

be

be then sown, and the roller be passed again across the line of its former movement, the job will be finished in the best manner possible ; but the common method of harrowing is a very bad one ; and trusting the seed without any operation to cover it, causes an inequality in the plant, for it is apt to fail where it does not fall into cracks.

Every man who is, or can be in the habit of this husbandry of paring and burning, should determine to sow turnips (or coleseed) on all the land that may have been pared and burnt after potatoe-planting : many farmers are not at all solicitous upon this point, because they are so very eager to sow white corn wherever it is possible to do it with any expectation of a crop : they put in oats on all early pared ; and wheat on all that is done afterwards ; but the practice is erroneous, nor can I see the gain by it ; for as to two crops of white corn, it is absolutely to be prohibited ; and as to three or four, such management is that of a barbarian : it arises from such execrable conduct, that there are many landlords prejudiced against this excellent husbandry. If oats or wheat are taken for the first crop, turnips or cole must be the second ; therefore, it is not easy to understand what the motive can be.

TURNIPS ON OLD GRASS.

Where a man is not allowed to pare and burn, or cannot do it for some other reason, he should be reminded that this crop succeeds well when sown on one earth upon old turf ; but it should be
ploughed

ploughed with a skim-coulter; then well worked, but shallow, with the scarifier, and the seed harrowed in. I have seen very good crops thus gained; and have had them myself even without scarifying. This is much better husbandry than putting in oats first, which should follow the turnips.

TURNIPS WITH RAPE-DUST, &c.

A practice very lately introduced, and immediately adopted and extensively applied by that great patron of modern improvements, Mr. Coke of Norfolk, is to drill rape-cake dust from the same machine, and at the same time with the turnip-seed; for which purpose a machine has been very successfully made on Mr. Coke's principle, by Mr. Burrel of Thetford. Rape-cake is a very common manure in Norfolk; when spread for turnips, it has been usual to sow it grossly powdered five or six weeks before the sowing the turnip-seed; but Mr. Coke has found at Holkham, that by means of grinding it to perfect powder, there is no necessity for any space of time between sowing the manure and the seed; and this may be probable enough. It is, however, a point in which comparative experiments would be valuable. By thus delivering the manure and the seed into the same pipes and shares of the machine, a ton does six acres instead of three. In whatever way the question of the time of application may be decided, still the importance of the machine remains the same; for, with various other manures, there is no
question

question yet made of the propriety of delivering
them at the same time as the seed. This is the
case with bone-dust, soot, coal and wood-ashes,
dried and threshed pigeons' dung, powdered night-
soil, and many others. Soot must, however, be
mixed with some rougher powder, to prevent the
cups smoothing their way in it, and by pressure
preventing the delivery. These are all cheap ways
of manuring for turnips ; and, as the seed and the
manure are deposited in close contact, the plants
receive immediate benefit, and obtain that quick
growth, in their early state, which enables them
best to escape the fly.

SWEDISH TURNIP.

A second, or perhaps a third sowing of this very
valuable plant, should take place in this month :
see the directions in May. And if the former
sowing was eaten up by the fly, the land should be
well scuffled and fresh sown.

CABBAGES.

Upon your cabbage-lands you should pursue the
same maxims as above laid down for turnips, only
in ploughing the manure in, always throw the land
on to the ridge, and set the plants in a single row
on the top of each : so the dung is covered up in
the ridges, and the plants in a proper situation for
profiting by it to the utmost. As to the distance
of the rows, you must be guided absolutely by the
richness of the soil : if you find the plants join
from row to row, when at four feet, then you have
proof that they should not be planted nearer ; but,

z   if

if they no more than join on three-feet rows, then
you would lose in the crop if you gave a greater
distance : two feet, from plant to plant, is the
proper distance.

When the manure is spread and turned in, the
proper way of planting will be to send women or
children in with bundles of plants, to drop them on
the tops of the ridges, at about two feet distance.
They will lay ready for the men, who may then
plant almost as fast as they can walk ; but, if they
have to get, carry, and set the plants, they will not
be able to do near the work they might with better
contrivance. The rows at four feet may be planted
at five shillings an acre. It is a rule among the
cabbage-planters in husbandry, never to water the
plants, let the season be as dry as it may, insisting
that it is entirely useless. Upon this I shall venture to remark, that in most years, if the land is in
fine tilth, and well dunged, this may be right, as
the expence must be considerable ; but I should
apprehend that, in very dry seasons, when the
new-set plants have nothing but a burning sun on
them, that watering would save the lives of vast
numbers, and might answer the expence, *if a pond
is near*, and the work done with a water-cart.

There is one use in cabbages, which appears not
to have met with the attention it merits : it is the
planting on those lands where turnips have failed.
A late-sown crop of those roots comes not often to
a profitable amount ; but cabbages planted on the
land, without any fresh ploughing, would turn out
a bene-

a beneficial crop for sheep late in the spring: in
all probability (unless on very light, sandy, or
lime-stone soils), of greater value than the turnips,
had they succeeded.

No farmer can entertain too high sentiments of
the necessity of gaining crops of green winter food:
the importance of having such food for his cattle,
and not depending totally on hay, is one of the
clearest axioms in the whole range of husbandry.
His profits will be amazingly lessened: his loss in
the want of manure felt severely for many years,
and on farms not abounding with hay, his expence
in buying it, or his loss in selling his cattle, will
be equally great. But, besides these accumulated
evils, there is another of a different nature, which
he should consider well: it is the change of his
course of crops. After either turnips or cabbages, he
sows spring corn, and with that spring corn clover.
On some soils, the grass is left but one year, in
others two, and in others, mixed with ray-grass,
&c. longer. The lay is ploughed up, and wheat
at once harrowed in. This is compendious, cheap,
and yet excellent husbandry; for the duration of
the grass is a constant fund of profit, at scarcely
any expence, and the preparation for corn is
carrying on in the most beneficial manner. But,
if the turnips fail, and no cabbages planted,
what is the consequence? Why, the farmer sows
wheat on the fallow, in hopes of a good crop, to
pay him for so much tillage as the land has received. This introduction of that grain at once

z 2   breaks

breaks the whole arrangement of his farm, and he is forced either to begin again, or to pursue that pernicious husbandry of sowing two crops of white corn running. He must either fallow for turnips again, or take a crop of barley, and then turnips : thus is he thrown out of his clover, though as important a crop as any on his farm, and launches into a series of tillage, that cannot but prove very expensive to him, without repaying the benefit that the clover-course would have done.

For these reasons, when the turnips fail, and cabbages are not planted, the land should be laid up in winter for barley, and the clover sown with it, which will turn out far more profitable than throwing in wheat.

The cabbages planted in April, and hand-hoed and horse hoed in May, should now have the second of each of those operations given : a hand-hoeing the middle of the month, which must cut up all weeds, and break the earth well of the narrow slip on which the plants were left. Towards the latter end, the double earth board plough should go in the intervals, splitting the ridge thrown up in May, and returning it to the rows. These operations will be of great utility to the crop.

The cabbages drilled in April where to remain, will demand much attention this month to keep them at a proper distance, the tops of the ridges well hand-hoed, and the intervals shimmed, that they may be gradually reducing to a fine state of pulverization. In all horse-hoed crops, these works should

should be particularly attended to while the plants are young, for when they are much branched out, the instruments cannot perform their work with any thing like the same effect.

CABBAGE FALLOW.

If a drought happens in the month of June, and the preparation of the fallow be not very forward, a farmer may be caught with his clods not sufficiently reduced to form the ridges : in this case, there is a tool so effective, that the following minute of my own practice may be worth attention.

In preparing my fallows for cabbages this year, (1793), I found a use in the Norfolk drill-roller, which I had not discovered before : I had got a ten-acred field, by heavy rolling, and harrowing, and repeated tillage, into pretty good order, and the dung well buried ready for the plants : this field was finished about the 9th of June : at the same time another ten acres were preparing; but here, instead of a large two-ox roller covered with lead, I ordered my bailiff to try the drill-roller, which requires four oxen : the effect was very great indeed : the clods, from a long series of continued dry weather, were large in spite of five ploughings, and much rolling and harrowing ; but once going over with this most effective tool, cut and bruised them to atoms, so that the land then ridged up for dung is in excellent order.   I have seen very powerful spiky rollers used, but not with equal effect. I should, however, observe, that no tool in the world can be expected to operate, if a fallow be

not

not prepared properly; that is, ploughed once
before winter, and cross-ploughed the moment the
north-east winds enable the teams to go on to the
land : this year I ribbled one field across, and clean-
ploughed one : the latter is by far the most effective, and leaves the fallow rougher and more exposed to drying winds. Rain is of great use in
pulverization, for a common pair of harrows, at
the right moment, will reduce the roughest land
to a fine state ; but there is a vast difference to the
temper of the land between reducing it while quite
dry, and while wet or even moist : in the former
case, friability is entirely preserved ; in the latter,
the *tread* of the horses is a pressure, which forms
that which will become a fresh clod, and of the
worst sort. It may be accepted as a rule in tillage,
to work wet and strong soils when quite dry; and
of all barbarous management, nothing is so abomi-
nable as to let dry seasons pass, in order to plough
such lands when rain comes. This Norfolk drill-
roller is so effective an instrument for the purpose,
that a strong land farmer should not be without
one ; in such a season as this has been, no common
tool will reduce land ; and for cabbages, the time
of rain must be used for setting out the plants, and
not lost for pulverizing the land, in order to be
planting afterwards in dry weather, or, what is as
bad, watering at an expence too great to bear.

CARROTS.

The carrot-crop will, in all probability, require a
hoeing this month, about the latter end of it. It
should

should be performed with common hoes, and the men who execute it, should take good care to kill all remaining weeds, and wherever they left the carrots double before, to set them out to the proper distances. This being the third hoeing, the land should be left in such order, as to require no more work, or, at least, nothing more than once going over it the latter end of August, to cut up straggling weeds, which may by that time have arisen.

POTATOES.

Another hand-hoeing must be given the potatoe crop, which should be so effectually performed, as to preclude the necessity of any succeeding ones ; because the plants will be too much grown to be hoed without damage in the operation. The crops planted in rows for horse-hoeing must have the second this month, given with the double mold-board plough : it must split the ridge before thrown up, and lay it equally to the rows.

MADDER.

Another hand-hoeing must be given to the madder-crops this month, in which the labourers must be extremely attentive not to damage the crop ; for the branches will be grown considerably, and they are so remarkably brittle, that the least rough usage breaks and damages them : they must not use longer than six or eight-inch hoes. The latter end of the month, the first horse-hoeing should be given. Put two horses, one before the other, in the swing-plough, and turn a furrow on each side *from* the plants, which will consequently throw up

z 4   a ridge

a ridge in the middle of each interval, and so it
should be left till the next month.

LIQUORICE.

This month the liquorice-plantation must be
hand-hoed again. Let the work be carefully performed with small hoes ; but the plant not being
nearly so brittle as madder, it will not require so
much nicety in the management.

HOPS.

If tying the binds to the poles was not finished
last month, it should be done early in this ; which
is also a busy season for cultivating the intervals in
the various methods practised in different hop
districts. About Midsummer, hops at Farnham are
pruned by cutting off the spare vines : these are the
perquisite of the work-people, who keep them for
hay in stacks to feed their cows, if they have any,
if not, they sell to those who have.

FLAX.

Weed the young flax : this is an expensive operation ; but the crop depends on it ; it must therefore be effectually performed.

LUCERNE.

The lucerne, drilled in the spring, will now want
a very careful attendance. It will not be advise-
able to horse-hoe it the first year, because its great
tenderness will not bear any accidental evils that
may arise in the operation ; but the hand-hoes
should be kept diligently at work ; the land kept
througnout this month perfectly free from weeds,
and the surface well broken by the hoes, to keep

it

it from any degree of binding. While the men are hoeing, they should never omit to stoop and pluck out such weeds with their fingers, as grow among the plants in the rows : this is highly necessary ; for, if they are left, they will injure the young lucerne much. Whoever cultivates this grass, must absolutely determine to spare no expence in the eradication of weeds. There is no plant will bear the neighbourhood of weeds so badly, and especially while it is young. If the hand-hoes are applied in time, and often enough, the expence will not be great ; but if, through saving, you defer it till they are gotten much a-head, the crop will either be lost, or the expence of cleaning enormous.

The old crops of drilled lucerne will be ready for cutting this month.

SAINFOIN.

The latter end of June, the sainfoin crops will, in general, be ready to mow : they should always be made into hay ; for no grass in the world answers so well for that purpose. It is a common thing to gain two tons per acre on dry land, that with any other crop would yield none at all : and the after-grass is extremely valuable, much more so alone, than the former value of the land.

Making the hay is the simplest of operations : never offer to shake out the swath, for it should be moved as little as possible, in order to preserve the leaf; leave the swaths long enough according to the weather, to be nearly made, then turn them over,

over, and leave them till ready to cock or cart. If
the crop is very great, or the weather unfavourable,
other turnings may be necessary, but be not too
busy.

CLOVER.

The latter end of June, the clover crops will be
ready to mow. In many situations it will not be
advisable to feed any more of it than can be dispensed with, the hay paying so much better.

In the making of all artificial grasses into hay,
particularly clover and sainfoin, it should be observed to act quite differently from the making natural grass. The latter is strewed about soon after
mowing; but the former should lie in swath a day
or two, then turned carefully, and lie a day or two
longer. In good weather, this makes it sufficiently.
It may then be got into cocks, in which it should
remain about two days, and then carted to the
stack. The whole is a very easy and cheap process.

MEADOWS.

The very early or rich meadows, and the highly
manured upland pastures, about great cities, will be
ready to mow in June. In executing the work,
observe particularly, that the labourers cut as close
to the ground as possible : grass never thrives well
that is not mown quite close, and the loss in the
crop of hay is very considerable ; for one inch at
bottom weighs more than several at top. In the
making it into hay, you will be a loser, if you have
not many hands ready for the work. It should be
                                          shaken

shaken out directly after the scythe ; wind rowed, that is, raked into rows. before the evening, shaken out again next morning, and in the afternoon got into grass-cocks : these should be opened the morning following, and got into the great cock by night ; by which time the hay will be well made, if no rain comes ; but, in case of bad weather the process will be more tedious. If successive rains come, so that the hay is damaged, and you are fearful of its turning out unprofitably, by all means salt it as you stack it ; a peck strewed in layers on the stack to a load of hay : it will have a very great effect in sweetening it, however bad it may be, even to blackness ; and it has been found by experiment, that horses and horned cattle will eat damaged hay salted, which they would not touch without that addition.

The following is the process in Middlesex :

This branch of the rural art has, by the farmers of Middlesex, been brought to a degree of perfection altogether unequalled by any other part of the kingdom. The neat husbandry, and superior skill and management that are so much, and justly, admired in the *arable* farmers of the best cultivated districts, may, with equal justice and propriety, be said to belong, in a very eminent degree, to the *hay* farmers of Middlesex, for by them may very fairly be claimed the merit of having reduced the art of making good hay into a regular system ; which, after having stood the test of long practice and

and experience, is found to be attended with the most desirable success. Even in the most unfavourable weather, the hay made according to the Middlesex manner, is superior to that made by any other method under similar circumstances. It is to be regretted that this very excellent practice has not yet, except in a very few instances, travelled beyond the borders of the county. But as it most justly deserves the attention and imitation of farmers in other districts, I shall, for their information, endeavour minutely to describe the method in which the Middlesex farmers make their hay.

In order that the subject may be more clearly understood, I shall relate the particular operations of each day, during the whole process, from the moment in which the mower first applies his scythe, to that in which the hay is secured either in the barn or in the stack. Before I enter more immediately on this task, I would just premise a few observations, viz. when the grass is nearly fit for mowing, the Middlesex farmer endeavours to select the best mowers, in number proportioned to the quantity of his grass, and the length of time it would be advisable to have it in hand; which having done, he lets it out as piece-work, or to be mown by the acre *.

* Each man mows from an acre and a half to an acre and three quarters per day : some there are who do two acres per day, during the whole season.—*J. M.*

About

About the same time he provides five haymakers (men and women *) to each mower. These last are paid by the day, the men attending from six till six; but the women only from eight till six; for an extra hour or so in the evening, when the business requires dispatch, they receive a proportionate allowance.

The mowers usually begin their work at three, four, or five o'clock in the morning, and continue to labour till seven or eight at night; resting an hour or two in the middle of the day.

Every haymaker is expected to come provided with a fork and a rake of his own; but when the grass is ready, and labourers scarce, the farmer is frequently obliged to provide both; but for the most part only the rake.

Every part of the operation is carried on with forks, except clearing the ground, which is done with rakes, and loading the carts, which is done by hand.

Having premised so much, I now come to the description of the business of the

First Day.—All the grass mown *before* nine o'clock in the morning is tedded (or spread), and great care taken to shake it out of every lump, and to strew it evenly over all the ground. Soon afterwards it is turned, with the same degree of care and attention; and if, from the number of hands, they are able to turn the whole again,

---

* Including loaders, pitchers, and stackers, and all others.— J. M.

they

they do so, or at least as much of it as they can, till twelve or one o'clock, at which time they dine. The first thing to be done after dinner, is to rake it into what are called *single* wind-rows * ; and the last operation of this day is to put it into grass-cocks.

Second Day.—The business of this day commences with tedding all the grass that was mown the first day *after* nine o'clock, and all that was mown this day before nine o'clock. Next, the grass-cocks are to be well shaken out into staddles (or separate plats) of five or six yards diameter. If the crop should be so thin and light as to leave the spaces between these staddles rather large, such spaces must be immediately raked clean, and the rakings mixed with the other hay, in order to its all drying of an uniform colour. The next business is to turn the staddles, and after that to turn the grass that was tedded in the first part of the morning once or twice, in the manner described for the first day. This should all be done before twelve or one o'clock, so that the whole may lie to dry while the work-people are at dinner. After dinner, the first thing to be done is, to rake the staddles into *double* wind-rows † ; next, to rake the

---

* That is, they all rake in such manner, as that each person makes a row, which rows are three or four feet apart.—*J. M.*

† In doing which, every two persons rake the hay in opposite directions, or towards each other, and by that means form a row between them of double the size of a single wind-row. Each of these double wind-rows are about six or eight feet distant from each other.—*J. M.*

grass

grass into *single* wind-rows ; then the double wind-rows are put into bastard-cocks ; and lastly, the single wind-rows are put into grass-cocks. This completes the work of the second day.

Third Day.—The grass mown and not spread on the second day, and also that mown in the early part of this day, is first to be tedded in the morning ; and then the grass-cocks are to be spread into staddles, as before, and the bastard-cocks into staddles of less extent. These lesser staddles, though last spread, are first turned, then those which were in grass-cocks ; and next, the grass is turned once or twice before twelve or one o'clock, when the people go to dinner as usual. If the weather has proved sunny and fine, the hay which was last night in bastard-cocks, will this afternoon be in a proper state to be carried * ; but if the weather should, on the contrary, have been cool and cloudy, no part of it probably will be fit to carry. In that case, the first thing set about after dinner, is to rake that which was in grass-cocks last night, into double wind-rows ; then the grass which was this morning spread from the swaths, into single wind-rows. After this, the hay which was last night in bastard-cocks, is made up into full sized cocks, and care taken to rake the hay up clean, and also to put the rakings upon the top of each cock. Next, the double wind-rows are put

---

* It seldom happens in dry weather, but that it may be carried on the third day.—*J. M.*

into

into bastard-cocks, and the single wind-rows into grass-cocks, as on the preceding days.

Fourth Day.—On this day, the great cocks just mentioned, are usually carried before dinner. The other operations of the day are such, and in the same order, as before described, and are continued daily until the hay harvest is completed.

In the course of hay-making, the grass should, as much as possible, be protected both day and night, against rain and dew, by cocking. Care should also be taken to proportion the number of haymakers to that of the mowers, so that there may not be more grass in hand, at any one time, than can be managed according to the foregoing process. This proportion is about 20 haymakers (of which number 12 may be women) to four mowers : the latter are sometimes taken half a day to assist the former. But in hot, windy, or very drying weather, a greater proportion of haymakers will be required than when the weather is cloudy and cool.

It is particularly necessary to guard against spreading more hay than the number of hands can get into cock the same day, or before rain. In showery and uncertain weather, the grass may sometimes be suffered to lie three, four, or even five days in swath. But before it has lain long enough for the under side of the swath to become yellow (which, if suffered to lie long, would be the case), particular care should be taken to turn the swaths with the heads of the rakes. In this state it will cure

so

so much in about two days as only to require being tedded a few hours, when the weather is fine
previous to its being put together and carried. In
this manner hay may be made and stacked at a
small expence, and of a good colour, but the tops
and bottoms of the grass are insufficiently separated
by it.

There are no hay-stacks more neatly formed, nor
better secured, than those of Middlesex. At every
vacant time, while the stack is carrying up, the men
are employed in pulling it, with their hands, into a
proper shape; and, about a week after it is finished,
the whole roof is properly thatched, and then secured from receiving any damage from the wind,
by means of a straw rope extended along the eaves,
up the ends, and near the ridge. The ends of the
thatch are afterwards cut evenly below the eaves
of the stack, just of sufficient length for the rain-
water to drip quite clear of the hay. When the
stack happens to be placed in a situation which
may be suspected of being too damp in the winter,
a trench of about six or eight inches deep is dug
round, and nearly close to it, which serves to convey all the water from the spot, and renders it perfectly dry and secure.

The Middlesex farmers are desirous of preserving
the green colour of their hay as much as possible,
though a lightish brown is of no disservice to it.
Hay of a *deep* brown colour, occasioned by its
having heated too much in the stack, is said to
weaken the horses that eat it, by promoting an

A a   excess

excess of urine, and consequently it sells at a reduced price *.

In the making of hay, some attention should be
paid to the quality of the soil, and the kind of
herbage growing on it. *The hard, benty, hay, of
a poor soil, is in little or no danger of firing in the
stack ; and should,* therefore, *be put very early together, in order to promote a considerable perspira-
tion, as the only means of imparting a flavour to
such hay, which will make it agreeable to horses and
lean cattle :* it will be nearly unfit for every other
sort of stock.

*It is the succulent herbage of rich land, or land
highly manured, that is more likely to generate heat
sufficient to burst into flame,* as it has sometimes
done : of course, the grass from such land must
have more time allowed in making it into hay.
This the Middlesex farmers are perfectly aware of ;
and, when the weather proves moderately drying,
they make most excellent hay. But when very hot
or scorching, they, as well as most other farmers,
under similar circumstances, are sometimes mistaken. In such weather the grass becomes crisp,
rustless, and handles like hay before the sap is sufficiently dissipated for it to be in a state fit to be
put into large stacks. But if that be done when
it is thus insufficiently made, it generally heats too
                                              much.

---

* Observation. If you would make your hay come out of
the stack of a fine colour, and the beauty of the flowers to
                                              appear,

much, sometimes becomes *mow-burnt*, and in some cases, though very rarely, has taken fire *.

The great quantity they have in hand at the same time, makes it extremely difficult to carry the whole just at the moment it is sufficiently made ; although it is certainly of considerable consequence that it should be so, *in order to its yielding the greatest possible weight,* and preserving its best quality ; as, every minute after that precise time, it continues to lose, both in weight, and in its nutritious properties, by evaporation †. Even the difference of an hour, in a very hot drying day, is supposed to occasion a loss of 15 or 20 per cent. on the hay, by its being carried beyond the point of perfection, and frequently even a greater loss is sustained.

The expences are, per acre, as under, viz.

	£.		
Mowing, 3s.; beer, 6d. -	£.0	3	6
Making and stacking, - -	0	9	6
Pulling the stack, and laying the hay pulled out, upon the stack, -	0	0	6
Carry forward -	£.0	13	6

appear, the hay you have shaken out of bastard-cocks to prepare for carting, should be cocked in the heat, and remain till the next morning ; then turn and open the cocks, for the air to take away the damp that is collected, which otherwise would heat in the stack, and of course the beauty of the colour would be done away.

* Hay stacked in a barn in the same state, would not heat too much ; and as to firing, no such thing was ever known.—*J. M.*

† Every thing we smell affords proof positive, that more than watery particles evaporate.—*J. M.*

Brought

Brought forward　-　£.0 13 6

Horses, harness, and carts,　-　　-　0 2 0

Straw for thatch, 3s.; thatcher and la-
　bourer, 6d.　　　　　-　　　-　0 3 6

Total expence per acre in the stack-yard is 19s. or 20s.
—*Middleton.*

WATERED MEADOWS.

These come in for mowing this month. Mr.
Boswell directs, that as soon as the hay is cleared,
cattle of any sort (no sheep) should be turned in
for a week to eat the grass out of the trenches,
and what may be left by the mowers. Then the
water should be worked on them, care being taken
to let it only dribble over every part as thinly as
possible, this being the warmest season of the year.
The first watering should not last longer than two
or three days, before it is shifted to another meadow : there will soon be an after-grass of such a
rich and beautiful verdure as will astonish a specta-
tor not accustomed to it ; and the quantity and
quality will be beyond conception, compared with
the state the lands were in before they were watered.
Mr. Boswell further cautions his reader to guard
by all means against keeping the water too long
upon the meadow in warm weather. It will very
soon produce a white substance like cream, which
is prejudicial to the grass, and shews it has been
upon the ground too long already ; but if permitted to remain a little longer, a thick scum will
settle upon the grass, of the consistence of glue,
　　　　　　　　　　　　　　　　and

and as tough as leather, which will quite destroy it.

FEEDING AND MOWING.

Relative to the application of grass, there are some common opinions, which I heard so often canvassed, or rather asserted, in discourse, that I gave a particular attention to them on my own farm. It has been said more than once, that mowing land exhausts it more than feeding ; and that pastures should be alternately fed and mown upon the same principles that arable lands are fallowed. I have remarked the effects of both on several of my fields, and also on my neighbours, and therefore can speak to it from better authority than mere conjecture. Several grass-fields on this estate, and some of my own, have been mown every year as long as the labourers remember : I have a minute of 22 successive crops of hay in one field, and yet neither that, nor any of the rest, shew more signs of being exhausted than others on similar soils that have been fed. Here are fields that have been constantly mown, separated only by a hedge from others that have been often fed ; the soil and treatment in other respects alike ; and yet the one are as good as the other ; nor are the few crops taken from the fed lands better than those from others mown. I have fed parts of fields, and mown parts, and the year following mown the whole ; nor could I perceive a difference.

Why is feeding thought to be so beneficial, as to rank with a fallow ? upon what principles ? it can only be the appearance of a large burtben of hay at once upon the ground, that constitutes so strong an idea of a crop ; and the notion of cattle in feeding manuring the land greatly.   As to the product, it is, probably, nearly the same when fed as when mown ; only the eating as fast as it grows, prevents the *quantity* appearing : the argument is therefore reduced to the manuring received from the cattle in feeding.

That this is not of much consequence from great cattle, I think there is reason to suppose.   In the first place, it is not laid on in one body, so as to occasion a fermentation in the soil.   In the next, it is dropped at an unfavourable season, summer.   It is also in such irregular quantities as to do mischief. Great cattle, while they dung, stand still, and drop the whole in one spot ; no grass is there to be found for a twelvemonth ; and when it does come, it is often rank, and left uneaten, occasioning loss, unless the scythe follows : and the quantity of grass thus hurt for a season is not inconsiderable.

Perhaps the treading of heavy cattle is hurtful to the grass ; the surface of the ground is too compact and bound without such an addition.

But the land receiving little benefit from feeding is not the only point : I conceive that a crop, such as we mow for hay, if cut early, is of benefit to it, from being at once on the land.   The thick shade in the summer breeds a fermentation, opens and oosens the surface ; of which any one may be convinced,

vinced, who examines the surface of two grass fields, one fed and the other mown ; and it must be a benefit to loosen the soil for the roots and fibres that are in general so bound and matted. These, I apprehend, are the reasons for the fact observed. But extend the argument, and suppose the hay converted to dung in the farm-yard, and then carried on to the field in proper quantities, and at a proper season, it is clear enough (all expences carried to account) which method will have the advantage.

THE TEAMS.

Continue to soil your horses and oxen in the stables, or under sheds, upon lucerne mown every day or two, and take care to have great plenty of litter, to spread under them, for treading into dung. They will raise immense quantities of most valuable manure with this management, and at the same time be kept at a much cheaper rate than if turned into any kind of pasture.

HORSE-HOEING.

The drilled crops of pease and beans, must be horse-hoed at least once in June. If they had received a first horse-hoeing in May, then this of June must reverse it : throw the earth back again to the rows, splitting the ridge in the middle of the interval. In these works of horse-hoeing, the plough should not be carried nearer the rows of corn than four inches : even at that distance, some of the corn will be apt to be buried.

FALLOWS.

The fallows, whether for wheat or barley, if in common management, should this month receive a stirring; by which the crops of weeds, that have arisen since the land was ploughed and harrowed fine, the latter end of April, or the beginning of May, will all be turned in and destroyed.

But in the more modern management, it is not necessary to give any ploughing this month; the weeds are better destroyed by the broad shim, or by the scufflers, which should work till just before harvest, and then one earth if wanted will be effective.

BUCK-WHEAT.

This crop bearing to be sown so late is, in many cases, a most valuable circumstance. By means of it, you have time to get the land into extreme good order, and quite free from seed-weeds. If the stubbles are broken up in October, he must be an indolent farmer that cannot get his land fine and clean by the middle of June.

BUCK-WHEAT AFTER TARES.

This is a very beneficial system, which was first explained by the Rev. Mr. Mosely, of Suffolk, and it is so much deserving of attention, that I shall insert his own account of it.

The excellent Norfolk method of managing light lands I generally adhere to, viz. turnips, barley, clover, and wheat; but finding, from a failure of clover in my two last crops after barley, that the succeeding ones were not equal to my expectation,
I de-

I determined to try something as a substitute for that excellent preparation. Tares, I was aware, were frequently sown, and excellent crops of wheat have succeeded; but, as there were near three months between the time of cutting tares and sowing wheat, I thought that something might be done in the interim, in order, not only to keep the land clean, but to improve the succeeding crop.

It was necessary to consider what would answer this end, that would not be attended with considerable expence; buck-wheat claimed the preference, as it was of quick growth, and had been recommended as a strong and lasting manure. I therefore determined to try the effects of it, and have reason to think that my expectation was not too much raised; for although I cannot with that certainty ascertain the real produce of the land as I can wish, as a considerable quantity of the wheat has been destroyed by vermin, yet still have I had the satisfaction of lodging in my granary as much as I usually have done in the common method of husbandry. The loss I sustained was, indeed, very considerable, from such small animals as mice, for there was not a rat in the barn, and will be a standing memorial to me for threshing my corn in the proper season. It was computed at one fourth of the whole crop. But, even deducting the loss, and allowing the increase to be equal to former years, will it not be right sometimes to alter the usual course, and substitute a preparation equally profitable as clover for the farmer's grand crop, wheat?

The

The land upon which this experiment was made, was light, and produced excellent turnips and barley, but seldom more than a moderate crop of wheat: 20 bushels per acre were as much as might be expected in a good season.

But although I cannot speak with precision in regard to the wheat crop, yet I can thus far affirm, that the additional profit from the rye, as spring feed, which succeeded the wheat, was more than equal to the original price of the buck-wheat. How long the effects of this manure will continue, I cannot possibly say, but, from the luxuriance of the rye, should not have made the least doubt of its operative qualities to the ripening that crop. The expence is trifling, for you cannot find any manure even for a single crop, equal in all respects to this for five shillings, which is, in general, the price of two bushels, and is sufficient for one acre.

But a material advantage there certainly is from two vegetable crops, the one immediately suc-ceeding the other, in cleaning the land, for although the rye was sown as soon as I could conveniently plough after the haulm was carried off, yet, upon breaking up the land after the rye was fed off, it was much cleaner than it was after the last fallow.

I wish I could have drawn a more accurate con-clusion from this experiment, as I find that it is the first that has been made in this manner; and would not have troubled you with this, had it not been by your particular desire, it being impossible to as-certain precisely the loss I sustained, consequently

mere

mere presumption to offer any thing as certain from it.

I hope hereafter to be more accurate, as I have six acres which have produced this season 12 waggon loads of tares, and are now sown with buckwheat, to be ploughed in the end of this month (June) as a preparation for wheat.

The field contained near six acres, including borders, and the produce was 29 combs 2 bushels of clean wheat, so that it may reasonably be set at 5 combs per acre, which is a much larger crop than I expected.

SHEEP.

In this month, the flocks of stock sheep are regularly managed : they live on the commons and sheep-walks, with little change or trouble. The stock intended for fatting, such, for instance, as wethers bought in April or May, and intended to be sold fat from turnips or cabbages the following winter, should be well kept.

WASH AND SHEAR SHEEP.

The first object in this work is to provide a convenient place for washing. It is common for men to stand in the water for it, by which they sometimes get bad colds and rheumatic complaints, and must besides be well supplied with gin : so disagreeably situated, they hurry over the work, and the wool suffers; a stream or a pond offers the requisite opportunity for doing it well, and at the same time comfortably to the men. Rail off a portion of the water for the sheep to walk into by a sloped

a sloped mouth at one end, and to walk out by
another at the other end, with a depth sufficient
at one part for them to swim ; pave the whole :
the breadth need not be more than six or seven
feet ; at one spot let in on each side of this passage, where the depth is just sufficient for the
water to flow over the sheep's back, a cask either
fixed or leaded, for a man to stand in dry ; the sheep
being in the water between them, they wash in perfection, and pushing them on, they swim through
the deep part, and walk out at the other mouth,
where is a clean pen, or a very clean dry pasture
to receive them.   Of course there is a bridge rail-
way to the tubs ; and a pen at the first mouth of
the water, whence the sheep are turned into it,
where they may be soaking a few minutes before
being driven to the washers.

Shearing is a business very ill performed in many
parts of the kingdom, so that it is probable that
one or two ounces of wool are left on an average
on all the sheep in it, which is mischievous to the
next growth ; for wool is in this respect like grass,
it will not thrive well if it be not cut close.   This
bad clipping arises much from longitudinal cutting.   The improvement of clipping circularly
round the body of the sheep began in Lincolnshire, and thence passed into Leicestershire.
The Earl of Egremont was assiduous in introducing
it into Sussex, the Duke of Bedford into Bedford-
shire, and Mr. Coke into Norfolk, so that gradually it will spread over the kingdom.   Possibly the
motive

motive for this mode of shearing was to add to the beauty of the animal. I know not whether there are shearers who can cut as close in a different direction, but I never saw it done. In winding the wool there are some absurd acts of parliament which operate : confidence in the common course of business ensures good washing, and so it will fair winding ; for he who once sells dirt clots and sand for wool, will find the loss when he deals again.

In common clipping, the pay per score is 2s. 6d. to 3s. 6d. for washing, clipping and winding, labour being at 1s. 6d. a day.

THE FLY.

Sheep that are kept in inclosures, and especially in a woodland country, should be examined every day, lest they be fly-struck : in twenty-four hours it may be almost past cure.

" As a preventive of the fly, the midland shepherds use curious applications, especially to the lambs. Train-oil is found to be efficacious ; but it fouls the wool, and makes the sheep disagreeable to touch. An ointment made of butter and the flowers of sulphur seems to be in the best repute *

Insects certainly have their antipathies, and to find out those of the sheep-fly is an interesting subject of inquiry.

---

* The butter being melted, a sufficiency of brimstone is stirred into it, to form an ointment of a pretty firm consistency. In application, a piece the size of a small walnut is rubbed between the hands, and these drawn along the backs of the sheep.

The

The method of destroying maggots here, is effectual, and if applied in time, simple and easy. Instead of cutting the wool off the part affected, and scraping off the maggots with the points of the shears, the wool is parted, and the maggots picked out with a knife, or otherwise dislodged, without breaking the coat ; and a small quantity of white lead scraped from a lump, among the wool ; which being agitated, the powder is carried evenly down to the wound. Too much discolours the wool : a little prevents any further harm from the maggots that may be left among the wool, driving them away from the wound ; and, at the same time, is found to promote its healing. In well shepherded flocks, which are seen regularly twice a day, there is no such thing as a broken coat."
—*Marshall.*

FOLDING,

This is a capital month for the sheep-fold, with those who still practise it. Now you may fold the cabbage and turnip-land, which are the crops that will soonest succeed the operation : the general rule for all manuring. Give each sheep a square yard in the fold, and go two nights on the same land.

THISTLE THE WHEATS.

In this month, the crops of wheat, if any thistles have arisen in them, should be weeded. The best manner of performing the work is with a small hand-hook. It should not be deferred longer than the beginning of June, or damage will be done to the crop by the treading.

DIG MARLE.

This is a good season for marling land: one of the most important works that can be done in husbandry. All farmers, that have marle under their fields and do not make use of such a treasure, are to be condemned.

In some countries, it is the common manure; and almost every where to be found when dug for ; in such places, the farmers have nothing to do but to resolve on the undertaking : they all acknowledge the expediency of the work, and seldom dispute the great profit of it ; but, in other parts, the knowledge of marle is very confined. It may perhaps be discovered half a century before it comes into general use. In tracts of waste land, or sheep walks and warrens, let at a shilling, or two shillings and sixpence an acre, marle being discovered, and rendering such land capable of yielding noble crops of turnips, clover, and all sorts of grain and pulse; the uncommon effect, and the amazing advantages made by it are so striking, that the use spreads fast. But, on the contrary, when marle is found under richer soils (inclosed countries, for instance, of ten or twelve shillings an acre), the case is different : it will not make such an improvement as on the poorer lands ; and, as great fortunes are not suddenly made by the use of it, the farmers in some districts will not be persuaded to use it with any spirit, possibly, not at all : they think that a rent, comparatively higher than the other tracts, will not allow of their spending such sums about it :

it : that they will not reap *equal* profit is undoubted ; but why not accept of twenty *per cent.*
advantage ? Should they reject it, because they
cannot command fifty ? If tenants are backward
in making use of marle on lands of ten or twelve
shillings an acre, their landlords should set them
the example, and shew that the work will answer
well.

Marle is of various sorts, and lies in various
strata : in some places, it is a soft, fat, soapy substance ; in others, it is hard as chalk, which are
called stone-marles : sometimes you find it white,
sometimes grey ; also blue, yellow, and a dark
brown. In some counties you have shell-marle,
which is composed of nothing but decayed shells.
The depths at which it lies are various : sometimes
only three feet from the surface, at others ten or
twelve, and in some places so deep, that it will not
answer to get it at all. The strata are also of different thickness, from two feet to twelve feet deep ;
but the general circumstances in which all true
marles agree, and which denote them to be real,
are the effervescence with acids, and the falling in
water : the crackling in the fire is a good sign, but
not alone determinate.

If it is uncertain what strata are under a farm,
it is ever advisable to use the screw-borer, to discover what soils are within reach. By means of
that instrument, you discover, at a trifling expence,
if there be any marle at command.

The best way of conveying it on to the land,
                                        if

if it does not lay very deep, is to open a sloping mouth, sinking the pit gradually, wide enough for a cart to drive in and out ; and, when you come to the marle, to work it away circularly, and to keep the pit ten or fifteen feet deep, by which means the expence of filling the carts will be much lessened. The expence of marling, when it is thrown in this manner into the cart, will be, upon an average, threepence to threepence-halfpenny per cubical yard, the filling and spreading ; and about fourpence-halfpenny for the teams, carts, and drivers : in all, eightpence per load, or cubical yard, or three pounds six shillings and eightpence per hundred loads. This will be a proper quantity for an acre of land : the benefit will last for twenty years, and the land always be the better for it.

DIG CLAY.

Where marle is not to be had, clay, in many places, is to be found at a moderate depth. This manure has few of the properties by which marle is to be known ; but yet it works wonderful improvements on many soils. In some light lands it has been preferred by many very good farmers to indifferent sorts of marle ; and this preference has been the result of attentive experience.

But the great point concerning clay is not so much the comparison with marle, as the use of it where no marle is to be had. On all light sandy soils it should be used with a confidence of success ; for the precedents of its good effects are so numerous, that we cannot have a doubt of its excel-

B b   lence.

lence. About sixty or seventy loads an acre, at the
same expence as of marle, will work an improvement great enough to shew how much mistaken
those men are, who think nothing but the finest
marles worthy of attention ; and upon heavier
soils, such as wet loams, brick-earths upon clay,
and loose hollow soils, that want a firmer texture,
clay is an excellent manure ; but there are vast
tracts of such land, that cover very fine veins of
clay, and yet the farmers know nothing of the use
of it. It is much to be regretted that their landlords do not give them a juster idea, by being at
the-expence of claying some small fields, until the
benefit of the improvement becomes conspicuous.

DIG CHALK.

Chalk is a manure common in many parts of the
kingdom, and this month is a very proper season
for digging it. A distinction is here to be made
between the chalks that are of the fat soapy kind,
and those hard ones that are flakey and different.
The first ought always to be ranked among the
marles, for such they really are ; but the latter is
properly chalk, and are of excellent use on many
soils . they work a great improvement on light
sands and light loams : they have in many places
been used with great success on gravels ; and on
clayey loams and clays they do extremely well,
mellowing them greatly, and bringing them into
much better order for ploughing, and much earlier
in the spring, which, on such soils, is always a
matter of consequence.   The expences of this improvement

provement are the same as of marle or clay, being
sometimes dug and thrown directly into carts, and
at others drawn up in buckets through shafts.
These variations are not of such importance as to
exclude the propriety of the improvement, even in
the most expensive countries.

EMPTY PONDS.

This is a proper season for emptying ponds, and
cleansing rivers ; for, being early in the summer,
you will afterwards have an opportunity of turning
the mud over, and thereby sweetening it, and laying it into the proper state for bringing on to the
land.   This is a part of husbandry too much neglected by many farmers ; but advantage should
always be taken of it by a good husbandman, when
he is lucky enough to succeed a great sloven ; for
then he will probably find all the ponds, &c. full of
rich mud.

It is improbable that pond mud, especially if
there is a stream into the water, should ever fail of
proving a good manure, when judiciously used.
The method ﹍f managing it, which has been found
the most beneficial, is the following :

As soon as the mud is dry, and hard enough to
spit, turn it over, and, in about three weeks or a
month after, mix it with an equal quantity of
chalk or marle : either bring the chalk to the mud,
or carry the mud to the chalk.   If lime is cheap
and plenty, it will be an excellent management to
add about one-fourth the quantity of mud in lime.
Let the whole be mixed well together, and remain

until September, when it should be turned over again, and spread upon pasture or meadow land in October. This is husbandry that will pay well.

RAPE OR COLE-SEED.

This plant may be sown now, in the same manner, and upon the same preparation as turnips.

RAPE OR COLE FOR SHEEP.

This crop is sown, when intended for sheep-feed, all through June and July ; but for seed, the first week in August will do : upon fen and peat soils and bogs, and black peaty low grounds it thrives greatly, and especially on pared and burnt land, which is the best preparation for it. In many respects the culture is the same as for turnips, only double the quantity of seed, as the crop is not commonly hoed. Two quarts an acre : but some sow three ; and I have heard of a gallon being used.

RAPE FOR SEED.

The Flemish culture much deserves attention. They sow in a seed-bed for transplantation.

Sow the seed thick, setting it out on an oat-stubble, after one ploughing. This is so great and striking an improvement of our culture of the same plant, that it merits the utmost attention ; for saving a whole year is an object of the first consequence. The transplanting is not performed till October, and lasts all November, if no frost ; and at such a season there is no danger of the plants not succeeding : earlier would, however, surely be better, to enable them to be stronger rooted, to withstand the frosts, which often destroy
them,

them, but the object of the Flemings is not to give
their attention to this business till every thing that
concerns wheat-sowing is over. The plants are large,
and two feet long ; a man makes the holes with a
large dibble, like the potatoe one used on the
Essex side of London, and men and women fix the
plants, at 18 inches by 10 inches; some at a foot
square, for which they are paid 9 liv. per manco
of land. The culture is so common all the way
to Valenciennes, that there are pieces of two,
three, and four acres of seed-bed often met with.
The crop is reckoned very uncertain : sometimes it pays nothing; but in a good year, up
to 300 liv. the arpent (100 perches of 24 feet),
or 8l. 15s. the English acre. They make the
crop in July ; and, by manuring the land, get
good wheat.

SOILING.

Soiling on lucerne, tares, clover or chicory,
should go on through the whole of this month.
In some soils, and situations, and seasons, it may
not be possible to do any thing in it in May, but
now these plants will every where admit it. The
mowings should be daily, and attention paid that
the food be not left loaded in carts, or given in the
racks or cribs in such quantities as to ferment,
which presently renders it unpalatable, and consequently refused by horses and cattle, much waste
ensuing. If the number to be fed be not so great as
to demand a one-horse cart load for every bait,
it will be proper to have an *ass car* for this purpose,

as it is very material that all which is brought home should be immediately distributed to the stock. A good farmer will have been attentive to secure as ample a provision of litter as possible ; if he has not reserved his wheat-stacks to be threshed at this season, which usually gives the best price, at the same time that it provides for littering at a season the best calculated for making dung. That summer is that season, there are several reasons for admitting.

Those farmers who have given particular attention to the state of farm-yard manure, as it is made in winter and in summer, and to the efficacy of both, can scarcely have failed to remark, that the superiority of the dung arising from any sort of stock, *commonly fed*, in summer, is very great to such as is made in winter from stock no better fed. The manure yielded by fat hogs, and by beasts fed on oil-cake, is of such a quality that the season does not demand attention ; but with all other stock I have great reason to believe, from many observations, that a farmer should make as large a reserve of straw. stubble, &c. for littering in summer; as possible.

Cattle, when soiled upon any kind of green food, as tares, clover, chicory, lucerne, or grass, make so large a quantity of urine as to demand the greatest quantity of litter ; the degree of this moisture in which their litter is kept, while the weather is hot, much assists a rapid fermentation, and great quantities of carbonic acid and hydrogen

are

are generated. The winter's cold, with superfluous moisture by rain or snow, has a contrary tendency; the manure is, comparatively speaking, weak and poor.

When I view the common spectacle of a large yard spread with a thin stratum of straw or stubble, and a parcel of lean straw-fed cows wandering about it, I think I see the most ingenious way of annihilating litter, without making dung, that the wit of man could have invented. Burning such straw upon the land before sowing turnips, would be an application far superior.

Cows thus managed, are amongst the most unprofitable stock that can be kept on a farm. With the best food and management, their dung is inferior, but thus kept on a wide expanse of thin litter well drenched in rain and snow, running to ponds and ditches, they destroy much, but give little.

When a farm is rich enough to summer-graze oxen, large or small, oil-cake feeding to finish, or wait for markets, is often profitable, and the litter sure to be converted into excellent manure; but when the grass-lands will not permit this system, a farmer cannot possibly be too sparing of litter in winter. Hogs form an exception, but I know not another.

It is a fact, that stock not in fattening condition make good dung in summer, but they do not in winter.

If, on experience, it should be found by others as it has often been by myself, that litter of all kinds is converted in summer to better dung than common

B b 4   winter

winter keeping can effect, the vast importance of raising amply various crops for soiling, acquires a fresh interest in the farmer's system. He will be sedulous to cover his fallows with tares, clover, and chicory, and apply a breadth of his very best land to lucerne; he will ever take care to have too much rather than too little, as an increase of his hay-stacks can in few cases prove any evil; and as these crops prepare for corn at the same time that they furnish support for cattle, horses, and swine, when dung is best made, they tend, in every way, to keep a farm in heart.

LONG AND SHORT DUNG.

Many experiments have been made, not only by garden farmers in the vicinity of Wimbledon, but also by Mr. Paterson, comparing long fresh dung with such as is well mixed and rotten, and the re-sult has been very generally in favour of the long dung.

DAIRY.

" I take it that oftentimes in very hot weather, the milk in a cow's udder, much agitated by driving, or running about, is in a state not very far different from that carried in a churn, which frequently makes the great difficulty in what is called bringing the cheese, or fixing the curd in the tub or pan : I have often heard dairy-women say, that it is sometimes very difficult to make it come at all, and instead of one hour (the time very commonly given by dairy women, in bringing the cheese) that it will frequently not come in three, four, or five hours,
and

and then in such an imperfect state, as to be scarce
capable of being confined either in the cheese-vat
or press, and when released from the press, will
heave, or puff up, by splitting or jointing, according as the nature or state of the curd happens to
be.   Whenever people find their cows in this situation, which in hot summer evenings must often
happen, especially where water is scarce, or in
grounds where there is very little shade; then it is,
that making use of a little cold spring water before
earning, or rendling, is useful, as that will make
the runnet take effect and the milk coagulate much
sooner.   It often happens, in some dairies, that the
work is quite at a stand : the dairy-woman not
knowing how to hasten the coagulum, or coming
of the cheese, thinks of putting more runnet in
to forward it ; but the nature of runnet being such
as will dissolve the curd in part coagulated, if
more be put in, disturbs the whole, and prevents
its becoming curd at all, or in a very imperfect
state, remaining in the whey, in an undigested
state that will neither turn to curd or cream, and a
principal part of the richest of the milk is then
cast away with the whey.   Cold water, with a little
salt (as hereafter recommended) will, in a great
measure, prevent this difficulty.   One great point,
or thing to be observed in first setting off, or rendling the milk, is carefully to observe the state of the
milk as to heat or cold : the grand medium, or state
it should be in when you put the runnet into it, is
what may be properly understood milk-warm ; if you
find

find it to be warmer than that, it is recommended
to put some fresh spring water into it, in such
quantity as will reduce it to the milk-warm state:
a quart, two, three, four or more, according to the
quantity of milk to be so cooled: many people may
think water will hurt the milk or impoverish the
cheese; experience shews it will not, but is a means
of the runnet more immediately striking or operat-
ing with the milk. I would recommend the use of
a thermometer, to shew the degree of heat milk
bears. I doubt not one may be constructed on a
very easy plan, that will cost a very little money,
and it will be well worth while to be at a small
charge to regulate a fault, of putting milk together
too hot, which is of more ill consequence than
people are aware of.—*Twamley*.

" Sometimes, if cheese be laid cool when first
made, or coming from the press, is dried outwardly
by means of a harsh cool air, when at the same
time the inside of the cheese remains in a moist
state, though the coat is hard and dry, when that
cheese is exposed to heat, either by lying near a
hot wall, or near tiles in hot weather, or by the
immediate heat of the sun, it will be drawn up
round, in the same manner, and by the same cause,
that a board is made round, or coffered up, by the
heat of the sun: rank cheese very often heaves,
from the cause before given, that makes it rank.
Cheese is very apt to split, or divide in the middle,
by being salted within, especially when people
spread salt across the middle of the cheese when
                                            the

the vat is about half filled, which curd, though in a
small degree separated by salt, never closes or joins,
and is much easier coffered up or drawn round than
other cheese ; especially thin cheese made in what
we call Gloucester vats, being round or rising in
the bottom, and the slider or cheese-board that is
laid over it, made convex also, in order to make
the cheese thinnest in the middle, that it may dry
quick for early sale. Then, if salted within, and
being laid soft on the shelf to dry, as it bears only
on the edge all around, it is almost sure to split,
and it is often seen ; scarce a cheese in some dairies
of this form but what do split. Salting a little in
the milk is greatly preferable, for these dairies in
particular.—*Twamley.*

" It is a fact well established, that the season has
great influence on the quality of cheese ; especially
on the defect more immediately under notice. In
1783, a dry hot summer, scarcely any dairy could
make good cheese. In some dairies more than
half the make was hollow, and even in the best
dairy I had an opportunity of examining, numbers
were " eyey ;" while in a common season, and
more especially in a cool summer, the same dairy
has scarcely a defective cheese.

" In North Wiltshire, an experienced, and very
intelligent dairy-woman observed, that when the
" crazey" (the crowfoot) is in full blow, she finds
her cheese particularly inclined to heave; while a
dairy farmer of the highest class in the same district, has observed that when the creeping trefoil,
white

white clover *(trifolium repens)* has been in full blow,
and in particular abundance, he has heard the loudest
complaints of the licentious disposition of the cheese.
It is not probable that any one species of plants
is the sole cause of the disorder. Almost every
cheese has its peculiar flavor, and its different degree of acrimony. Nothing is more likely to give that
almost caustic quality which some cheeses are possessed of, than the common and bulbous crowfoots :
not only their flowers, but their leaves, are singu-
larly acrid. On the other hand, there are several
circumstances which render it probable that a re-
dundancy of the creeping trefoil tends to aggravate
the disorder. Dry seasons, by keeping the grass
short, give it an opportunity of spreading. Manure is well known to encourage it ; sometimes in
a singular manner. Sheep-feeding pasture grounds
produce a similar effect, partly owing perhaps to
the blade grasses being kept short ; and in part to
the soil being meliorated by a fresh manure ; and it
has been observed that a suit of cow-grounds,
which have been occasionally fed hard with sheep,
are very difficult to make cheese from : while a few
sheep among cows may, by picking out the clover,
be serviceable to the dairy."—*Marshall.*

PARE AND BURN.

The men employed in this business should be
kept steadily at work throughout the whole of this
month ; if heavy rains impede the drying and
burning, let it be remembered that the paring may
probably go on the better for it, so that whatever
the

the weather may be, this operation, which is of such essential importance in many improvements, need not stop.

STATE OF WHEAT CROPS.

The young farmer will now be naturally led to watch the progress of his wheat crops: no accurate judgment can be formed till this month, which will enable him to make various observations which a man of any curiosity will not omit. It is remarked by a late writer, that wheat which has carried a green and flourishing countenance throughout the winter, often loses its verdure in the spring, and assumes a yellow sickly aspect. In the spring of the year 1780, the forward sown wheat was so much affected by the cold weather in the months of April and May (it having been one of the most backward springs I ever remember), as to become exceedingly yellow, and was interspersed throughout with innumerable patches of different tints, which patches, wherever they appear, are always accounted a certain and infallible prognostic of a bad crop of wheat; it having been remarked that the fields where these patches abound, do seldom if ever recover: though it is otherwise in fields which have not these patches in them, since with kindly weather in June, the corn on these latter fields often surmounts the mischief occasioned to the blade by the vernal cold, according to the old proverb current among the farmers, and expressed in their homely lines, " I came to my
wheat

wheat in May, and went right sorrowful away : I came to my wheat in June, and went away whistling a merry tune.   After a dripping summer, bread corn is generally dear, as there is no weather so inimical to the wheat on the ground as wet, especially on the deep rich lands, where the largest crops are raised ; and even on poor chalky soils, it is matter of doubt with me, whether a wet summer be not rather in- jurious than beneficial to the wheaten crop, though such moist weather may haply increase the growth of straw.   But although this reasoning generally holds good, yet I have sometimes known the crops of wheat turn out very prolific after a wet summer. The year 1777 was one of the wettest that could have been remembered, and the spring had been uncommonly wet and chilly, so that the farmers, from the great abundance of straw, and from an observation of the unkindly state of the air through- out the summer, expected that their wheaten crops would turn out to bad account, and that conse- quently this grain would fetch an advanced price in the ensuing winter, but the event falsified their prediction, and the public were served with bread at a reasonable rate throughout the winter, the price of wheat never exceeding 42s. or 43s. per quarter.   This unexpected fertility was occasioned, as I conceive, by a kindly and favourable season at the blooming time, for in this year the wheat was very backward in coming out of the hose, and dur- ing the time it remained in blossom, the weather

was

was most favourable for that purpose, being, in truth, the only part of the summer unaccompanied with rain or wind. From these observations may be drawn the following corollary: that when the wheat hath a good blooming time, though the rest of the summer, both antecedent and succeeding this period, may have been unkindly; yet so much depends on the kindly state of the air at the blossoming season, that little danger need be apprehended to the crop from the weather in any other part of the summer. On the other hand, though the summer months may in general have been such as to promise a good crop of wheat; yet should a wet and unkindly season intervene while the corn is in bloom, the produce will not be analogous to the general state of the weather during the greater part of the summer months, but to that particular prevalence of it at the time when the wheat was in bloom, a time whereon seems to depend the future welfare of this and every other vegetable.

A series of easterly winds at the blooming season is often highly prejudicial to the wheaten crop: in the year 1771, the weather having been such as above described, there appeared a very capital defect in the wheat after the blooming season was past: on opening the chests of the ear, were perceived several small maggots, resembling in size and colour, the male blossoms of the wheat, and for which I at first view mistook them. These maggots lay in a cluster within the chest, and adhered closely to the nib of the seed: within some

of

of the chests the corn had attained nearly to half
its growth with these maggots preying upon its surface.　On a closer inspection into the nature of this
malady the succeeding year, I could clearly discern
the maggots adhering to the female blossoms, and
in whatever chests these maggots were found, the
male blossoms, which in a kindly state of the air
are suspended without the chests, and are connected to the female blossoms by very slender fila-
ments, and by this economy, apparently convey
the fructifying quality to the female blossoms, were
in close contact with the latter within the chests,
amongst which the maggots effect their lodgment,
and, as I observed before, bear, on a superficial
view, a strong resemblance to the male blossom, but
on a closer inspection, are found preying on the
female blossom, and covered by the male : from
whence it seems reasonable to conclude, that these
maggots are the produce of a small fly, which settling on the male blossom whilst it is performing
its office, may there deposit its eggs, which being
instantly conveyed to the nib of the seed, are succeeded by a progeny which are the maggots in
question.　These eggs may possibly retard the male
blossoms from completing their office, and prevent
their flying off, which they ought to, when the
female blossoms become impregnated, and this may
be the reason why the maggots are always found
adhering to the male blossoms.

Of the male blossom of wheat it is to be remarked, that if by wind or other accident it becomes

comes disunited from the ear, a succession of bloom still continues to supply its place, and this a second or third time, which displays the wisdom and goodness of Providence, in thus securing to us this necessary part of our aliment from the variety of accidents to which it is liable in this critical period of its vegetation ; and for this reason stormy weather at the blooming season is not of such very fatal consequence to the growing crops of wheat as many people imagine. But in a wet or clouded atmosphere, the danger is much greater ; for by this density of the air, the chests are so intimately closed as to prevent the male blossoms from escaping out of the hose, and hence ensues a corruption within the ear, which occasions, as hath been remarked, the evil mentioned.

A mild and open winter is by no means kindly for the growing crops of wheat, not only from the blade having by such weather been encouraged to push forward with too much celerity, and thus becoming winter proud, as before remarked ; but for another reason, namely, that the weeds will be apt to spring up in great abundance, and meeting with no resistance in their growth, spread over the surface and become a formidable enemy ; and should a dripping summer succeed, the mischief accruing from a weedy crop will be still increased to a very high degree.—*Bannister.*

FLAX.

The flax crop will this month want a careful weeding by hand : it should be done with atten-

tion, not to go on to it in wet weather, and to beat
it down as little as may be.

HEMP.

Some writers have recommended weeding hemp
in June ; but on all soils proper for this plant it is
unnecessary, the crop will get the better and destroy
all weeds : if these get ahead among it, it is a sure
proof that the soil has been improperly chosen.

PLANT HOLLY.

No plant makes so good a hedge as holly; if preserved with any attention in its infancy, it will in a
few years be impenetrable to man or beast.   It
often fails from being planted at an improper season ; for there is not the least certainty of any
success except by planting about Midsummer. The
plants should be from six to nine inches high, and
well rooted ; they should not be let into the sloping face of a bank, but on a level tablet left for
that purpose, and well defended on both sides, to
keep both sheep and hogs from it.

TRAVELLING.

If our young farmer has any relation, friend, or
confidential bailiff that he can trust his farm to
for ten days or a fortnight, let him now take his
nag for a summer tour, to view some farms in well
cultivated countries, and to introduce himself to
the conversation of his intelligent brethren, from
whom he will be sure to learn something useful.
In this month are the sheep shearings of his Grace
the Duke of Bedford and of Mr. Coke ; he cannot
do better than be present at one of them, as he
will

will there meet with able cultivators from every part
of the kingdom, and may learn where best to direct
his steps, whatever may be his object : and this, let
me remark, is no inconsiderable proof of the great
national utility of those meetings. I have met
farmers both at Woburn and at Holkham, who
were in the progress of such journies, were properly and usefully inquisitive, and without doubt
received no slight advantage from the knowledge
thus gained. This is a good season for a journey;
the corn of all sorts shews itself to advantage;
the turnip season is in full operation ; lucerne is
mowing for soiling ; the marle carts are at work ;
the lime-kilns active, and most of the works of a
farm either in operation or effect.

LIME.

The lime-kilns ought to be in full work in this
month, and there is no better time for carting and
spreading it. At this period the proper land to
spread it on is the turnip fallows, which now being
in full tilth, if the surface be well harrowed after
receiving the manure, the union of it with the soil
will be intimate ; it should be left some time before
ploughing for slacking and re-imbibing the carbonic acid driven from it by the act of burning.
Modern chemists are much inclined to attribute
great effects to this acid in the business of vegetation ; the point is by no means fully elucidated,
and does not very well accord with the very small
benefit derived from lime when laid on certain poor
soils ; but as there are many others on which the

good

good effects of lime are unquestionable, and our
farmer may be supposed to have satisfied himself
by previous experiments, he will bestow the expence only where he is sure of a re-imbursement.
Where fallowing for wheat is the system pursued,
lime is also spread on these fallows throughout all
the summer months.   A bushel to a rod is a very
good dressing : and even half the quantity well
applied has a considerable effect.   On waste lands,
such as fresh drained bogs and mountainous
moors, the greater the quantity, even to 500 or
600 bushels per acre, the greater is the effect, and
probably the profit also.

SPRING TARES.

It is exceedingly good husbandry to sow spring
tares in this month, and a quart of cole-seed over
the same land in order to have a very wholesome
and nourishing food for weaned lambs in autumn.
This is a practice on the South Downs in Sussex,
from which great benefit is derived. The breadth
of land to be thus applied, will depend on the other
articles provided for the same application.   The
land first soiled or fed of winter tares, may be
ploughed for this purpose, and thus two beneficial
crops gained in one year.

BEES.

This is the principal season for swarming ; a careful attention should therefore be paid to the hives,
that the swarms be not lost.

HOGS.

The principal stock of swine may now be feeding

ing in clover or chicory : but if due attention be
paid to the great object of raising manure, our
young farmer will supply them very amply in their
yards with these plants, or with lucerne or tares ;
but plenty of litter should be given ; they will pay
well for whatever attention is bestowed of this sort.
The most profitable litters of the whole year are
those of the sows which pig in June ; every one
should be carefully reared.

HOE BEANS.

The bean crops must be well attended to throughout this month, and the horse and hand-hoes kept
at work. Let the young farmer remember, that
this crop is his fallow for wheat, and must on no
account be neglected. He has, of course, Berkshire shims of various breadths in the cutting
plates, adapted to the spaces he has allowed as intervals, and also to the height of the beans, that
he may use them at any time without damage to
the plants. For these operations he has nothing
to fear but a very wet season, which much impedes
all these necessary works ; he ought therefore to
make so active a use of every dry time, that he
may not be forced to lose much labour in doing
work twice which is better executed at once when
the weather is favourable.

WARPING.

This singular operation begins in June.

The husbandry which I am about to describe
under this title, is one of the most singular improvements I have any where met with, and far

exceeding any other that has been heard of. It is
practised only in Lincoln and Yorkshire.

The water of the tides that come up the Trent,
Ouze, Dun, and other rivers which empty themselves into the great estuary of the Humber, is
muddy to an excess ; insomuch that in summer, if
a cylindrical glass, 12 or 15 inches long, be filled
with it, it will presently deposit an inch, and sometimes more, of what is called warp. Where it
comes from is a dispute : the Humber, at its
mouth, is clear water ; and no floods in the countries washed by the warp rivers bring it, but, on
the contrary, do much mischief by spoiling the
warp. In the very driest seasons and longest
droughts, it is best and most plentiful. The improvement is perfectly simple, and consists in nothing more than letting in the tide at high-water
to deposit the warp, and permitting it to run off
again as the tide falls : this is the aim and effect.
But to render it efficacious, the water must be at
command, to keep it out and let it in at pleasure ;
so that there must not only be a cut or canal made
to join the river, but a sluice at the mouth to open
or shut, as wanted ; and that the water may be of
a proper depth on the land to be warped, and also
prevented flowing over contiguous lands, whether
cultivated or not, banks are raised around the fields
to be warped, from three or four to six or seven
feet high, according to circumstances. Thus, if
the tract be large, the canal which takes the water,
and which, as in irrigation, might be called the
                                        grand

grand *carrier*, may be made several miles long ; it has been tried as far as four, so as to warp the lands on each side the whole way, and lateral cuts made in any direction for the same purpose ; observing, however, that the effect lessens as you recede from the river ; that is, it demands longer time to deposit warp enough.

But the effect is very different from that of irrigation ; for it is not the water that works the effect, but the mud, so that in floods the business ceases, as also in winter ; and it is not to manure the soil, but to create it. What the land is, intended to be warped, is not of the smallest consequence : a bog, clay, sand, peat, or a barn floor ; all one ; as the warp raises it in one summer from six to sixteen inches thick ; and in hollows or low places, two, three, or four feet, so as to leave the whole piece level. Thus a soil of any depth you please is formed, which consists of mud of a vast fertility, though containing not *much* besides sand ; but a sand unique. Mr. Dalton, of Knaith, on Trent, sent some to an eminent chemist, whose report was, that it contains mucilage, and a very minute portion of saline matter ; a considerable one of calcareous earth : the residue is mica and sand ; the latter in far the largest quantity : both in very fine particles. Here is no mention of any thing argillaceous ; but from examining in the fields much warp, I am clear there must be clay in some, from its caking in small clods, and from its cleansing cloth of grease, almost like fuller's earth. A con-

c c 4   siderable

siderable warp farmer told me, that the stiffer warp
was the best; but in general it has the appearance
of sand, and all glitters with the micaceous particles. So much, in general, as to the effect; the
culture, crops, &c. are circumstances that will best
appear, with others, in the following notes, taken
on the spot.

Mr. Webster, at Bankside, has made so great an
improvement by warping, that it merits particular
attention. His farm, of 212 acres, is all warped; and
to shew the immense importance of the improvement,
it would be necessary only to mention, that he gave
11l. an acre for the land, and would not now take
70l. an acre; he thinks it worth 80l. and some
even 100l.: not that it would sell so high at' present; yet his whole expence of sluices, cuts, banks,
&c. did not exceed 2500l. or 12l. per acre; from
which, however, to continue the account, 1500l.
may be deducted, as a neighbour below him offers
5l. an acre for the use of his sluice and main cut,
to warp 300 acres, which will reduce Mr. Webster's
expence to 1000l. or about 5l. an acre. Take it,
however, at the highest, 12l. and add 11l. the purchase, together 23l. an acre; if he can sell at 70l.
it is 59l. per acre profit. This is prodigious, and
sufficient to prove that warping exceeds all other
improvements. He began only four years ago.
He has warped to various depths, 18 inches, 2 feet,
$2\frac{1}{2}$ feet, &c. He has some that, before warping,
was moor-land, worth only 1s. 6d. per acre, now
as good as the best. Some of it would let at 5l.

for

for flax or potatoes; and the whole at 50s. He has
20 acres that he warped three feet deep, between
the beginning of June and the end of September, and
18 acres, part of which is three feet and a half deep.
He has applied it on stubbles in autumn by way of
manuring; for it should be noted as a vast advantage
in this species of improvement, that it is renewable
at any time; were it possible to wear out by cropping or ill-management, a few tides will at any
time restore it. As to the crops he has had, they
have been very great indeed; of potatoes from 80
to 130 tubs of 36 gallons, selling the round sorts
at 3s. to 3s. 6d. a tub; and kidneys at 5s. to 8s.
Twenty acres warped in 1794, could not be ploughed
for oats in 1795, he therefore sowed the oats on
the fresh warp, and scuffled in the seed by men
drawing a scuffler; eight to draw, and one to hold;
the whole crop was very great: but on three acres
of it measured separately, they amounted to 14
quarters one sack per acre. I little thought of
finding exactly the husbandry of the Nile in England. I had before heard of clover seed being
sown in this manner on fresh warp, and succeeding
greatly.

He warped 12 acres of wheat stubble, and sowed
oats in April, which produced 12 quarters an acre.
Then wheat, 36 bushels an acre. His wheat is
never less than 30.

Six acres of beans produced 30 loads per acre,
or 90 bushels: one acre, measured to decide a
wager,

wager, yielded 99 bushels. Has had 144 pods from one bean on four stalks; and tartarian oats seven feet high. One piece warped in 1793, produced oats in 1794, six quarters an acre: white clover and hay seeds were sown with them, mown twice the first year: the first cutting yielded three tons of hay an acre; the second one ton; and after that an immense eddish. Warp, Mr. Webster observes, brings weeds never seen here before, particularly mustard, cresses, and wild celery, with plenty of docks and thistles.

Flax, 40 to 50 stone per acre.

A sluice for warping, 5 feet high, and 7 wide, will do for 50 acres per annum; and if the land lie near the river, for 70. Costs from 400l. to 500l.

Mr. Nicolson at Rawcliff, takes the levels first; builds a sluice; if a quarter of a mile or half a mile, 60 acres may be done the first year; the drier the season, the better. The clough, or sluice, 400l. eight feet wide, and five or six feet high: a drain 14 feet at bottom, and as much more at top, 30s. to 40s. an acre, of 28 yards: banks four to eight feet high, and expence 7s. to 20s. an acre of 28 yards. Begin at Lady-day till Martinmas; but all depends on season; the depth will depend on circumstances. If a landlord warp, it should be deep at once; if a tenant, shallow and repeat it, as good corn will grow at six inches as six feet; at three inches great crops; the stiffer the warp the better. Some seasons, sow corn the year after.

Warp

Warp is cold, and if deep takes time : a dry year best : great seeds. Crops ought to be, beans 20 loads ; oats 10 quarters ; wheat 10 or 12 loads ; never barley. After six years potatoes, and good flax : he makes it worth 40l. to 50l. an acre.

Mr. Wilson's idea of warping is very just : to exhaust the low lands in favour of the hills ; then to warp six inches deep, to exhaust that to make the hills ; then to warp again ; and by thus doing to keep the warp land in the highest order, and at the same time work a great improvement to all the higher grounds.

Note, by a Commissioner employed in Warping.— " Warp leaves one eighth of an inch every tide, on an average ; and these layers do not mix in an uniform mass, but remain in leaves distinct.

" If only one sluice, then only every other tide can be used, as the water must run perfectly off, that the surface may incrust, and if the canal be not empty the tide has not the effect. At Althorpe Mr. Bower has warped to the depth of 18 inches in a summer.

" Ten quarters an acre of oats, on raking in the seed on warp, the more salt in it the better ; but one fallow in that case necessary, to lessen the effect, or it hurts vegetation."

A very great object in this husbandry of warping, is the application of it in other districts. They have much warp on all the coast from Wisbeach to Boston, &c. and though a long succession of ages has formed a large tract of warp country, called there

there *silt*, yet no attempts that I have heard of have been made to warp artificially there.

Should any proprietor into whose hands this Calendar may fall, or even any farmer, living near a muddy river, consider well the position of his ground, and try the amount of the subsidence of the water in a cylindrical glass jar, for a treasure may be near him without his knowing any thing of the matter.

PLOUGH IN GREEN CROPS.

This is a sort of manuring which has many advocates and some enemies, resulting probably from their having drawn conclusions from trials on different soils and under different circumstances. The probability of success is greater when the vegetable is ploughed in at Midsummer than at Michaelmas, as the warmth of season must considerably aid the fermentation. Whatever the plant may be, whether rye, tares, or very early sown buck-wheat, it should be ploughed down with a skim-coulter plough, which is the only means of turning it so completely in as to be quite concealed from the eye; and this operation should take place at least three weeks before sowing turnips; when that seed should be very lightly harrowed in.

MOUNTAIN IMPROVEMENT.

The improver of moors and mountains should take care to employ hands enough in summer for executing all the works belonging to that season : in such situations the winter is usually very long and very severe ; few works can then go on except

quarrying

quarrying for walls and lime, and digging drains, and in some cases only the former; a good use should therefore be made of all the summer months for paring and burning, building, walling, &c. &c.

HIRE HARVEST MEN.

At Whitsuntide it is usual for farmers to make an agreement for their harvest : see the Calendar for August. But the young cultivator should now have it in his mind.

BURN DRY WEEDS FOR MANURE.

Our young farmer may perhaps want to be reminded, that spreading any sort of dry vegetable substance on the land, and setting fire to it previously to harrowing in, or drilling turnip-seed, is one of the most powerful manures that can be used. There are situations where fern from wastes, warrens, &c. may be collected in almost any quantity : if he has it in his power to preserve more than he wants for littering, he should save it carefully for this use. In the fens of Cambridge and Lincoln, it has been long a custom to burn oat and other stubbles (of reaped crops), and the effect resulting from it was probably the origin of a practice which I first heard of in the latter county ; that of burning straw for this purpose.

The most singular practice which I ever met with in manuring, subsists on the Wolds : it is that of spreading dry straw on the land, and burning it. At Lord Yarborough's I first heard of this custom. His Lordship's tenant, Mr. Richardson,

a very

a very good and intelligent farmer, gave me the account, having long practised it with success. The quantity is about five tons an acre. At Great Lumber he straw-burnt a piece in the middle of a field preparing for turnips, and on each side of it manured with ten loads an acre of yard-dung, and the burnt part was visibly superior in the crop. In another piece the same comparative trial was made in 1796, for turnips, which crop was much the best on the burnt part; and in 1797, the barley equally superior. On another farm he had at Wold Newton he did it for turnips, then barley, and laid with sainfoin ; and the burnt straw was better in all those crops than yard-dung. Burning gorse in this manner returns great crops, but the expence is too high. He is clearly of opinion, that it is the warmth from the fire that has the effect, and not the ashes ; for the quantity is nothing, and would blow away at one blast. It is proper to observe, that they do not value straw used in feeding cattle, at more than 4s. or 5s. per ton.

Mr. Mallis, of Lumber, is of the same opinion, and thinks four ton enough : he never knew that quantity fail for turnips.

This straw-burning husbandry I found again at Belesby. Mr. Lloyd, who, as I should observe, is an excellent farmer, thinks that it takes six tons per acre, which will last longer in its effect, and beat the dung which that straw would make ; and

in

in general lasts longer than common dunging. Keeping much cattle, he cannot practise it, but highly approves it.

In discourse at Horncastle Ordinary, on burning straw, the practice was much reprobated; yet an instance was produced that seemed to make in favour of it. Mr. Elmhurst, of Hazlethorpe, burnt twelve acres of *cole-seed* straw on eight acres of the twelve, and the effect was very great, and seen even for twenty years : he sowed wheat on it, four bushels an acre, and had five quarters : the four acres upon which nothing was burnt much the better land, yet the crops on the burnt part were by that made equal to the rest. But in another similar experiment for turnips, Mr. Rancliff observed the result, and the effect, though good, lasted only for one crop. Mr. Kirkham, who was in company, gave it as his opinion, that as cattle would not eat stubble, it might be beneficial to collect and stack that, and before turnip sowing burn it.

The Rev. Mr. Allington, of Swinop, has cut and carried gorse, and spread it on other land, and burnt it in May for a manuring for turnips; but has done it twice, and it answered very well; but of course it is to be noted, that this is done only when it cannot be sold for faggots, which sell at 8s. per hundred ; so that the expence would be 4l. an acre, as 1000 are produced per acre, and he burnt the produce of one acre upon another : the effect was great in the turnips; the barley was
better

better for it ; but he has not attended to it in the
seeds, because hard stocked with sheep. He has
burnt on the land for turnips, the long straw dung
from the surface of the farm-yard, and he had
better turnips there than where the dung was laid.
This has been the case in two experiments he has
made.

A general practice through the mountains of Gas-
cony, and almost to Bayonne, is that of manuring for
raves, a sort of turnip, with the ashes of burnt straw.
I observed several fields quite black ; and demand-
ing what it was, my guide told me of this common practice here : afterwards I saw them strewing
straw thickly over land, part of which had been
already burnt on. They do this on a wheat-stubble ;
but not thinking that stubble enough is left, they
add much wheat-straw, and setting fire to it, burn
the weeds as well as the straw, and clean as well as
manure the land. With such quantities of fern
on all their extensive wastes, I asked why they did
not burn that, and keep their straw ? The reply
was, that fern makes much better dung than straw,
so they burn the straw in preference. As soon as
the operation is over, they plough the land, and
harrow in rave-seed. One large field, thus treated,
I saw ploughing for that crop. They both hoe
and hand-weed the raves, and have them sometimes very large ; many as big as a man's head.
Use them for oxen.

SAINFOIN.

When the plants of sainfoin are thin on the
ground,

ground, it is a very judicious practice to suffer the crop to remain the first year for seed, which will thicken the swath in the succeeding crops. The seed will probably be worth 5l. per acre; the straw is good horse-fodder; the plants are not at all damaged at present, and their number greatly increased for the future.

JULY.

FARM-YARD.

ALL, or much of the compost in the yard being carted on to the land, you may now, if you have leisure, begin to cart in the layer of turf, ditch-earth, chalk, marle, clay or peat, upon which you are to fodder. There is no necessity of performing this work in July; but, as it may probably prove a leisure time for the teams, it is mentioned as a business that should be in hand, as a prevention of their standing still. It should be executing from this time till the end of September. As the importance of it is very great, being the source of the most material improvements on a farm, it should be resolved on early.

The management of the farm-yard must, however, depend on the system pursued relative to using dung in a long or in a rotten state. If in the latter, the layer of earth that was spread at bottom for absorbing urine and the draining of the dung has been mixed by turning over, and may be supposed to have been carted on for turnips. But if the dung is taken fresh from the yard, the earth in that case is to be examined, as it should remain till well impregnated, and this may not have been effected at present, in which case it will be left for a longer duration and successive soiling

on.

on. Whenever it has absorbed its due portion, so that white soils, as chalk and clay marles, are become blackish, then is the time for removing them with the dung, and they will be equally beneficial.

TURNIPS.

Now is the time for hand-hoeing turnips; a work perfectly understood in many parts of the kingdom; but so much neglected in some, that it will be proper to enlarge a little on the method of performing it.

Supposing turnip-hoers to be scarce, or to demand extravagant prices, or none to be had, order some hoes to be made by your blacksmith : the iron part nine inches long, and three or four broad, neatly done and sharp : put handles five feet long in them. So provided, take your men into the field, and yourself with a hoe should accompany them : make them hoe the crop boldly, and not be afraid of cutting too many up. Direct them to strike their hoe round every plant they leave, and fix upon the most vigorous and healthy growing ones. By this means they will leave the plants twelve inches asunder; for, their hoes spreading at every cut nine inches, they cannot spoil your crop by not cutting freely. This work must be done by the day, and you must attend the men well, to see that they cut the land pretty deep, so as to kill all the weeds, and also such turnips as they strike at. In about a fortnight after, send them in again to rectify former omissions, in which time

they

they must break all the land again with their hoes,
cut up the remaining weeds, and wherever the turnips were left double, thin them. The men will be
aukward at this work the first year; but, by degrees, they will be able to do it well, and by mixing
new ones among them every year, the art will not
be lost.

In countries where turnip-hoeing is commonly
practised, the work is generally done by the piece:
four shillings an acre for the first hoeing, and two
shillings and sixpence, or two shillings, for the
second, were common prices: but now it is in
some districts five -shillings or six shillings for the
first hoeing, and four shillings for the second.
When done by the piece, the farmer's principal
attention should be to see that the work is well
done; for, in all these operations, the men are extremely apt to run over their work in a slovenly
manner, aiming only at making good earnings: the
farmer should see that they cut up all the weeds,
and leave the turnips every where single. The
crop must have two hoeings, which should leave
it *perfectly* clean, and the plants at regular distances.

If the turnip-hoers are not to be procured in
number sufficient to execute the work soon enough,
the plants should be well harrowed, which will
thin and keep them from running thickly together
in bunches. It is common for the men to bargain
with the farmer to have a harrowing given before
they hoe.

DRILLED

DRILLED TURNIPS.

But wherever hand-hoeing turnips is not well understood, and men for it easy to be had, the Northumberland system should by all means be practised. The rows on the tops of the ridges, as described last month, are hoed by women in great perfection ; they should be set out in good land a foot asunder ; and on weaker soils at nine inches. I have known women send in their children before them to thin the plants with their fingers, leaving them at the distance required, and follow themselves with six-inch hoes for cutting the ground, and making very good work. This is a great improvement, much deserving the attention of all farmers who live in counties where the labourers are ignorant in common turnip-hoeing, or where they are scarce, or apt to impose in the prices demanded. The crops are as good, and in the opinion of many, much better than broad-cast ones.

SOOT TURNIPS.

In 1803, my son had a crop of turnips drilled in the Northumberland manner, which, as soon as the young plants were seen in the rows, he sooted at the rate of twenty bushels per acre, throwing the soot by hand out of a seed-lip, in a stream as near as might be along the row of plants. They escaped the fly, and were the only turnips in the neighbourhood that did so. It was a thought of his own, for he never read the following passage in Ellis :

" Turnips

" Turnips sooted about twenty-four hours after they are up, will be entirely secured from the fly." *Practical Farmer, or Hertfordshire Husbandman,* 1732, p. 86.

COLE-SEED.

This plant may be sowed through all this month, which is indeed the principal season for it. The preparation of the land is exactly the same as for turnips; and it has equal success with turnips when sown on pared and burnt land, which secures a better crop generally than any other method of manuring. Two or three, and even four quarts an acre of the seed are sown. It is not common husbandry to give it any hoeing. Upon peat soils pared and burnt, (as in the fens of Cambridge and Lincoln), it is reckoned much superior to turnips for feeding and fattening sheep; and usually sells, if a good plant, at 50s. an acre.

COLE SEED WHERE TURNIPS FAIL.

The first, and even second sowings of turnips, may have failed by the end of July; in this case, some farmers prefer sowing cole-seed rather than turnips a third time.

CABBAGES.

The crop planted in April or May must be looked to this month. As they were both hand and horse-hoed in June, perhaps they will not want any more culture till August; but this depends on the season : if the weeds grow, let them be killed; for the best rule in this matter is, to hoe
suffi-

sufficiently to keep the crop perfectly clean, and to horse-hoe whenever the intervals have been bound by rains or otherwise.

The crop planted last month must be hand-hoed before the middle of this : in which work you should be attentive to cut up all the young weeds that grow near the plants, and break all the land on the tops of the ridges ; but the men need not hoe the sides of them or the furrows, as the plough in horse-hoeing will cut them much better. Some fresh earth should also be drawn to each plant, earthing it up as it were. The first horse-hoeing should be given soon after ; in which operation the plough should take off a furrow from the ridges on each side, and throw up a small ridge in the middle of each interval, which will let the air into the earth on which the plants stand, and pulverize and sweeten it. The cabbages will be left on a narrow slip of earth, ready for the second hand-hoeing, which will be given with great ease.

This work must, however, be done with much care and attention, for if the plants are left in too small a space, and the sun be powerful, they will suffer : the stripe of earth the plants are left in should be nine inches wide ; and, if the weather is very hot, a furrow turned back again, at least on one side, as soon as may be. Afterwards the horse-hoeing should be given with the shim of three shares ; one *low*, for cutting the bottom of the furrow in the intervals, and two others, four inches higher (being drawn up at pleasure through the

block),

block), for cutting the sides of the ridges without removing too much earth from the plants. And this tool followed after a time by the double mould-board plough, to sweep out the furrows and round up the ridges.

The cabbages drilled in April where to remain, must now be horse and hand-hoed ; and in the latter work, whenever it is executed, well pulverized earth free from weeds should successively be drawn to the stalks of the plants.

WELD.

This is the season for pulling weld, which is done by women, and bound in small sheaves ; these are set to dry, which takes from one week to three. It is then stacked, and is the better for a sweat. In three months it is bound in fresh sheaves of two stones (14lb.) each, and is then saleable. A good crop is from 15 cwt. to a ton : the price varies greatly : sometimes to 16l. and 20l. a ton ; at others, 4l. or 5l.

POTATOES.

The crops of potatoes, planted in rows, must have a third horse-hoeing this month. The common way of ploughing backwards or forwards *every* time of horse-hoeing, is not well adapted to this crop ; for cutting the roots, when the plants are in full growth, hurts the crop, and you destroy runners that would produce potatoes. For this reason, the third horse-hoeing should be given first with the shim, which cuts and loosens the earth, without turning it over, or forming any ridge. Some

Some of them work with many little triangular shares, some with single flat ones, and others only with coulters; but any of them that cut up fresh moulds at the bottom of the furrows, will answer the purpose. A double mould-board plough (a common instrument in some counties), should follow the shim in about a week; and, striking the furrows, throw up all the loose earth against the ridges, banking them up. There is a great use in this operation; for it throws up fresh earth for the roots to shoot into, which is preferable to taking it away from them, after they have advanced at all in growth.

CARROTS.

The carrot and parsnip crops will want a hoeing this month; which should be given while the weather is dry. These operations are never neglected, but the farmer is sure to lose a guinea for every shilling he saves.

BEANS.

The horse-hoed crops of beans must be attended to very carefully: and they are now so high, that if a horse-hoeing is given this month, it must be very carefully done. Whether the shim or double mould board plough be used, it must be drawn by a whipple-tree as short as permits the horse to work, and hung on to a springing fixture at the beam-end, in this form:

by

by which means the whipple is raised, that if it
does brush the beans, it is so high in the stalk,
that they bend easily to it without suffering damage ; but the higher it is thus raised the better.
I have seen them work in Kent, when men from
other counties thought it impossible.   In this state
of the crops, the block of the shim is in a position
longitudinal with the rows, otherwise the ends may
break the stalks.   In common, however, the only
horse work wanting this month is earthing up.
Weeds are never to be left, the hands and hand-
hoes are ever to attack them.

LUCERNE.

The lucerne will be ready to cut again this
month : if it was drilled for horse-hoeing, the intervals must be directly horse-hoed the contrary
way to the last.   In respect to hand-hoeing, the
best rule will be to do it according to the growth
of weeds : there is no necessity for it while the
rows continue clean : the weeds that arise among
the plants in the rows, should be plucked out, and
particularly all grasses, which are the greatest enemies to this crop.

But if the lucerne was sown according to the
directions given in this Calendar, that is, either
broad-cast, or in drills at nine inches, probably
nothing need be done in cleaning this month ; as
one or two scarifyings in the year, will keep it
sufficiently clean.

BURNET.

This month the crops of burnet left for seed
will

will be fit for mowing : the seed is apt to shed, if care is not taken in mowing it. It is best threshed in the field, like turnip or cole-seed, and the straw made into hay. It yields very great crops of seed ; and some persons have asserted, that it is as good for horses as oats ; but no satisfactory trials of due continuance have been made on it.

The following notes deserve attention :

My burnet, though very green and beautiful all the winter, made no great progress till the middle of April following, when I thought it absolutely necessary to feed it. I did so ; but I did it too late, and kept my cattle upon it too long, from the middle of April to the 20th of May. This was a very great mistake : the burnet plants were now headed for seed, and the stock fed chiefly upon the heads, which greatly lessened my quantity of seed, as well as retarded the growth of the plants. I turned ewes, lambs, and calves into the field, and they all fed very greedily upon the burnet. From what I had heard of Mr. Rocque, I very much expected them to scour ; but there was not the least appearance of it, and the cattle throve accordingly.

The 6th of July I began to mow, the weather being favourable : six men and four boys threshed and cleared the seed in seven days. I had 200 bushels of very fine clean seed, as many sacks of chaff, and seven loads of hay, from a field of seven acres and a quarter.

Satisfied that 200 bushels of seed would be

more

more than I should be able to dispose of, I was not anxious after another crop, being rather desirous of seeing what it would perform as a pasture. Accordingly, in about 10 or 12 days after the field was cleared, I turned seven cows, two calves, and two horses into it ; they all throve very remarkably, and the cows gave more, and we thought a richer milk than in any other pasture : I really expected, as burnet is so strong an aromatic, that the milk would have had a particular taste, but far otherwise, the milk, cream and butter were as fine, if not finer tasted than any from the best meadows. I am satisfied, that there is no better pasture for cows, whether milched or barren, than burnet. The weather was now extremely droughty, all our pastures were burnt up, yet the burnet flourished and grew away as if it had a shower every week. My stock of cows, horses and calves, before mentioned, pastured in it almost continually till Michaelmas : by the middle of November it was grown so considerably, that I have again turned in six head of cattle ; and if the weather be not severe, I am of opinion, it will maintain them till Christmas.

The burnet straw, or haulm, is, after the seed is separated from it, a very useful fodder for horses, cows, calves, and sheep : the chaff is of good value, if mixed with any other, however ordinary, chaff. I have fed all the above mentioned stock with it promiscuously together in one field ; putting the

haulm

haulm into racks and the chaff into troughs, and if the haulm was chopped with an engine, it would still be of much more value.

Burnet, I am fully persuaded, will prove a very great acquisition to husbandry on many accounts, but more particularly for the following reasons.

Burnet is a good winter pasture, consequently it will be of great service to the farmer, as a constant crop he may depend upon, and that without any expence for seed or tillage, after the first sowing; whereas turnips are precarious and expensive, and when they fail, as particularly this year, the farmer is very often put to great inconveniences to keep his stock.

It affords both corn and hay. Burnet seed is said to be as good as oats for horses. I know they will eat it very well : judge then the value of an acre of land which gives you at two mowings ten quarters of corn and three loads of hay.

The seed indeed is too valuable to be put to that use at present ; though it multiplies so fast, that I doubt not but in a few years the horses will be fed with it.

It will bear pasturing with sheep.

It makes good butter.

It never blows or hoves cattle.

It will flourish upon poor light, sandy, stony, shaltery, or chalky land.

Burnet, after the first year, will weed itself, and be kept clean at little or no expence.

The

The cultivation of burnet is neither hazardous nor expensive : if the land be prepared, as is generally done for a crop of turnips, there is no danger of any miscarriage, and any person may be supplied with the best seed at 6d. per pound.

I make no doubt but that burnet might be sown late in the spring, with oats or barley. A gentleman in my neighbourhood did so last summer, and it succeeded very well. I should think a buck-wheat season, which is sown the last of all corn, would suit it very well ; but of this I have no experience, and could wish to have the experiment tried. A pea field drilled, in rows, and kept clean, would make an excellent season for burnet, as the pea crop would come off soon enough to prepare the land with two ploughings by the middle of August, after which time I should not chuse to sow it.

It very frequently happens, that every farmer who sows many acres with turnips, has several worth little or nothing : the fly, the black catterpillar, the dry weather, or some unknown cause, often defeating the industry and expence of the most skilful farmer. When this happens, as it too often does, I would by all means advise him to sow it with burnet, and in March and April following he will have a fine pasture for his sheep and lambs.—*Communications to Dr. Templeman.*

MOW GRASS.

All meadows and pastures, not mown in June, should be cut this month. Hay-making being in
many

many seasons such ticklish work, and so extremely
expensive, the farmer should take care to manage
it with as good contrivance as he can. To have a
plenty of hands is a material point; for, if good
use be not made of favourable days, the work will
certainly be unprofitable. In order to this, the
farmer should have some other work always in
readiness for his people, in case the weather is too
wet for hay-making. For men, he may have com-
post-hills to turn over and mix, borders to grub or
dig up, carting manure, &c. Women he may employ in stone-picking, weeding, &c. When many
hands are kept, this management will save much
useless expence. In the making the hay, the getting it at last on to the large cock should never be
omitted.

Many farmers only run it up in broad rows, and
load from them on to the waggons; but it is better
to employ all the hands in cocking it: for, if the
cocks are large and well made, the hay will take no
damage in them, even in very heavy rains; and,
by all the men being so employed, much the more
will be secured.

HAY.

Mr. Ducket's method of trying the heat of his
hay-stacks well deserves noting. He thrusts a scaf-
fold bolt, or other stout and long iron bolt into it,
to give an easy admission to a gun rod, with a
strong worm at the end of it, with which he screws
out a sample, and discovers not only the heat, but
the colour of the hay: if the stack wants air, he
makes

makes many of these holes, which give vent to the
heat, and answer the purpose of a chimney.   The
preceding summer was so favourable for hay-making,
that, according to custom, much hay was spoiled by
hurrying together too quick, and many stacks fired.
Experience should convince men, that there is more
danger in a fine year than in a bad one.

THE TEAMS.

All this month, the horse and ox teams should
be soiled daily with lucerne, in the house or yards;
but if in the latter, they must have water always at
command, and also sheds for shelter; and if the
farmer does not provide plenty of litter for treading
into dung, he neglects the principal part of his profit.   Lucerne is the best plant for this purpose,
and an acre of it will go much farther than of any
thing else.   Chicory is good, so is clover; and
tares, mown every day, will answer well in the
same use.   In want of these give natural grass;
but any of them are better, with plenty of litter
for dung, than turning the horses or oxen into the
field.

THE FALLOWS.

Have an eye to your fallows this month, and do
not follow the example of those farmers who totally neglect them for works of hay and harvest.   A
farmer carries on his business very unprofitably, if
he does not keep men and horses enough for all
works : it is unpardonable to suffer the fallows to
be over-run with weeds.

A ploughing well-timed, just before harvest, is
certainly

certainly of much consequence in fallowing, a
work in which well-timed ploughings are of more
consequence than the mere number of earths given.
It is necessary in such a work to suppose this business of fallowing, but the modern well informed
husbandman will, after his first year, but rarely have
recourse to them.

FOLDING.

Where folding is the system, it should this month
be followed with unremitted diligence : the lands
usually fixed on for this purpose are the wheat fallows, which is very judicious in those farmers who
have no crop sown between the turnips and wheat;
but let the attentive, accurate husbandman lay it
down as a rule, ever to fold those lands first, which
will be first sown.   During this month, he should
fold such fields as are destined for August-sown
grasses, of whatever sort, or tares : if the manure is
left long before sowing, the benefit reaped by the
crop will not be nearly so considerable.

WEAN LAMBS.

Before this month goes out the lambs of the
flock should be weaned ; in this business they are
much earlier in Sussex than they are in Suffolk.
Clover in blossom is, of all other food, the most
forcing ; sainfoin rouen excellent ; and if the farmer has neither, he ought to have made a reserve
of a sweet good bite of fresh grass for them.   It is
essential that due provision be made before-hand.

DIG MANURES.

Do not let the marle, chalk, mud, or clay carts,

stop this month : it is a very proper season for the work, and should be pursued with spirit, while the season admits it, on all soils : I say *on all soils*, because in winter, wet or heavy ones must not be carted on. These manures, though expensive at first, are cheap in the end ; for they last many years. In all works of carting, attend particularly to the employment of your team : use as few horses as possible. For this purpose, the small three-wheeled cart is well adapted : one horse is sufficient for two of them : one loading while the other is driving away, by means of the third wheel, which supports the weight of the cart and load, instead of the fill-horse in large carts : they do not hold more than fifteen bushels ; such will do for winter-carting on grass-lands, without poaching. If the draft is not distant, three or four men will thus be employed by one horse, which is an excellence that no other machine can boast. Now let any attentive cultivator reflect on the importance of an odd horse performing much of the carting of a farm, while the others are going regularly on with their tillage or road-work. Whoever will consider this comparison in the proper light, will be sensible that it is an economical way of carrying on business.

MADDER.

In case the season in the spring proved so unfavourable to planting madder, that the work was delayed until the last week of May, or the first of June, the fields so planted should now be horse or hand-hoed, as most wanting. The best way is to

use

use the shim : not for turning a ridge against the
rows, as the plants will yet be too weak for that
operation, but merely to loosen the earth of the
intervals, thereby to kill the weeds, and prepare the
soil for throwing up against the rows by a succeeding operation.　Hand-hoeing and weeding should
depend on the number of the weeds that arise
among the plants.　Let the cultivator of madder,
through the whole process of the crop, remember,
that he must be to the full as accurate as a gardener : his soil must be rendered but little inferior
to a dung-hill : all weeds must be for ever eradi-
cated; not one must injure the plants: his land
must always be kept perfectly loose and well pulverized ; for a crop that depends merely on the
quantity of the roots, can never thrive to profit in
land that is bound, or in an adhesive state.

CUT PEASE.

Forward white pease will be fit to cut early in this
month.　If the crop is very great, they must be
hooked; but if small, or only middling, mowing
will be sufficient.　The stalks and leaves of pease
being very succulent, they should be taken good
care of in wet weather : the tufts, called wads, or
heaps, should be turned, or they will receive damage.　White pease should always be perfectly dry
before they are housed, or they will sell but indifferently, as the brightness and plumpness of the
grain are considered at market more than with hog-
pease.　The straw also, if well harvested, is very good
fodder for all sorts of cattle and for sheep; but if

it receives much wet, or if the heaps are not turned, it can be used only to litter the farm-yard with.

BARLEY.

Some of your barley will probably be ready for mowing towards the latter end of this month, particularly the Fulham sort, which is frequently cut the middle of July, a fortnight before any other sort, though sown at the same time, and on the same land. This early mowing has several advantages ; many weeds are cut before they seed, which, in a fortnight longer, would shed, and consequently injure the ensuing crops. The trouble and attention of harvest is lessened : for a part, at least, of the barley crop may be in the barn, before other farmers, who do not use this sort, begin to mow.

WHEAT.

August is the principal month for cutting wheat, under which head I shall treat of it more particularly. I mention it at present merely to consider the conduct of many sensible farmers, who are fond of cutting their wheat, at least ten days before it is ripe. There is reason to think this practice a very good one : the corn is left in the field longer than common, to finish in that manner the ripening : the advantage is the fineness of the grain. If you are desirous of carrying to market a sample of wheat that shall exceed all others, it must be thus harvested ; and I have heard more than once several very attentive farmers assert, that they lose nothing in measure by this management. It is at least worthy the trial of all good husbandmen, were it only for the convenience

nience of somewhat dividing their harvest : the last fortnight in August is so busy a time, that many of them scarcely know how to get in their corn, upon account of all sorts then requiring attention at once.

MILDEWED WHEAT.

Be very attentive to the wheat crops this month : they are every where liable to this fatal distemper, which admits but of one cure or check, and that is, reaping it as soon as it is struck. The capital managers in Suffolk know well, that every hour the wheat stands after the mildew appears is mischievous to the erop. It should be cut, though quite green, as it is found that the grain fills after it is cut, and ripens in a manner that those would not conceive who have not tried the experiment, which I have done many times ; reaping so early, that the labourers pronounced I should have nothing but hen's-meat. They were always mistaken, for the sample proved good, while others, who left it longer, suffered severely. The fact is now pretty generally known and admitted.

BUCK-WHEAT.

I have known this crop succeed well and yield largely, when sown so late as the first week in this month; and it is a very valuable circumstance, that a man can have so long a period for tillage, and then raise a crop which certainly classes with ameliorating ones, and which prepares well for wheat.

PARE AND BURN.

Wherever there is an improvement going on of

any extent, this work should never stop while the weather permits it to be continued. In the spring for potatoes; then for turnips; now for cole-seed; and, when that is over, then it may go on for wheat, or on mountains for rye. An improver should not let the hands thus employed go to any other work. The stoutest and most skilful hand will not be able to pare (without the burning) more than an acre in a week, even where the work is smooth and free from impediments.

HOGS.

During this month the stock of swine may be supported on clover, chicory, and lucerne; for sows that have pigs, and for weaned pigs, the early sown lettuces on rich warm land will be ready, and prove very useful. Garden beans planted for this purpose are also applicable to the use of all sorts of swine. This is not a month of difficulty for this animal; and the young farmer should take care that his dairy-wash is accumulating in his cisterns for sows and weaned pigs, for a time when they may want it more than at present.

FAT OXEN.

Careful graziers make it a rule, however extensive their farms may be, to ride round and see every beast in every inclosure, at least once a day. Fences demand perpetual attention, and high-fed cattle are apt to break their bounds if this article suffers through neglect.

Beasts that are soiled in stalls or yards, have, through all this season, plenty of food, supposing
a proper

a proper succession of those crops which have been often mentioned for this use.

WARPING.

As this capital improvement, where yet known, goes on only in summer, the farmer should of course keep his works active every tide, and never lose one through neglect, or from having his sluices, &c. out of order.

MANURING NEW LAYS.

This part of the management will not be found essential if the land be laid down in the courses prescribed; it however will at all times be found very beneficial. The best time for it is in August or September, if done the first year, being then a year old, when a moderate dressing will much promote the thickening of the herbage. But upon soils rather unfavourable to grass, on which the success is at all doubtful, I should prefer (if it can be done but once) to delay it to the period when new lays are apt to fall off, that is in August of the third year, if fed; but if mown, immediately after clearing off the hay, which is the best time of all others for manuring grass-land.

Top-dressings of soot, sifted ashes, malt-dust, and other bodies which will wash in the first heavy rain, should be sown the end of February or beginning of March.

SHUT UP ROUEN.

There is scarcely a more important object in the range of common farming management, than that of converting rouen after-grass, after-math, whatever

it

it may be called, to the greatest profit. If it be consumed in the general manner by feeding soon after the fields are cleared of hay, or in the autumn, the value is small, rarely amounting to more (except in watered meadows) than from 7s. 6d. to 15s. an acre; and the reason of this low value is, that food is usually plentiful at this season; but kept for ewes and lambs, and other stock, in the very depth of the winter, and in spring, when food is scarce, and if turnips fail, greatly so, it is of such a value, that whoever once makes the trial of it will never fail to value it highly. By all means let the young farmer make as large a reserve as he can possibly spare, for when his neighbours in spring are much distressed for want of food, and perhaps hundreds of sheep and lambs dying around him, his will be well fed, and himself, in this respect, on velvet.

AUGUST.

AUGUST.

HARVEST-MEN.

THE agreements with harvest men, in various parts of the kingdom, are extremely different, and even in the same place there are many variations, some farmers pursuing one method, and some another. A common way in some parts, is to agree with the men for all, by the acre; to reap or mow, turn, shock, make, cart, stack or barn, drive, &c. &c. to do all the business of the harvest, in short, at so much per acre : this is a very good way; but it requires a man to be almost as watchful as day-work : for a very strict eye should be had to the *manner* in which every thing is done; that the men do not cut the corn at improper times; that they take proper care to turn it after rain, and to get it perfectly dry into the barn. A pretty sharp attention will be requisite to all these points, and many others. On the other hand, when the work is done by the day, month, or week, it requires constant attention, early and late, to see that the men work their hours; and that upon *carrying*, in dubious weather, they work as long as they can see, unless the dews are heavy; for it is a maxim in most countries, that men are not to talk of *hours* in harvest, but to do whatever they are ordered.

In many counties, it is the custom to board the harvest-

harvest-men, and in some they are fed at an extra-vagant rate : I would by all means advise the eco-nomical farmer to vary this matter, if possible, un-less the men really work at a great rate, and stick to it early and late ; but, if the custom is rooted so deeply, that they will not give it up, then it is an object of attention to make the expence as mo-derate as possible, which must be by a previous plan of fatting a beast or two, and a few sheep for the purpose ; and also by providing whatever else may be consumed.

For many years, that is to say, till the scarcities, I put out my harvests to the men at 15 acres per man, and 4s. per acre for spring corn, and 5s. per acre for wheat, beans, and pease ; three bushels of malt per man instead of beer, and from 5s. to 7s. 6d. per man in lieu of *earnest*, dinner, gloves, and *hawky*, or harvest-home supper, at which rates the whole harvest came to about 3l. 10s. per man : it rose to 5l. 5s. and then even to 6l. and 7l. 7s. a man, so that at present, in some parts of Suffolk, the expence is not less than 10s. 6d. an acre, every branch of labour included. And this is much lower than it is in some counties. In the fens of Lincoln and Cambridge, where cottage building has by no means kept pace with improvements, I have known 10s. 6d. a day and ten pints and an half of ale given ; and 27s. per acre for reaping oats. There the strangers who go to assist in the harvest will let themselves only for the day ; they are found at four o'clock in the morning on certain bridges, and the

the bargain is made for the day, according to weather and competition. As the price of labour in
common with other expences of farming must
eventually regulate the rent, landlords are blind
to their interest in not building cottages : unite this
with the baneful custom of not giving leases,
which prevent farmers building, and the folly must
be seen in its true colours. In order to bring the
harvest business together, I have treated this matter here, but the young farmer is to remember that
the harvest-bargain is usually made long before this
time ; Whitsun-Tuesday is the common day for it in
Suffolk.

WHEAT HARVEST.

Now is the time that the farmer gives the first of
his attention to that golden crop, WHEAT. Having
been a year at least, perhaps a year and half, or two
years, in gaining it, he is now anxious to get it safe
within his barn. Bad weather now greatly injures
his profit : he must have many hands at work to
make the best use of fine seasons, or he will gain
the name of an *afternoon farmer.*

There are two ways of cutting wheat, reaping
and mowing : the first is the common practice, used
time immemorial, and by far the better.

The low reaping called *bagging* is preferable to
mowing ; they cut thus near London nearly as low
as the scythe.

Reaping is a work often put out by the acre to
the men, and it may be done as well so as most
works :

works; but it is necessary to observe, that they do not cut or bind in improper weather, and that they make the sheaves no larger than proportioned to the quantity of weeds, and the ripeness of the corn. In the forming them into shocks or stacks, there is, in some counties, an art of making them in such a manner, that they shoot off the water, and are kept tolerably dry in wet weather, without being laid so close, as not to dry with the sun and wind : it is a good practice, and deserves imitation. I have, on another occasion, mentioned the practice of covering the shocks of wheat-sheaves between Sandwich and Dover with cloths and mats. Mr. Boys informs me that mats are more commonly used, and that the practice is found to improve the sample of wheat, so that the Dover bakers give a clear preference to corn thus treated. The mats cost 7 d. each. Some farmers leave their corn standing so long, that it is ripe enough to *cut and carry*, as they call it : that is, they cart home the sheaves as soon as they are bound : but this will only do for very clean crops.

In a farm-yard, where there are teams enough, carting the wheat crops requires three waggons : one loading in the field, one unloading, and one upon the road going backwards and forwards : five or six horses are sufficient for them, and two men to pitch, two to load, one to drive, and two to unload ; in all seven : which make good dispatch.

But the use of one horse carts is very superior, whatever

whatever the number of horses; let each be in a well-formed cart, and much more ground will be cleared than the waggons can effect.

In some counties, it is common to stack all the wheat, if they stack any thing; and they are certainly right in the practice. No rats can get into a stack, if it is built on a floor, raised on posts, in the common manner; and, wheat, being in sheaves, admits the cut ends of the straw all to be laid outwards, so that the grain is defended from every injury, from external attack: whereas any corn that is not bound up, is subject to some little damage. Wheat is also found to carry a finer countenance out of a stack than a barn: the admission of the air gives it a brighter colour. In getting a stack into the barn for threshing, difficulties sometimes arise: a whole one should be got in at once, it being very dangerous to leave a broken stack exposed all night. It must also be done in dry weather, which in winter the farmer may wait for in vain some days, and thereby find inconvenience. Some of these evils would be remedied, and at all times a great expence saved, if a window was cut in the side or end of the barn, and the stack built against it, near enough to lay some short planks from one to the other, and so do the whole by hand, throwing from the stack at once into the barn. These are points that should be considered at harvest, when the stacks are built.

STACKS FOR THRESHING-MILL.

The invention of this excellent machine has not
been

been attended with one half of the advantages
which might have flowed from so useful a discovery,
for want of combining the use of it with the
various connected circumstances of the farm-yard.
This business of stacking corn, for instance, must
receive an entirely new arrangement in consequence of building a threshing-mill. By means of
no other additional expence than that of an iron
rail-way, and placing the stacks on frames resting
on block-wheels, two feet diameter, a very considerable annual expence in labour is saved in carting stacks to the barns, in loss of corn, and in
waiting for weather, as well as in the saving of
threshing by flails, and all the attendant evils of
pilfering and leaving corn in the straw. This is a
material object which cannot receive too much
attention from both landlord and tenant.

BARLEY, &c. HARVEST.

The barley crops should generally have good
field room, laying five or six days after mowing :
they will improve, and, if a heavy shower of rain
comes, it will not diminish the farmer's profit : it
will make the grain swell, and measure more per
acre ; for maltsters reckon much on their profit in
such dry harvests, that the barleys receive no rain
after they are mown. But ever observe, that barley; oats, &c. be quite dry when you cart them :
corn is always greatly damaged from being carried
in damp or moist : a heat is contracted in the mow,
the grain much discoloured, and the straw spoiled.
While barley lays on the swath, if much rain comes
it

it is apt to sprout. In the wet harvest of 1801, this crop in Norfolk presented a most melancholy spectacle ; three or four wet and very warm days made it grow to such a degree, that when the swaths came to be turned, they looked as if feathers had been strewed along every swath. Many thousand acres were thus damaged : those farmers escaped best who *lifted* the swaths before they were dry enough to turn ; they raised them lightly from the ground with forks to let air in ; a practice worth recommending. After the fields are cleared, they are raked with an instrument generally called a *dew-rake*, from its being used in the dew of the morning : a man draws it by a broad leather strap. This is a bad contrivance ; the work goes on slowly, and, being hard, the men often neglect doing it well, and much corn is left in the field. Instead of it, there is in some counties a machine, called a *horse-rake :* a rake ten or twelve feet long, drawn by one horse. This machine expedites the work greatly, at the same time that it does it much better. The use of it should be universal ; for one will work against twenty men, as I have experienced ; and the price is not above five guineas and an half complete.

Barley and oats in some countries are reaped, an excellent custom where they cut low enough ; for it is not with these as with wheat, which yields a crop of stubble ; if reaped with spring corn, what is left in the field is lost to the farm-yard. But by reaping, some of the evil of a wet harvest is remedied,

medied, provided the sheaves be made small enough.

BUCK-WHEAT.

This is a difficult crop to harvest ; for the least improper treatment makes it shed the seed in the field, to the great loss in product : if ripe, it should be mown only in the dew, and left to dry in the field ; and, if it stood but a few days too long, it must also be carted in the dew, or it will shed in carting. The grain being black, the colour of the sample is not a matter of consequence.

It is only the very early sown crops, however, that can be ready in any part of this month ; it is not commonly ripe till the end of September or beginning of October.

PEASE.

All strong crops of hog-pease must be hooked, and not mown, and care should be taken to turn the heaps after rain ; for the stalks and leaves are so succulent, that the straw will presently spoil if it is neglected. If they are stacked, great care must be taken to thatch the rick immediately, and to do it perfectly well ; for a little wet getting in will be of great damage to the pease.

BEANS.

Beans are always reaped and bound in sheaves, like wheat, and being generally late in harvest, and extremely succulent, they require being left a good while in the field ; and for the same reason, they should be tied in small sheaves. In binding, there are variations : the bands are made in some places

of

of wheat straw ; in others of yarn twine, which will last two years, if the threshers are careful to save them.   Beans do well in a stack.

TURNIP AND RAPE-SEED.

Crops of turnip-seed, and rape or cole-seed, are extremely various, uncertain, and subject to many misfortunes ; they must be conducted with great spirit, or the loss will probably not be small.   The principal point is to make good use of fine weather; for, as they must be threshed as fast as reaped, or at least without being housed or stacked, like other crops, they require a greater number of hands, in proportion to the land, than any other part of husbandry.   The reaping is very delicate work ; for, if the men are not careful, they will shed much of the seed.   Moving it to the threshing-floor is another work that requires attention : the best way is to make little waggons on four wheels, with poles, and cloths strained over them : the diameter of the wheels about two feet ; the cloth-body five feet wide, six long, and two deep, and drawn by one horse : the whole expence not more than 30s. or 40s.   I have, in large farms, seen several of these at work at a time in one field.   The turnip or rape is lifted from the ground gently, and dropt at once into these machines, without any loss : they carry it to the threshers, who keep hard at work, being supplied from the waggons, as fast as they come, by one set of men, and their straw moved off the floor by another set ; and, many hands of all sorts being employed, a great breadth of land is finished in a day. All is stopped by rain, and the crop much damaged : it is therefore of very great consequence to throw in as many people as possible, men, women, and boys, to make the greatest use of fine weather.

SOW RAPE.

The seed, when intended for a crop to reap, should be sown the beginning of this month, or the end of July. The preparation is the same as that for turnips.

GLEANING.

The custom of gleaning is universal, and very ancient : in this country, however, the poor have no right to glean, but by the permission of the farmer ; but the custom is so old and common, that it is scarcely ever broken through. It much behoves the farmer, in some places, where it is carried on to excess, to make rules for the gleaners. and not suffer them to be broken under any pretence whatever.

The abuse of gleaning, in many places, is so great, as deservedly to be ranked among the farmer's evils : the poor glean among the sheaves, and too often *from* them, in so notorious a manner, that complaints of it are innumerable. Make it therefore a law, that no gleaner shall enter a wheat field until it is quite cleared of the crop : this is the practice in many places, and great advantages are found from it. But, upon this plan, always desist from turning any cattle into the field, until the poor have gleaned it ; for, if a use is made of keeping

ing them out while sheaves are there, merely for
an opportunity of turning hogs and other cattle in,
it is double dealing, and a meanness unpardonable.

FARM-YARD.

At the leisure time of harvest, such as the wet
days, when the team cannot carry corn, and while
all the harvest men are employed in reaping and
mowing, if the works of tillage do not require attendance, let the horses and oxen be kept to earth-
cart, to form the bottom layer in the farm-yard,
carrying marle, chalk, turf, ditch-earth, or pond-
mud : the quantity in proportion to that of the
dung which you expect will be raised.

TURNIPS.

The second hand-hoeing of the broad-cast turnip crops must be given some time this month, nor
should it ever be omitted on account of works of
harvest.   In counties, where turnip-hoeing is a
common business, there is no difficulty in this, as
men enough are always to be had.   In some places,
many of them make it their business to hoe all harvest through, earning more at it than by other
field-work.   But in countries where hoers are scarce,
a farmer should always consider his turnip crops
when he agrees with his harvest-men, and hire a
sufficiency to set them to hoeing as regularly, when
the turnips want it, as to reaping when the wheat
is ready.

Look well to your drilled crops : both the horse
and hand-hoeings must be given whenever weeds
arise, or the land seems to be growing adhesive.

WHEAT AMONGST TURNIPS.

Mr. Walker, a considerable farmer at Harpley, in Norfolk, invented and executed on a large scale, one of the most singular practices that I have met with : that of hoeing in wheat seed at the second hoeing of his turnips on sand land. The wheat got up well, and was not damaged by the sheep feeding the turnips, but, on the contrary; if fed in a dry season and not too late, improved ; by this method he got crops of three quarters an acre without the expence of a shilling in tillage. I viewed some of them with much pleasure. A singular idea, that may be applicable with great profit on certain soils and in certain cases. He practised it so extensively as to lessen the number of his horses in consequence of it.

CABBAGES.

The beginning of this month, the second horse-hoeing should be given to the Midsummer planted crop of cabbages : the earth thrown into a ridge, in the middle of each interval, by the first, should now be split by the double mould-board plough, and thrown half to one row, and half to the other : this earth, which has been some time exposed to the weather, well be in fine order for the young fibres of roots to spread in ; nor should it be stirred by the succeeding operations : for the cabbage is a plant of such a luxuriant growth, that the roots have power to follow the well pulverized land thus thrown up, and the cabbages will certainly be of a size proportioned to the quantity of food the

roots

roots command. Care should also be taken to keep the tops of the ridges perfectly clean from weeds by the hand-hoe : none should be suffered to grow ; for on this part of the management much depends.

SOW CABBAGE-SEED.

This is the season of sowing for those crops that are transplanted in April. Plough a piece of well fallowed land until it is as fine as a garden ; then manure it amply with very rotten dung, and turning it in, harrow in the seed ; a pound of seed to every three acres of the intended crop.

But a preparation superior to ploughing and dunging, is that of paring and burning a thick coat for plenty of ashes, and adding a thin dressing of very rotten dung ; turn them in together and roll in the seed and bush-harrow. The plants thus escape the fly and slug.

Having on former occasions mentioned the great importance of this crop, the less is necessary at present ; still, however, I must urge our young farmer to determine to have as many August-sown cabbages as he can want for cattle, sheep, and swine, from the first of October to the last of December. The size they come to is superior to spring-sown plants, but they will not, in general, last longer than December. The use is however so great ; so exceedingly valuable for autumnal fatting of oxen, feeding cows, fatting wethers, feeding hoggit lambs, and supporting the whole herd of swine, that one may, without hazard, assert the farmer who doe,

F f 3   not

not make a provision of them, to be negligent in a
very material point of his business. I beg leave to
refer the reader to my *Northern* and *Eastern Tours*,
and to very many registers in the *Annals of Agriculture*, for abundant information on this subject.

DRILL CABBAGE-SEED.

Cabbage-seed may be drilled on ridges where it is
to remain the last week in this month, or the beginning of September, as directed in the Calendar
for April; but the reader is to observe, that the benefit of the practice has not that superiority which
attends the April drillings, by which a transplantation in June is avoided. When the seed is sown in
August, that operation takes place in a much safer
season.

POTATOES.

The potatoe crops in rows must be hand-weeded
if necessary; but it is now probably too late to
horse-hoe. If the intervals are weedy, or bound
at all, or the plants not sufficiently earthed up,
run the shim through them, but with much care,
which will cut up weeds, and loosen the earth:
after which the double mould-board will strike them
clean, and throw the earth against the rows, banking them up: the running roots and fibres will
follow such new thrown-up earth, and increase the
crop.

LUCERNE.

The lucerne will be ready again for cutting: if,
unfortunately, you have any drilled with too wide
intervals, attend well to the state of the land, and
take

take care to keep it in loose, well-pulverized order, and perfectly clean from weeds. But with crops put in as they ought to be, that is, broad-cast, or the rows at nine inches, nothing will be wanted of this sort.

SAINFOIN.

It will be now time to turn into the sainfoin fields that were mown in June ; but you should be cautious of feeding it with all sorts of cattle indiscriminately. Sheep, if kept too long on it, bite out the heart by eating into the bulb.

DIG MANURES.

This is a work which should never stop for hay, harvest, or any thing else, if the farmer has money in his pocket, and his plan is thus to improve his farm : the sooner the work is done, the longer he has the benefit.

FOLDING.

If this practice is pursued the fold should never stop in this month : the flock will bear it. Remember the general rule of folding the land that will be first sown.

HOGS.

This is a common month for the sows to bring their second litter of pigs ; and, if the farmer has not had the forecast to provide plenty of wash in his hog cisterns, he will find the disadvantage. Clover will not do for sows and pigs ; they must be fed on the skim-milk, butter-milk, and cheese-whey, that have been collecting together through the preceding months, while the dairy was at its

height ;

height ; bran, pollard, barley, buck-wheat, or pease ground into meal, and small quantities mixed in it.

Lettuces now come into use, and are of excellent service to the sows and pigs ; and may be deemed necessary if the dairy is small ; and in every event to tend to saving corn. He who keeps many swine cannot be too attentive to providing such articles of food as shall save corn feeding. Hogs are reckoned, when kept in great numbers, an unprofitable stock ; but it is merely for want of making a due provision of crops for them ; a few acres of each sort will carry a great herd of swine : but let no gap occur between the finishing one crop and beginning another.

CARROTS.

About the latter end of this month the carrot crop should be examined. It will require a slight hoeing, not an expensive one ; but just to cut up the few weeds that may be supposed to have arisen since the last hoeing. If the former hoeings have been well performed, only a hand-weeding will do.

PULL HEMP.

The time of pulling is about the beginning of August, or, more properly speaking, 13 weeks from the time of sowing : the leaves turning yellow and the stalks white, are signs of its maturity ; the male and female hemp are pulled together : indeed, when the crop is thick, it is impossible to separate them. The expence of pulling is generally estimated at

1s.

1s. per peck, according to the quantity originally
sown.

When it is all taken up, and bound in small
bundles, with bands at each end, to such a bigness
as you can grasp with both hands, it is conveyed to
a pond of standing water (if a clay pit the better),
where it is laid bundle upon bundle, direct and
across, thus, ▦ this is termed a bed of
hemp, and after it is piled to such a thickness as to
answer the depth of the water (which cannot be
too deep *), it is loaded with blocks and logs of
wood, until all of it is totally immersed : after remaining in this state four or five days, as the weather shall direct, it is taken out and carried to a
field of aftermath, or of any other grass that is
clean and free from cattle ; the bundles being un-
tied, it is spread out thin, stalk by stalk ; in this
state it must be turned every other day, especially
in moist weather, lest the worms should injure it.
Thus it remains for six weeks or more : then it is
gathered together, tied in large bundles, and kept
dry † in a house till December or January.   bundle

In the fens the male and female, or femble and
seed-hemp, are frequently separated.   This may

---

* This deserves experimental inquiry : watering hemp is a
partial rotting through fermentation : the vicinity of the atmosphere must, for that purpose, be necessary. The best hemp
ponds I have seen have not exceeded the depth of five feet.

† It might do as well stacked, if kept perfectly dry.

arise

arise from their hemp being coarser, and the stalks
larger. To attempt it, says a manufacturer, in
Suffolk, would be, I think, unprofitable, if not
impracticable.

Hemp, when left for seed, is seldom water-
retted, from the additional trouble and expence;
but it would be better if so done. It is generally
stacked and covered during the winter, and is
spread upon meadow-land in January or February.
If the season suit (particularly when covered with
snow) it will come to a good colour, and make
strong coarse cloths. It is much inferior to hemp
pulled in proper time, and water-retted.

PULL FLAX.

This also is the season for pulling flax : it is
bound in small sheafs, and conveyed to the steeping-
pit, where it remains about ten days on an average,
and is then grassed. To name the time of the
chief works on this crop is sufficient : flax *draws*
the land, and returns no more to it than hemp.
I cannot advise the young farmer to have any
thing to do with it. If from singular circumstances he is inclined to try it, he should procure a
man accustomed both to the culture and dressing.

SET STOCK LAMBS.

Fairs for the sale of lambs in several sheep districts take place in August : and it is, upon the
whole, as proper a season as any other, when the
whole are collected, to draw into different parcels,
is a convenient moment for separating that portion
which is meant to be kept for the farmer's own
                                                    use.

use. The common management of a flock is to
sell a certain number of crones every year, and to
keep that number of the very best ewe lambs to
supply their place in the flock : and, in making
this selection, the farmer or his shepherd usually
(whatever the breed may be) rejects all that manifest any departure from certain signs of the true
breed : thus, in a Norfolk flock, a white leg, and
a face not of a hue sufficiently dark, would
be excluded, however well formed : in the same
manner a white face on the South Downs ; in
Wiltshire a black face, would be an exclusion, or a
horn that does not fall back ; in Dorsetshire a horn
that does not project, &c. &c. and where the produce is annually sold lean, there is reason in all
this ; for customers who have been used to, and
prefer certain breeds, as having paid them well,
are apt to be fastidious when they purchase. Some
farmers in this selection look chiefly at size, always keeping the largest frames : but this is probably erroneous, unless they *keep* very high. It
connects with a question by no means ascertained,
whether sheep do or do not eat a quantity of food
proportioned to their weight ? In general, it is a safer
rule to chuse a well-formed lamb, or that indicates
the probability of making a well-formed ewe, rather than to select for size. The attention that is
to be paid to wool, in the breeds that produce the
carding sort, will depend on the price to be received : if the farmer lives in a district where the
price

price of the year is given equally to all flocks,
there is little encouragement to lessen quantity for
the sake of quality ; retaining, however, in idea,
the fact that both are attainable, that it is very
common to see coarse *breeched* sheep with light
fleeces ; and those of a fine quality heavy in
weight.   The Spanish fleeces, which are finer than
any other, are heavier than those of our finest
woolled sheep.   With combing wool, the importance of the fleece depends still more on price ; we
have seen it 8s. a tod ; and it has lately been 36s.
Quality is of very little consequence indeed compared with quantity, and when wool sells high, no
prudent breeder will set his stock without being
governed considerably by this object.

The high prices at which new Leicester,
and new South Down rams let and sell, has
opened a field of speculation in sheep-breed-
ing.   It is sufficient to remark, that this spirit
of breeding, whether it shall prove durable or
not ; whether much money shall or shall not be
made in it in future, is not what any prudent man
beginning business will adventure in but with
great caution : men of such immense fortune are
now taking a lead in it, and are in many respects
doing it on such liberal principles, that the wisest
conduct of such farmers as I may be supposed to
address, is to take proper opportunities of converting their experiments to their own (the farmers)
profit.   Leave the expence to them, but when you
                                                    can

can convert the profit to your own advantage. In setting a stock of lambs, therefore, you may mark a score of the best, for a future ram to be picked up when opportunity offers; or, better still, to send to the tup of some ram-letter that takes them in at a reasonable price per head. By every year selecting five or six per cent. and by every year covering that number by a ram better than any of your own, the flock must be on the improving hand, and this may be done at a very small expence.

This moment of setting the stock lambs is, that of adding to, or diminishing the number of a flock, by keeping more or fewer than the crones sold. This is a very material part of the business: on a farm with a given stationary sheep-walk, it is probably regulated by circumstances that rarely change: but, on inclosed farms, where the sheep are supported by fields alternately in grass and tillage, variations may easily be supposed, and the question of hard or light stocking, that is, of close feeding or a head of grass, then comes in to decide the number kept. If the produce or profit per head is looked to, the conduct to be pursued is evidently to stock lightly; but, if the return is looked for in corn from fields laid down for refreshment by rest, then close feeding is a very material point, and the number kept will depend on it. With all the grasses, &c. that do not decline from age, the more sheep you keep, the more you may keep, and the more corn you will reap when such are ploughed;

a cira circumstance too important to be forgotten. But the young farmer will remember, that upon this system he must not have a *show* flock, or let *the vanity of a farm* have the least influence with him: if in this way he will have something to talk of, a score or two of pampered favourites, the fewer the better, for they may cost him more than they are worth.

SELL LAMBS.

Having set his own stock, he drives the rest of the ewe lambs and all the wethers to the lamb fair; and it will be satisfactory to him, in ascertaining comparative prices, as well as for knowing the progressive state of his flock, to weigh and register the weights of all. Let him also acquaint himself by proper inquiries among his neighbours, of the expences which ought to be incurred in driving, hiring grass or stubbles, shepherd and assistants, board, hurdles, &c. &c.

If the fairs for the sale of lambs are later than August, as in September, and even to Michaelmas, as in some districts, great care must be taken to keep them in forcing food, as in spring tares, early sown rape, good grass of the right degree of bite, &c. &c. in order to promote their growth and increase their value. But to sell in August is more beneficial.

KEEPING ROUND.

This term is not very expressive of its meaning, but common among farmers, for describing a different sheep system, that of so proportioning the number

number of ewes as to keep all their produce for the butcher. If he is in this system, his principal object is to consider wether lambs as fatting stock, and feed them accordingly from weaning to the knife. His crones the same from this period.

LAYING DOWN LAND TO GRASS.

This is the best season of the whole year for this very important operation, and no other admissible for it on strong, wet, or heavy soils. Spring sowings with corn may succeed, and do often, but that they are hazardous I know from forty years experience. In all my trials I never failed with an August sowing but once, and that was with crested dog's-tail gathered too early by the women, knowing the intention too soon.

Preparation.—I suppose this to have been either a fallow or winter tares sown very early, and mown in June for soiling, which may give nearly two months in the very heat of the summer for tillage: such a preparation is one of the most effective for cleaning land that can be; and it may further be supposed that the land had been favoured in the course of crops previously to the year of laying down.

Seeds.—These should be varied according to the soil, as in the following table :

Clay.

Clay.	Loam.	Sand.	Chalk.	Peat.
Cow-grass.	White clover.	White clover.	Yarrow.	White clover.
Cock's-foot.	Ray.	Ray.	Burnet.	Dog's-tail.
Dog's-tail.	York white.	York white.	Trefoil.	Cock's-foot.
Fescue.	Fescue.	Yarrow.	White clover.	Rib.
Fox-tail.	Fox-tail.	Burnet.	Sainfoin.	York white.
Oat-grass.	Dog's-tail.	Trefoil.		Ray.
Trefoil.	Poa.	Rib.		Fox-tail.
York white.	Timothy.			Fescue.
Timothy.	Yarrow.			Timothy.
	Lucerne.			

In regard to the quantities per acre of these plants, this must necessarily depend on the means of getting them. In situations where women and children are fully employed, it may be difficult to procure large quantities, gathered by hand : in such places a man must be content with what can be bought. *Crested dog's-tail* is so very generally to be thus procured, that I cannot but suppose it, in a good measure, at command. However, without adverting to this point, I may remark, that from the lands which I have laid down to grass to a considerable extent, and in which I have used every one of these plants largely, except the poa, and that on a smaller scale, I am inclined to think that the following quantities may be safely recommended:

Clay.

Seeds.			Substitutes.	
Cow-grass,	-	5 lb.		
Trefoil,	-	5 lb.		
Dog's-tail,		10 lb.	Yorkshire white,	2 bush.
Fescue,	-	1 bush.	Timothy,	1 lb.
Fox-tail,	-	1 do.	Do. 4 lb.; or, York white, 1 bush.	

Loam.

White clover,	-	5 lb.	
Dog's-tail,	-	10 lb.	Ray, 1 peck ; Rib-grass, 4 lb.
Ray,	- -	1 peck.	

Fescue,

Seeds.			*Substitutes.*
Fescue,	-	3 pecks.	Yorkshire white.
Fox-tail,	-	3 do.	Timothy, 4 lb.
Yarrow,	-	2 do.	Cow-grass, 5 lb.

Sand.

White clover,		7 lb.	
Trefoil,	-	5 lb.	
Burnet,	-	6 lb.	
Ray,	-	1 peck.	
Yarrow,	-	1 bush.	Ray, 1 peck; Rib, 4 lb.

Chalk.

Burnet,	-	10 lb.	
Trefoil,	-	5 lb.	
White clover,		5 lb.	
Yarrow,	-	1 bush.	Ray, 1 bush.

Peat.

White clover,		10 lb.	
Dog's-tail,		10 lb.	York white, 6 pecks.
Ray,	-	1 peck.	
Fox-tail,	-	2 do.	Rib, 5 lb.
Fescue,	-	2 do.	Cow-grass, 4 lb.
Timothy,	-	1 do.	

But here I must observe generally, that if the land, thus laid to grass, be intended for sheep, it is not an object of very great consequence to sow only the finer grasses; as close feeding, after the first year, will make any grass named in these lists fine, and sweet, and productive; but this effect depends altogether on its being constantly fed close; that is, all seed stems being prevented from rising. Every good farmer is sensible of the necessity of this with *ray-grass*; but most

G g   unac-

unaccountably does not extend a similar concern to other grasses. I have laid down above 200 acres chiefly for sheep ; and I have stocked the fields so early in spring, and so thickly, as just to keep down the seed stems : the *cock's-foot, oat-grass,* and *Yorkshire white,* with this management, have proved sweet feeding grasses, not at all rejected, even in fields where the flock had a choice.

Sowing.—The even distribution of the seeds being of much importance, a calm day should be chosen for the sowing, and particular attention given to prevent the seedsmen mixing too many together : let the farmer remember, that the ex- pence of going often over the ground, is nothing on comparison with the benefit of having each sort equally distributed.

Successive management.—In this respect, no other attention is necessary than to keep every sort of stock out of the field most assiduously through all the following autumn and winter. Some writers direct manuring ; but this should have been done previously to sowing the winter tares, if done at all. After sowing, none is admis- sible but such dressings as may be sown by hand.

CONVERSION OF POOR LAYS.

There are, on many farms, tracts of barren lays, from moss, poverty, neglect, and bad herbage, upon which a very great improvement may be made by a single ploughing in August. For this purpose, a strong four-horse plough must be used with a skim-coulter ; then go over it twice in different

rent directions with the scarifier, so as not to disturb the flag; harrow it once, and immediately
sow a quarter of a peck of cole-seed, two bushels
of cock's-foot, and one bushel of Yorkshire white
per acre, adding some of whatever seeds may be
procured at the moment cheaply.   Leave it unfed
and untouched till the March following, in which
month, and through April, load it well with sheep;
the use will then be very great ; keep sheep feeding it heavily through the year ; the cole will be
killed, and you will have a pasture worth treble
what it was before.   The expence is small, and
the improvement rapid.

SHUT UP ROUEN.

The time for shutting up after-grass for use the
following winter and spring, will depend on the
richness of the soil : directly from the scythe is
the proper period for lands of moderate fertility,
that let from 12s. to 25s. an acre ; but, in fields
of greater richness, from 25s. to 35s. August is a
better month, feeding till then quite bare : and, on
still richer lands, September may do.   On the fine
salt-marshes of Lincolnshire there is such a spring
all winter, that two sheep an acre are fed without
any previous exclusion.   This husbandry cannot
have too much attention, for it is by far the most
certain dependence a man can have for his flock at
the most pinching period of the whole year.
Lands fed in the spring may be kept equally with
those mown.

SEPTEM-

SEPTEMBER.

WHEAT.

UPON all cold, wet, and backward soils, September is the best season for putting in wheat, provided other circumstances permit it ; such are principally the weather, for wheat should not be sown till rain comes in *tolerable* plenty ; and never in a dry season. Upon dryer and warmer soils, it is better to postpone this business till October. But in this case, let our farmer remember, that clever and other layers should be ploughed (if rain comes) in September, for it is a great advantage to have such layers remain unsown for three weeks or a month after ploughing. Another general observation is, that in proportion to the earliness of the sowing, may be a small deduction from the accustomed quantity of seed ; two bushels sown any time in this month, are equal to three in November.

SORT OF WHEAT.

These are numerous, and many of them known in different counties under different names, which necessarily causes some confusion in reports that are made on this subject. It is here necessary to notice but a few of the sorts.

1. Red lammas ; a red straw, red ear, and red kernel ; reckoned by many farmers the best of all the sorts hitherto known, and yielding the finest, whitest

whitest flour. There are also a yellow and a brown lammas.

2. Hoary white; white straw, ear and grain.

3. Bearded; productive on very poor, cold, wet land; but a coarse grain, and sells for an inferior price.

4. Clark wheat; red blossom, chaff, and straw, but white grain; a favourite sort in Sussex.

5. Hedge wheat; white: very productive.

6. Velvet; a distinct sort from the hoary white; it is a white wheat, and though not weighty, yields much flour; a very thin skin.

7. Cone wheat of various sorts, so called from the shape of the ear.

STEEPING THE SEED.

The modes of steeping, brining, and liming the seed, are innumerable; all are equally intended as precautions against the smut. I made several experiments on this object, from which it appeared, that steeping from twelve to twenty-four hours in a ley of wood ashes, in lime-water, and in a solution of arsenic, gave clean crops from extremely smutty seed, but a short time in those mixtures had a much less effect.

WHEAT AFTER FALLOW.

If there is one practice in husbandry proved by modern improvements to be worse than another, it is that of sowing wheat on fallows: all I shall therefore observe under this head is, to note that in some counties the fallows are ploughed just before harvest on to two bout ridges, ready to plough and

sow under furrow in the spraining method a seeds-
man to every plough which reverses the ridges.   In
others they lay their lands into ten or twelve furrow stitches or lands, and sow some under furrow,
some under the harrow. Ridges vary exceedingly, according to their wetness ; and in Kent they have by
means of the turn-wrest plough, no lands at all,
but a whole field one even surface. It would be
useless to expatiate on the circumstances of fallow-
wheat, which ought no where to be found. If
fallows be, or are thought necessary, let them be
sown with barley or oats, or with any thing but
wheat.

WHEAT AFTER BEANS.

Beans, if well cultivated, form the best preparation for wheat : I have seen in Kent a field of wheat
which followed four preparations, beans, clover, tares,
and fallow, and the first was superior to all the
rest ; next the clover, then the tares, and the
worst was after the fallow. If our young farmer
has a bean-stubble on which he intends sowing
wheat, he should be as early as possible in giving it
the due tillage ; this will depend on soil, for on
some it will be more advantageous to trust to the
shim, scarifiers, and scufflers, than to the plough.
If the land is very clean, the great Isle of Thanet
shim will cut through every thing, and loosen the
surface sufficiently to enable the harrows to leave it
as clean and fine as a garden, women attending to
pick and burn. If less clean, the Kentish broad-
share may do the work more effectively. In other
g   cases

cases the scuffler may be equal to the business.
When he has got the surface to his mind, he is to
consider whether or not he should plough it, which
is advisable if the soil be of a firm, solid, tenacious
quality, and if he does not intend to drill the
wheat : if he ploughs such a soil he may not have
any apprehension of root-fallen wheat, failing
roots from a loose bottom ; but he will bring up a
new surface that may drill with difficulty, whereas
that which has received the influences of the crop,
atmosphere, and of his late operations will be in
exactly the right temper for the drill to work in.
If the soil is of a more loose, friable quality, and
he should plough down the fine surface he has
gained, he will give the wheat too loose a bottom,
and he will run the chance of a root-fallen crop.
In all such cases, or in any that have a tendency to
this circumstance, he should determine not to
plough at all, but drill directly ; a method in which
he saves tillage, and has the probability of a better
produce. This is a new practice on strong land, but
I have seen such success in it as leaves no reason
for doubting the soundness of its principles. Mr.
Ducket, on a sandy soil, did it for years, and with
great effect. It should be remembered, that whatever other circumstances may influence the growth
of this grain, it loves a firm bottom to root in, and
rarely flourishes to profit when it is loose and
crumbly, nor will a depth of such mould do if the
under stratum in which it will attempt to fix its
roots, be from its quality repellent. The best

basis

basis is the cultivable earth firm from not having been lately disturbed. No such rules can be general, but the case happens oftener than some are willing to suppose.

WHEAT AFTER CLOVER.

Clover forms a very excellent preparation for wheat, yielding ample crops of the golden grain at a very light expence: so that, while the Norfolk husbandry of, 1. Turnips; 2. Barley; 3. Clover; 4. Wheat; is practicable on a farm, dry enough for turnips, and rich enough for wheat, a man may well enough be satisfied with his profit; but after many repetitions (and this system has been common for above fifty years), it was found that two evils gradually appeared, which were unknown at the first introduction of it. Turnips demanded ample manuring where they were once produced of larger size without any; and the clover became so subject to failures, that it was no longer easy to have it every fourth year. This created the necessity of variations (of which more in another place), but still retaining clover as the preparation for wheat. The husbandry, however, was very imperfect, from the mode of putting in the wheat, which was merely by the harrow, in which method (for the skim-coulter was unknown) the seed was too apt to fall into the seams of the furrows, and came up consequently among whatever grass and weeds might be in the land. The discovery of dibbling was a very capital improvement: in this method the seed was deposited in the

the centre of the flags, and the regular treading the land received, pressed down the furrow, and gave a degree of firmness not otherwise attainable. The success was great ; and had the labouring poor kept to that care and accuracy which they began with in dibbling, the practice would never have lessened ; but the great earnings they made gave a spur to their avidity, and they have both in Norfolk and Suffolk done it of late years in so slovenly and careless a manner, that drilling is every where coming in, instead of a practice esteemed by many of the most intelligent farmers as unrivalled when well performed. At present, it is thought in those counties the mark of a bad farmer to sow broad-cast wheat on clover.

The land having been ploughed a fortnight or three weeks*, it is to be well rolled down with a heavy roller, and then dibbled : here, as in all other cases, the chief attention is to be paid to the dibblers making the holes deep enough, and to the children dropping equally without scattering. It is then bush-harrowed. Six pecks of seed is enough for two rows on a flag in this month. But if only one row, still I would recommend as much seed to be put in. And another observation it is necessary to make, that if the land is known to be given to the mildew, an increase of seed on that account is right, whatever the soil or season ;

---

* On to drill-stitches, if that husbandry is at any time to be practised in the field.

by

by reason of the well known fact, that all thin crops suffer more from that distemper than such as are thicker.

In regard to drilling, the various directions given in the spring Calendars, relative to accurately ploughing the lands either for one stroke of the drill-machine, or for a bout of it, are equally applicable to drilling wheat. The operations are the same, and therefore to dwell on them needless; but, it should be remembered, that in ploughing all lays, the use of the skim-coulter is very great, and in some cases indispensable : an effective harrowing should precede the drill. The quantity of seed the same as in dibbling.

A very singular experiment of Mr. Ducket's, on preparing a clover lay for wheat, should here be mentioned : he had a field in which wheat rarely escaped being greatly root-fallen : not to lose sowing it with that grain, and at the same time to guard against the experienced malady, he scarified it repeatedly, till he had torn up the clover, and also gained tilth enough for drilling in ; then he collected the clover fragments, and carted them into the farm-yard to make dung, and drilled the field : the wheat having a firm bottom in an unstirred soil, escaped the disease, and yielded an ample produce : very singular husbandry, and admirably adapted to the peculiarity of the soil. The clover-bulb, which would have secured the dreaded looseness had it been turned down, made a large quantity of dung, and therefore was not
lost

lost to the farm, though the particular field was deprived of it. No saving in expence was here made, but an extraordinary one incurred; but it secured a crop where otherwise there would have been none.

WHEAT AFTER TARES.

A good crop of winter tares leaves the ground in such loose, putrid, friable order, that it is much better husbandry to sow turnips or plant cabbages on it, than leave it to receive tillage for wheat. However, if this management should take place, the land should not be ploughed at all, but left to consolidate at bottom, to become firm for the roots of the wheat to fix in, and the surface worked with the scarifiers or scufflers, according to its temper, just sufficient to keep it clear of all weeds, and in that state, drill the wheat without any ploughing. This I have practised with good success.

WATER-FURROWING.

A circumstance of much importance in the culture of wheat, but oftentimes strangely neglected, is water-furrowing: this work should be well and effectually performed on all lands, except those that are perfectly dry all winter through. The water-furrows should be ploughed as soon as the field is finished sowing, ploughing, &c. and then a spit should be dug out from the bottom of them, and laid on one side opposite the rise of the land, and the loose moulds shovelled out: the openings of all the furrows should likewise be cleansed, so that the water

may

may have an easy fall out of every furrow into the
water ones. The number of these must ever depend on the variations of the surface ; the only general rule being to make them so numerous, that
no water can stand on the land in the wettest weather. In bottoms of fields, or other places, where
there is a double slope of the land, it is necessary
to cut double water-furrows, about a yard or four
feet from each other, to take water from each
descent.

BUY IN SHEEP.

If there is not a regular flock kept upon the
farm, the annual purchase may be supposed to take
place at fairs towards the end of August, or all
through September : and the sort most commonly
bought is wether lambs, and I believe more profitably than any other. They used to be had for 15 s.
in the breeds of Norfolk, the South Down, and
others of a similar size ; but of late years, they
have risen to 20s. and even higher. There are
two systems of keeping them ; one is, to put them
to the very highest keep, and push them by every
means to sell as soon as possible ; the other, and I
believe the more advantageous method, is to keep
pretty well till March, and then to full keep, and
begin to sell in harvest, continuing till September
or Michaelmas, and then clearing all. In this way
I have often known the purchase money doubled,
besides the fleece. Sometimes much more is done,
but this may be looked for on an average of years.

BREED

BREED OF SHEEP.

In buying in the wether lambs mentioned in the preceding article, there will not probably be much choice in the vicinity ; and it is not commonly a profitable speculation to send into distant counties for breeds very different from those of the neighbourhood. At present, such has been the spirit of making these importations for the purpose of breeding, that there are not many districts where a farmer has a choice. The new Leicester first, and then the South Downs, have made remarkable inroads in various parts of the kingdom. The former come in competition with all the long-woolled breeds ; and the latter with all the short and middling woolled ones. When bred in sufficient plenty to be had as wethers, the new Leicester are *generally* to be preferred. The only doubtful exception I have heard is, *when wool sells high, upon very rich marsh land in Lincolnshire.* I know of no competition between the South Downs and other fine woolled breeds, in which the South Downs should not be preferred. In buying wether lambs for mountainous moors in the North of England, the black-faced long coarse woolled Scotch are the common sort, and I believe much superior to any other for regions of real and severe hardship.

In all this matter, the young farmer is to bear in mind, that for an annual stock, he is to discard all prejudices that are attended with expence ; these

must

must be transferred to the breeding systems, and *there* well considered before they are adopted.

CRONES.

It is a common system in many inclosed districts, to buy old crones in September, to put the ram to them in October, and to sell the lambs as they become fat for the butcher, and then to fatten the mothers, clearing within or about a year from the time of buying. This system is well enough where fences are very secure, and food very plentiful ; but in general it is inferior to wether lambs.

In buying any sort of sheep that are to be *wintered* on turnips, the young farmer should not calculate on more than ten to an acre of *very good* turnips : and, in providing stock for fattening, the best acres must not be expected to fatten more than seven or eight ; and middling crops not more than five or six ; due provision being made beside for taking the stock when turnips are done.

FATTING BEASTS.

You must now be very attentive to the state of your fatting beasts, and the remainder of their food ; see, therefore, that the cattle do not stop for want. A beast that is nearly fat must have plenty ; he is nice, and if he is at all curtailed in his *pasture*, will fall off. It is excellent management to have August-sown cabbages now ready for the fat beasts, and to carry them on to Christmas : grass declines after this month ; and if rouen is freely turned into in September, twenty to one but
the

the ewes and lambs will be distressed in March
and April: whatever grass is now used on the farm,
will pay far better by sheep than by feeding at
present. The soil, however, must in some measure govern this distribution : for all lands that
are subject to floods, or that have a tender and
poaching surface, should be left quite bare before
the heavy autumnal rains come. The farmer's judgment must be exercised in this, as in so many other
cases : dry sound soils answer best for kept rouen.

Remember that beef is cheaper at Michaelmas
than at any time throughout the year ; for grass-
fed cattle are then at once brought to market : this
should give the attentive grazier an idea of varying
from the common method : to sell only a part of his
cattle at this time.

In drawing off a lot or lots for sale, it is common
to sell the fattest, and keep on the ill-doing ones
for further exertions. If the food provided be not
costly, this to a certain degree is *admissible :* but if
the beasts are for cake or corn ; or the quantity of
other food rather limited, it is a very questionable
conduct. I would not give expensive food to beasts
which have proved themselves unthrifty, but on the
contrary, draw off for this purpose the most thriving ones in the lot : the contrary conduct has often
been the reason why all winter fatting has been so
heavily condemned. The moment a grazier is well
convinced that he has a beast that is an *ill-doer*,
the first loss is the best, and he should get rid of
him as soon as he can.

COWS.

COWS.

The dairy of cows must have plenty of grass throughout this month, or their milk will be very apt to fail. Lucerne, mown green, and given them in a yard, is the most profitable way of feeding : the product is so regular, that it is an easy matter to proportion the dairy to the plantation, and never be under a want of food ; for lucerne, mown every day regularly, will carry them into October ; and, although some persons have asserted that cows will not give so much milk thus managed, as when they range at large, and feed how and where they will, it is not a matter of inquiry ; because, if they give less, the quantity will pay more clear profit, than more in the other case : there may be some inferiority ; but the cows are kept on so small a quantity of land, that there remains no comparison between the methods, for profit.

THE TEAMS.

These must be kept at work. Wheat-sowing is a business in which we usually stretch a point, and make the ploughs do full work. Both horses and oxen should be kept this month to lucerne, &c. mown every day : they will work as well on it as any other food ; but, while they plough, they must have oats and chaff with it ; for no grass at this season of the year, is so nourishing as it was in the summer.

MANURE GRASS.

Dung should not be kept until it is rotten, that
it

it will wash into turf; because, by that time, it loses its virtue at a great rate, and, while in full fermentation, it is of such great utility to all land. There are many succedaneums for dung, or at least for making it go much farther on grass than on arable: the proper composts are chalk, clay, turf, ditch-earth, pond-mud, lime, ashes, soot, with some dung; all, or some of these mixed together, will be in order for spreading on grass lands, and will be highly sufficient to keep them in great heart, with but a small quantity of dungad The end of this month is a proper season for carrying such composts on: lay about 15 or 20 yards an acre. It is difficult to over-manure arable lands, but very easily done on grass; because large quantities do not wash in quick enough. Let the compost heaps be spread very regularly. A good farmer will manage to give his pastures, unless they are very rich, a dressing of compost every four years. Always add a pound of common salt to every cubic yard of compost, sprinkled in in turning over.

SCARIFY GRASS LANDS.

Scarifying grass is a new practice of some ingenious gentlemen, but not yet become common husbandry. It consists in cutting the turf with a plough of coulters, or with a ploughing harrow: so that the surface may all be cut or torn: this operation is on principles directly contrary to the common idea of rolling in autumn, which is done with design not only of levelling for the scythe, but also of pressing the surface as much as possible, for which the heaviest rollers are chosen, until some are worked that require six horses.

BURNET.

Observe not to let any cattle pasture your burnet fields after mowing, either for seed in July, or for a second crop of hay in August ; for the greatest peculiarity of this plant is to afford a full bite in March ; and, if you leave it six or eight inches high in October, you will find more the beginning of March, and in possession of the leaves it had in autumn ; for the winter's frosts have not much effect on it. Upon this caution, therefore, depends much of the advantage of burnet : some who have found fault with it, and asserted that it is unprofitable, have fed off the after-grass in autumn bare, and let their sheep and cattle get into it in winter. It is then no wonder the burnet does not answer the character given of it by others, who have managed in a different manner.

FERN.

Cut fern, called, in some places, brakes and brakens. This is most profitable work, and should never be neglected. Carry it into the farm-yard, and build large stacks of it for cutting down through the winter as fast as the cattle will tread it into dung ; also for littering the stables, ox-houses, cow-houses, hogsties, fatting-sheds, &c. By having great plenty of it, you will be able to raise immense quantities of dung, which is the foundation of all good husbandry ; and it is well known, that no vegetable yields such a quantity of salts as fern : from which we are to conclude, that it is well adapted to the making manure. The good farmer, in this work of bringing fern, should

should not confine himself to his own wastes, of which he may have none, but purchase it of his neighbours, if they are within a tolerable distance. It will answer exceedingly well : he need not therefore fear paying for it, as for refuse straw.

STUBBLE.

This month is the proper time for cutting the wheat and rye stubbles, and raking into heaps, for carting home to the farm-yard for litter, upon the same principles as fern is carted there. This is a business strangely neglected in most parts of the kingdom; but is nevertheless of great importance : the stubble left on the land is of little advantage as a manure, but it prevents the plough from turning in the land with neatness, but carted into the farm-yard it becomes an excellent manure. Any sort of litter there is valuable, and serves for the cattle treading into dung. In those parts of the kingdom where this use of stubble is common, the price for cutting and raking into heaps is from 2s. 6d. to 3s. 6d. per acre; a very small expence, compared with the great advantages that undoubtedly result from it.

HOPS.

This is the season for hop-picking. To name it is sufficient in such a work as this, for no prudent man would venture upon so operose a business as this article of culture, by means alone of such information as he could find in books. Mr. Marshall has treated the subject in detail; and many other writers have given information concerning it, from Reginald Scott to the present time.

PLOUGH FALLOWS.

Now, if you have leisure, let your ploughs turn up all sorts of stubbles : this is one of the material parts of husbandry, in which some farmers are greatly wanting : they form very mistaken ideas respecting this part of husbandry, suffering their lands designed for fallows, &c. to remain till after barley-sowing, before they break them up.

WATER-FURROWING.

Make it a rule to water-furrow all wet fields, as fast as the plough leaves them : this is an important work on autumnal ploughings ; for the dryness and health of the land depend on the cutting them with judgment.

LUCERNE.

The lucerne will yield another cutting probably this month ; but, at farthest, the first week in October ; after which the succeeding short growth is no object ; but, if it is cut the beginning of this month, there may be another the end of October. But, when the last is taken, manure the plantation with long dung.

Lucerne requires much manure : it will, on good land, yield very beneficial crops without any ; but, to be carried to the highest perfection, not only of *product*, but also of *clear profit*, it must have plenty.

SOILING.

This is a period in which some farmers, who understand soiling well, through the height of summer, are apt to grow negligent in it ; but if due preparation was made, by a right succession of cuttings of
                                                        lucerne,

lucerne, chicory, tares, &c. with a reserve, if wanted, of common grass of the right age, it should be steadily adhered to throughout the month.   There is plenty of food, it is true, in the fields, but this should be eaten by sheep, which should be cheaply maintained through the months of September and October.

SOW WINTER TARES.

This is the principal month in the year for sowing winter tares. The earlier they are got into the ground the better ; for the difference of forwardness in the spring, from only a week difference in the time of sowing, is sometimes great.   Three bushels an acre are the common quantity of seed, broad-cast ; but some sow only two or two and a half thus early. If they are drilled at six inches, two are enough.   Our young farmer will proportion the breadth of ground he ap. plies to this crop, to the circumstance of the quantity of lucerne or chicory he has for the purpose of soiling : if he has little or none of those plants, he must appropriate a good breadth to tares, for in such case he will find them very advantageous.

SOW WINTER TARES ON POOR PASTURES AND MEADOWS.

This very extraordinary husbandry was the invention, if I may be allowed the expression, of a very ingenious and excellent farmer, Mr. Salter, of Norfolk : wherever he improves poor meadows or pastures by spreading earth, clay, chalk, marle, gravel, &c. he harrows in winter tares on such manuring, and, if wanted, grass-seeds or white clover.   And it is a curious fact, which I have witnessed on his farm at Winborough, that the grass-seeds succeed better

where

where winter tares are sown than in any spots where by accident there happens to be none. The improvement of the old grass by the tares is very great, and the value of the tare crop is considerable in soiling, or in hay. This husbandry is new, extremely interesting, and much deserves imitation in many cases.

LAYING TO GRASS WITH WHEAT.

Next to laying down upon a clean fallow and sowing the grass-seeds in August, I believe as good a system as can be pursued is, to sow the seeds with wheat very early in September, provided the weather be favourable for putting in the wheat. I have had very good success in this method. The land must be either a fallow, or sown early with winter tares, and these mown for soiling, after which there is plenty of time through the heat of the summer for fallowing the land. The seeds are detailed in the Calendars for April and August.

FAILURE OF NEW LAYS.

If the recommendations given in this work be closely attended to, there is little reason to apprehend this ; however, as it is possible, from extremely unfavourable seasons, something should be said on it. Such a failure can scarcely happen to more than one or two of the seeds ; in this, or indeed in any case of failure, fresh seed should be sown in a moist time in the spring, and if a flock of sheep can be driven over the land, it will be a good way to cover them ; if not, it should take its chance, for a roller will not so well effect it, and a harrow cannot enter without mischief. If a very large fold (five or six yards to a sheep) be run over the field once in a place, and the seeds sown

before

before the sheep enter, success is almost certain. At least I have found the benefit of thus thickening new lays in seasons not peculiarly favourable.

Should, however, a total failure from any unforeseen cause take place, the better conduct will be, in fields that were sown in the spring, to clear the corn as early as possible, and ploughing once, harrow in fresh seeds immediately; this will succeed very well if they are got in in the month of August, or early in September; the sooner the better : and in this case the land should be very well rolled in October, in a dry season. If the failure happens in land sown in August, it should have three earths in dry weather in the spring, and the grass-seeds re-sown with buck-wheat in May : that is not a crop for clays and wet loams, but I have known it succeed well in a dry summer; should the season be wet, it will give little seed, and should be mown when in blossom for soiling cows. It is an ameliorating plant, never exhausting any soil, and therefore preserves in the land the fertility gained by the operations previous to the former sowing. And I may here generally observe, that grass-seeds of all sorts, and on all soils, never succeed better than with buck-wheat, of which not more than one bushel an acre should be sown. There is a district in Norfolk where buck is highly valued for this object. It is a profitable article of cultivation on the very poorest barren sands.

AUTUMNAL MANAGEMENT OF NEW LAYS.

This is a point of considerable consequence; and in proportion to the moisture of the soil. All trampling of cattle and horses is pernicious, for the soil, after a crop of corn, or after the tillage of a fallow, is very

tender,

tender, and affected by every impression : it is also bad to feed the plants, as I have found by much ex‑perience. The safe way is to keep every thing out through both autumn and winter. The profit of feeding is absolutely nothing, for the pasturage in the spring for sheep is of far more value, by reason of not eating it in autumn : at the former season it affords a most valuable and very early bite for ewes and lambs.

OCTOBER.

OCTOBER.

SEASON.

IT is necessary to remind the reader, that a circumstance takes place in respect to this month, which scarcely holds with any other in an equal degree : by *October*, is to be understood that period of good or tolerable weather which usually takes place before the change by rain, snow, or frost stops most field operations ; what is now directed to be done must often be executed in November : if the farmer cannot effect it in the first of these months he must do it in the second.

HIRING FARMS.

This is commonly the month for hiring and stocking farms, and moving from one to another. Upon such occasions, the farmer should have his attention awakened : he should be equally clear-sighted to the advantages of a farm and to the disadvantages, that he may be able to draw a balance between them, and compare that balance with the rent demanded. Let him remember, that he must equally discard a too solicitous prudence, which doubts every benefit, and a too daring courage, which overlooks or lessens real evils. It must be open to almost every person's observation, that many lose themselves in deliberating concerning a farm : they have so many mistaken rules of judging, that we sometimes see them reject farms which soon after are hired by others, and prove the

fortunes

fortunes of such : they are apt to take one false guide
in particular, the success of the last tenant. If a man
makes a good deal of money on a farm, or leaves it
for a much larger, numbers will immediately apply
with eagerness to get it ; but if a tenant or two
break, or are poor on a farm, most of the neighbours
consider little farther ; they attribute too much of
the ill success to the land, and avoid it, under an
idea that, without a fall of rent, no money can be
made on it.

Soil.—Let the farmer that is debating whether he
should hire a farm that is offered him, examine the
soil well, to be able to determine its nature, the
stiffness, moisture, exposure, levelness, slope, sto-
nyness ; what draining, manuring, fencing, &c.
will be wanted : let him see to the roads, distance of
market, prices of commodities, labour, &c. ; let him
fully acquaint himself with the state of tithes or gathering. He should know the poor-rates, attend to
the compactness of the fields, and consider well the
covenants relative to cropping ; for many such are
extremely detrimental to a good conduct of the land.

One general rule in hiring a farm should not be
forgotten—to fix on good land, and he can scarcely
pay too much for it ; but, for poor soils, the least
rent is sometimes too high to be consistent with profit. By *poor soils*, however, are not to be understood
such as have a command of lasting manures, that work
great improvements ; nor waste lands, which, under
that false denomination, are often found the most
profitable of all.

The sound, mellow, rich, putrid, crumbling, sandy
loams,

loams, are of all soils the most profitable ; such as will admit tillage soon after rain, and do not bake on hot gleams of sun coming after heavy rains, when finely harrowed : such land is better worth forty shillings an acre than many soils deserve five.

The next soil I shall mention is that of the stiff loam, which is nearest allied to brick earth ; this, till drained, is in general an unkindly soil, without plenty of manure. It is known in winter by being very adhesive upon walking over it ; is long in drying, even when little or no water is seen upon it : for which reason it is generally late in the spring before it can be ploughed. When quite dry, it breaks up neither so hard and cloddy as mere clay, nor near so crumbly and mellow as the good loam. If it is in stubble, it is apt to be covered with a minute green moss. There are many varieties of this soil, but all agree in most of these circumstances, and in being what the farmers call poor, cold, hungry land. When hollow-ditched, and greatly manured, it yields any thing ; but those who hire it should forget neither of these expences.

The gravelly soils are numerous in their kind, and very different in their natures. Warm, dry, sound gravelly loams, are easily distinguished in winter. They admit ploughing all winter through, except in very wet times ; always break up in a crumbly state of running moulds; and if a stubble, will dig on trial by the spade, in the same manner. If under turnips, you may perceive, by walking through them, that it will bear their being fed off.

The wet, cold, springy gravel, is a very bad soil ; it is known in winter by the wetness of it ; and in
spring,

spring, by its binding with hasty showers. It rarely breaks up in a crumbly state, or shews a mellowness under the spade. Very expensive drains greatly correct its ill qualities, but it requires a prodigious quantity of manure to fertilize it.

Some gravels are so sharp and burning, that they produce nothing except in wet summers; but such are known at any season of the year. Sands are as various as gravels, and are all easily discoverable in their natures. The rich, red sand, is, I believe, as profitable a soil as any in the world. It has at all seasons a dry soundness, and at the same time a moisture without wetness, which secures crops even in dry summers. The spade is sufficient to try it, at any season of the year.

The light sandy loam is, likewise, an admirable soil: it will bear ploughing, like the preceding, all winter long, and appears quite sound and mellow when tried with the spade. If it lies under a winter fallow, the best way to judge of its richness, is to remark the state of the furrows, and the degree of adhesion in the soil. Stiff land, being dry and crumbly, is a great perfection, and sand, being adhesive, is an equally good sign.

When, therefore, the farmer views a light sandy loam, whose sound dryness is acknowledged, he may presume the soil *is rich*, in proportion to its adhesion.

If it falls flat in powder, and has no adhesion, it is a *mere sand*. The white chalky *marm* is often cold and wet, will not bear ploughing in winter, unless the weather is very dry or frosty; runs excessively to mortar with a heavy shower when in a pulverized state.

state. It is a cold soil, of little profit, except with peculiar management : but answers best when dry laid down to sainfoin.

In general, let him lay it down as a maxim, that strong, harsh, tenacious clay, though it will yield great crops of wheat, is yet managed at so heavy an expence, that it is usually let for more than it is worth. Much money is not often made on such land. The very contrary soil, a light, poor, dry sand, is very often indeed in the occupation of men who have made fortunes. Some permanent manure is usually below the surface, which answers well to carry on : and sheep, the common stock of such soils, is the most profitable sort he can depend on.

All stiff soils are viewed to most advantage in winter : the general fault of them is wetness, which is in the greatest excess at that season of the year. If the fields are level, and the water stands in the land, notwithstanding the furrows are well ploughed and open, it is a sign that the clay is very stiff, and of so adhesive a nature as to contain the water like a dish. It is likewise probable, that draining may prove insufficient to cure the natural evil of such land. This kind of soil, likewise, shews itself in the breaking up of stubbles for a fallow ; a very strong draught of cattle is then necessary to work it. It breaks up in vast pieces almost as hard as iron. When it is worked fine, it will run like mortar, with a heavy spring or summer shower. These soils will yield very great crops of beans and wheat, &c. They must, like others, be cultivated by somebody ; but I would advise every friend of mine to have nothing to
do

do with them; never to be captivated with seeing
large crops upon the land; for he does not see at the
same time the expences at which they are raised.

Peat, bog, moor, and fen, in many variations are
very profitable; but the expences of improvement
demand a calculating head. The vicinity of lime or
marle is then of great importance.

In respect to grass lands, the marks for judgment
are different. These are best examined by attending,
first, to the circumstances in which they are most deficient : and then to such as are in their favour.
The more seasons grass, fields are viewed in, the
better; though any one is sufficient for a tolerable
judgment.

One evil attending these lands is, that of being
too wet; the signs of which can never be mistaken
or overlooked in any season of the year. In winter,
it is at once perceived by walking on it ; at all times
of the year by the herbage which generally abounds
on it, such as rushes, flags, and a great quantity of
moss ; and also by the colour of the grass, which is
mostly blue at the points ; sometimes of a dirty
yellow hue, and always coarse. If the soil is the first
described stiff clay, and the surface level, the evil will
be very difficult of cure ; if of the other sort of clay,
or stiff loams, draining will have great effects.

Grass fields on gravelly soils are, if the gravel is
sharp, very apt to *burn* in dry summers ; but they
give great and sweet crops in wet ones, provided the
land is a gravelly loam. An *absolute* gravel should
never be under grass. A farmer should not, however, regret having a pasture or two of this sort in
his

his farm, being of excellent use in winter for feeding sheep and lambs on with turnips, &c.

The low meadows, whatever the soil, on the banks
of the rivers and brooks, are in general good, but
often subject to the misfortune of being overflown
in summer, which not only ruins crops of hay before
they are cut, but carries them away, perhaps, when
just made.

Many grass fields on all soils, consist of so bad an
herbage as to be of little value. Made up of weeds,
and the worst and coarsest of grasses, if a landlord
will not allow such to be ploughed, the farmer should
minute the rent accordingly. This fault is visible at
all seasons.

A river that does not overflow, running through a
farm, is a very favourable circumstance, as it indicates
a probability of all the grass fields being well watered :
that is, for cattle.

Size.—Another matter of great import, in the hiring a farm, is the taking no larger a one, than the
sum of money a man can command will stock properly.
A common fault among farmers is, the hiring too much
land for their money : they are extremely eager to
farm as much as possible : the certain consequence of
which is the conducting the soil in an imperfect manner. In the neighbourhood of great cities and towns,
variety of manures are to be had, in some places
cheap ; but, if the farmers have not money, how are
they to make use of such advantages ? For these,
and other reasons, a farmer should not think of venturing on a tract of land which he cannot command ;
that is, farm as seems best to him.

Contiguity

Contiguity of the Fields.—Many farmers too often overlook this circumstance. If they attended to it, as much as their profit required, we should see landlords reforming their estates in this particular, more than many do at present. There is not a more expensive, perplexing circumstance in a farm, than the fields being in a straggling, disjointed situation. The disadvantages are numerous and striking.

Covenants.—Many landlords are very tenacious of the covenants which they have usually inserted in their leases; so that a man, when he approves a farm and agrees to the rent, may find the conditions of tenure proposed to him, such as are incompatible with his interest, his designs, and even with good husbandry.

The merit or reasonableness of covenants must be considered always, on comparison with the nature of the farm. It is for want of this consideration that unreasonable covenants are ever proposed.

These prohibitions are often foolish, but sometimes admissible : they must depend on local circumstances, to be well weighed by the farmer who hires.

Ascertainment of Rent.—This is a very important part of the business in hiring a farm ; but the other circumstances already detailed precede it, rent in a good measure depending on them. The principal point here necessary to touch on, is the combination of rent, tithe, and rates, in one sum. Knowing the capital intended to be invested, estimate the interest of it at not less than 10 per cent. and then calculate the expences and produce ; the former deducted from the latter, leaves that sum which the farmer can afford

ford to pay in these tnree species of rent. Deduct
further the tithe and rates, and the remainder is what
he can afford to pay to the landlord. If rent be valued in any other way, it must be erroneously and
deceitfully done, and no dependence can be placed
on it.

STOCKING FARMS.

The benefit to be derived from the occupation of
land, depends so much on the farmer commanding
the requisite capital, that it is extremely necessary
for the young beginner to be well advised on this
essential point. If he is fixed in business by some
experienced relation, he will not want the proper instruction ; but as many adventurers (as they may be
called) are every day making efforts to try their fortune in the culture of the earth, and many gentlemen
taking farms into their hands, sometimes without
due consideration of the necessary expences, it is
proper to minute a few observations on the subject.

Thirty years ago, the sum that was usually appropriated to stocking a farm, varied from 3l. to 5l. an
acre ; and it was a general idea, that the latter sum
was sufficient for any farm, part arable and part grass,
of no uncommon fertility. Rich marshes were, of
course, excluded in the calculation ; and light flock-
farms were often stocked for 3l. per acre. But these
matters are now greatly changed ; rents are much
increased ; tithes are compounded at a higher payment ; poor-rates are enormously risen ; all sorts of
implements comprehended in the article *wear and
tear*, are thirty or forty per cent. dearer ; labour is
in many districts doubled ; the prices of cattle and

sheep

sheep are greatly advanced; so that, at present, the
same farm which at that period would have been very
well stocked, and the first year's expences provided
for, at the rate of 5l. per acre, now demands 7l. to
8l. per acre. But it is to be remembered, that in all
such estimates it is necessary to suppose that every
implement bought in is new, and that the live stock
be good of the sort, and that the first year's expences be provided for, though a portion of the
crop *may* come in before the whole payment is made.
A man cannot be at his ease if he does not thus provide; nor will he be able to make that profit by his
business with a small capital which will attend the
employment of a larger. By *profit*, I would be understood to mean a per centage on his capital, which
is the only satisfactory way of estimating it. If, by
stocking a farm with 5l. per acre, he makes 7 or 8
per cent. profit; and, by stocking in the proportion
of 8l. per acre, he makes ten per cent. (and this difference will, I believe, often be found), it must be
sufficiently apparent that the loss by the smaller
stock is a serious evil. It will depend much on situation and local circumstances: the benefit of procuring
manures, or litter to make dung, may, in some
places, be very great, in others much less; but not
to be able to profit by every favourable opportunity
that may attend the spot on which a farmer is fixed,
must be highly disadvantageous. To irrigate land is
an expensive operation; but to omit or postpone it,
for want of money for the undertaking, is to lose
perhaps the capital advantage of a farm. Cases of
this sort might be greatly multiplied; and there is
                                                    not

not one that does not call on the farmer for an ample capital.

Of all farms, a warren is hired with the smallest capital; but there are marshes in Lincolnshire stocked at the rate of above 30l. per acre.

The annual expence ot many hop-grounds amounts to 30l. and the capital to above 60l.

If a farmer does not make ten per cent on his capital, he must either have a bad farm or bad management, or *the times* must be unfavourable. He ought to make from twelve to fifteen per cent. Some farmers make more, when corn is at a fair price.

SETTING THE FLOCK.

In stocking the farm, our young farmer will have to determine on the breed and system of sheep. I shall here give one caution, and that is, to be very careful that he do not enter into such expences in buying fashionable breeds, as may cripple his exertions in more necessary objects.

SERVANTS.

This is the time that farming servants are chiefly hired, and the attentive cultivator should consider well before he determines on the number or the quality of his servants. A considerable business requires the employment of a bailiff; and as such a servant may have material effects on the conduct of a businass, it is always right in a farmer to consider the nature of his own case well. If his farm is extremely large, if his culture is complex, or if he is absent a part of the year; in these circumstances, the employment of a bailiff may be absolutely necessary; but there are so many sorts of bailiffs, that a man may

qualify

qualify the measure almost into that of hiring a common servant. In a large farm, his business is to be perpetually on the watch on all the people, of whatever sort, employed ; consequently he must not work, which confines him to one place. This part of his employment renders it necessary that he should be of a rank something above the best sort of servants and workmen ; for if one from that class has the command given him, he will not be well obeyed. A bailiff should ever preserve a due authority over all the people employed ; and, for this purpose, his master would find it very useful to allow him to hire his own servants and labourers, or at least to give him liberty to turn any of them away.

Respecting market transactions, all buying, selling, bargaining, and receiving money, it is highly advisable for the master to do all business of that sort : it is dangerous to trust these servants too much : not for fear of their running away with money ; but numerous money transactions, of which it is impossible the master can have an entire check, have at least a tendency to give opportunities of dishonesty, which may have a bad effect ; and market meetings, for the transaction of this sort of business, are likewise too apt to hazard the sobriety of a bailiff.

As to other servants, the principal are the ploughmen ; for on them depends, in a good measure, the success of all crops. In a large business, it will be very difficult to have all good hands ; but a man should aim at it as much as possible; for a bad ploughman makes very indifferent work, but skims the land

in

in an irregular manner, and rice balks half he goes over.

If no bailiff is kept, you must be more attentive in hiring ploughmen; to chuse such as will be obedient without that round of murmuring and complaints so often heard from these men : if the people be not docile, you will find great difficulty in having the land managed in the manner you like best. Shepherds, hog-herds, cow-herds, driving-boys, and all other servants, are now hired; and as characters are scarcely ever given among farmers, it much depends on your quick judging of the accounts the fellows give of themselves; and every man is a physiognomist.

Some great farmers board their men-servants and boys with their bailiff : it is one way of lessening trouble, and with one bailiff in twenty may be a proper arrangement; but far better for the farmer to have all his people under his own eye: he ought to consider himself as answerable, in some degree, for their religious and their moral conduct; to keep them regularly at their church, and, as much as possible, to prevent all swearing and profane language, for he may depend on finding servants, thus kept to a decent, orderly, and sober conduct, proving much more useful assistants than an ill-regulated, profligate set : it is to be hoped that our farmer has higher and better motives, but in a mere worldly view, he will soon be convinced of this fact. If he keeps a bailiff of the better sort, and single, there are many advantages in having him eat at the farmer's table.

Is it more advantageous to keep many servants and

few

few labourers, or many labourers and few servants?
Twenty years ago I believe the latter was more profitable, but at present the reverse. The price of day
labour, and the difficulty of getting it, are increased
more than the wages of servants and the expences
of house-keeping : and there is an advantage, not a
trifling one, in the farm-house being made a market
for many inferior articles of the farm-produce. The
certainty of commanding hands is a great object.
However, much will depend on the local circumstances and population of the district; and much on
the due regulation of the farmer's family.

FAMILY ARRANGEMENT.

Many accidental circumstances, gradually bring into a certain train the common habits of domestic
life; but it would often be more advantageous to lay
down a plan to be pursued within doors as well as
without : such ideas may not always be practicable,
but the mere aim will not be without its use. Our
young farmer, on entering his farm, must necessarily
arrange his plan of life and house-keeping, a subject
which should not be wholly omitted, yet admits but
a few cursory hints that may possibly give a turn to
his reflections, and being properly worked on in his
mind, may possibly produce a beneficial effect on his
conduct. It is not every man that has the power
of thinking to any marked utility; but he whose
mind is inquisitive, may think to advantage on every
subject. A prudent economy, free from all sordid
avarice, will by every one be admitted as right; but it
demands some reflection on entering life, or a farm, so
to arrange every day circumstances, that they shall flow
from

from the plan adopted; or at least that such plan shall have no tendency to counteract. In regard to house keeping, the safest way is to assign a stated weekly sum for it, which should on no account be exceeded. An annual one for his own dress and personal expences; the same for his wife and young children. And he should, in prudence, keep the whole allotted expence so much *within* his probable income, as to possess an accumulating fund for contingencies, children, &c. &c. And if he expects the blessing of the Almighty on his industry, he will not forget the poor in such distribution : I do not mean by *rates*, but by *charity :* and this hint demands one observation : a very material evil attending the support of the poor by rates, is the natural tendency they have essentially to lessen, if not to cut up charity by the root; that they do this in many hearts cannot be doubted; but it is a horrible, and a national evil. Let our young farmer accustom his mind to very different reflections, remembering that what he pays in poor-rates he is forced to pay, and that it is a part of his calculation in stocking his farm : if he expects to prosper (but not from that motive *only*, or he might as well close his purse) let him so accustom himself to kind offices and assistance to his poor neighbours, whoever they may work with, as to gain a habit of reaping pleasure from his free benevolence.

In such calculations as I have hinted at, he may safely estimate his profit at 10 per cent. on his capital: from 4000l. his income derived from his farm ought to be 400l. a year. He should lay up 50l. and as much more as his better interest may permit. To ex-

pend

pend this in *extra* improvements, may be the most advantageous investment, provided he owns his farm, or
has a long lease, not otherwise.

To attend markets and a few fairs, is a necessary
part of a farmer's business; but to a young man it is
a very dangerous part; it is too apt to give the evil
habits of drinking and dissipation : evil company is
every where to be found, and many a farmer has
been ruined by a want of a careful selection of his ac-
quaintance, and by not avoiding the contraction of
habits which cannot be indulged with safety. As a
safeguard against all evils of this tendency, an ha-
bitual attention to the duties of religion, will have
more efficacy than all the philosophic morality which
so much abounds upon the tongues of many : by religion, I mean that of the national church, the most
excellent that has been any where established for the
instruction of the human species. He can have no
true friend that will not advise him to keep the Sab-
bath piously and strictly himself, and make his family
do the same : many a judge has traced the origin of
crimes that have brought labourers to the gallows, to
Sabbath-breaking ; and if the source of failures among
farmers were as well explored, they would be traced
to the same spring. Serve God on Sunday as you
serve yourself on Monday : if you are a pagan, a deist,
a moral philosopher, you are, to a certain degree,
in reason, answerable for the paganism, deism, or
moral philosophy of your children and servants; if a
christian, you are surely the same for their christianity :
you may gain by this, but cannot lose.

The fashionable sheep-shearings, farming clubs,
                                                  societies,

societies, &c. render another remark not absolutely unnecessary : a steady, careful old farmer may not be the worse for mixing a good deal in company of a much higher rank than his own ; but a young man with a small degree of animation may suffer by it. His eye and his mind become insensibly accustomed to objects and habits of living to which he was before a stranger ; to steer clear of all imitation is not a very easy task, but it is an extremely necessary one : if after an excursion which has carried him into great, and what is called good company, he returns home not quite so well satisfied with home as he was before, he has contracted a taint that may be worse than the scab among his sheep. The caustic of wholesome self-discipline becomes necessary. He should guard carefully against one of the most natural propensities, or *his* pleasures (partially assuming somewhat of the garb of business) will prove like the pleasures of so many other classes, treacherous dissipation, and lay a foundation for uneasiness and regret.

THE FARM ACCOUNTS.

In the arrangement of the business of a farm, this is an object of no inconsiderable importance ; the modes of keeping them are almost as numerous as there are farmers in the island. The most plain and simple method commonly used, is that of entering all payments on one side of a book, and all receipts on the other, and balancing when the transactions of the year are ended : and this method gives a tolerable idea of the single object of profit and loss. I say *tolerable*, for it is but a tolerable one.

FARM-YARD.

This may be the last month of cattle remaining abroad, and if so, the farm-yard should be in order to receive them.

Good and convenient yards are of such great importance to spirited husbandry of all sorts, that, in the hiring a farm, a man should attend to this point; but if he finds himself on a farm where it has been neglected, and that the advantageous circumstances of a new one more than balance the expence of alterations, let him determine to remedy the evil himself, which may generally be done at no great expence. Let him run a high, warm fence, about a piece of ground large enough for all his cattle, contiguous to the barns and other buildings. It will pay the expence of good pales very well; but a much cheaper fence is, to build a stack of stubble, fern, ling, or straw, about eight or nine feet high, and five or six wide, and to thatch it for preservation : no fence is so warm for cattle. This inclosure he must gravel or chalk at bottom, to keep it always firm, and hard enough to shovel up earth or dung. Throughout the leisure times of the summer or autumn, a layer one to two feet deep, of marle or chalk, turf, ditch-earth, peat, &c. should be spread in it; and upon that layer the cattle may be foddered with straw, hay, &c. all winter. Plenty of stubble, fern, or straw, constantly spreading as fast as they tread it into dung, or lie wet or damp : the stables, cow-houses, hog-sties, fatting-stalls, if any, should be cleaned on to it; and, if the farmer fats any beasts on turnips, he may give them in binns in such a yard; by which means the quantity

of

of dung he will raise will turn out immense, provided he has plenty of litter.

THE TEAMS.

About the latter end of this month, the horses must be put to dry meat ; that is, hay, oats, and chaff. The hay should be given them cut into chaff with straw : as to oats, if the horses are worked constantly, they should be allowed two bushels per horse per week, which will be no more than sufficient to keep them in good heart, and make amends for the loss of lucerne : with this food they may be worked regularly.

But this system of feeding is expensive, and there is a way to lessen the cost, which is by substituting carrots instead of oats, or, at least, instead of the greatest parts of the oats. If you apply the chief of your carrot crop to other purposes, still you should determine to allow a small quantity weekly to all your horses, for the mere purpose of keeping them in good health.

HORSES OR OXEN.

In stocking a farm, the question, whether to employ horses or oxen, or both, will necessarily demand the farmer's attention. If he lives in a country where both are common, he will probably from practice have fixed his ideas sufficiently for the regulation of his business ; but if he live where horses only are known, he may be inclined to try oxen, in which case some practical observations on the subject may not be useless, from one who has had many years experience of both, and of bulls also.

There are two cases in which oxen are certainly
more

more beneficial than horses : first, when a farmer lives in a district where there is a breed of cattle well adapted to work ; and, secondly, when his farm is so large that he can buy in a considerable' lot of cattle annually, at a small expence *per head*, and feel no inconvenience in turning out such beasts from the teams to fattening as do not work well. In both these cases I have little doubt of the superiority of oxen to horses. But in countries that do not possess a breed of cattle well adapted for work in the state of oxen ; and on small farms whence fairs must be attended perhaps at the distance of an hundred miles to purchase a few, and consequently at a great expence per head, and possibly without land for fattening any, the benefit will be very questionable. In such a case I should prefer the bulls of the country, which are every where to be procured probably much cheaper than oxen ; are broken in with but little difficulty ; which work well, and which will recover from fatigue sooner than any ox. This I believe from what I have experienced, and from all the information I have procured, is stating the question of the comparison of horses and oxen as nearly to the truth as it can be done, in few words. There are, however, some works in the business of a farm, in which horses are better than either oxen or bulls, and therefore it may be advantageous to keep a few horses,

The ox teams should this month be kept on straw and cabbages, and, in default of the latter, on turnips ; but cabbages are superior. Let them have good straw always in their racks.

The

COWS.

The dairy of cows, supposing the system of the farm to be field-feeding, are now to be taken into the yard, where their food must vary according to their state : the dry ones must be put to cut chaff, and those in milk, in another yard, to cabbages, which are found, on experience, to give no disagreeable taste to the milk ; but good chaff must be given with them. Young cattle should be put with the cows in milk, as they cannot be kept too well. On no account let any of these cattle out of the yards : they only poach and damage the grass fields. And let it be remembered, that the grass now to be had in meadows and pastures, suitable to kept rouen, is of far greater value for sheep in more pinching seasons.

Such cattle as have been in the yard or sheds, and supported by soiling, may now have their food gradually changed to dry meat, roots, or cabbages.

FATTING BEASTS.

This is the proper time to take the large fatting oxen, that have had the summer's grass, and put them to turnips, cabbages, or carrots : turnips with cut chaff will do ; but not near so well as cabbages or carrots : both which will fat a large ox as well as any food. You may either stall-feed them under cover, or let them be loose in a straw-yard, well littered in either case : and, if the latter, they should have open sheds to retire under at pleasure.

This is also the month for purchasing beasts of the smaller sort, for fatting on the same articles of food, particularly turnips and cabbages. It is this plan of appropriating the turnips and cabbages of a

farm

farm to fatting beasts throughout the winter, in a well-littered farm-yard, that converts the straw, fern, stubble, &c. into such quantities of dung, as improves the land more than any other method whatever.

The quantity of turnips and hay which stalled oxen eat, appears from experiment to be a ton of turnips, besides chaff or hay, in a week, for an ox of 75 stone, (14lb.) ; 12 cwt. a week for a cow of 32 stone, with variations of course.

HOGS.

Now also put full-grown hogs to fatten : a business profitable, particularly in respect to the improvement of a farm by dung. If he gets the market price for his pease, barley, beans, buck-wheat, &c. and saves carriage upon them, at the same time getting a fair price for his swine, lean, he certainly makes a considerable profit upon the whole transaction, though not an immediate one, as the mere fattener of hogs: but, what is of much greater consequence, is the raising of rich and most valuable manure.

The most profitable method of converting corn of any kind into food for hogs, is to grind it into meal; and mix this with water in cisterns, in the proportion of five bushels of meal to 100 gallons of water ; stir it well several times a day, for three weeks, in cold weather, or for a fortnight in a warmer season, by which it will have fermented well and become acid, till which time it is not ready to give. It should be stirred immediately before feeding. Two or three cisterns should be kept fermenting in succession, that no necessity may occur of giving it not duly prepared. The difference in profit between feeding

in

in this manner, and giving the grain whole, or only ground, is so great, that whoever tries it once, will not be apt to change it for the common methods.

Pease-soup, however, is an excellent food for hogs, and may, for what I know (but I have not sufficiently compared them), equal the above, especially if given in winter milk-warm ; but the expence of fuel and labour must be remembered.

For the general stock of hogs, cabbages are this month of incomparable use. Swine at this season are often very cheap, and it is of material consequence in that case, that the farmer be largely provided with a food, by means of which he can keep this stock for a better market. Without this plant, he cannot keep great stocks of swine to the best advantage.

PUT FAT SHEEP TO TURNIPS, &c.

This is the proper time to begin to feed off the forward sown turnips. A general rule, which ought rarely to be departed from is, not to begin to fat lean stock on this food ; they rarely pay for it. Sheep should be nearly half fat when they begin turnips ; nor will they feed to profit if lean. I have made the comparison repeatedly with the same result. Upon every soil that will bea it, the turnips ought to be eaten, by hurdling, where they grow, for to draw them, unless absolutely necessary, is most unprofitable management. A lean stock should follow, to eat what the fat ones leave.

MANURE GRASS.

If this was omitted after mowing, it should not be deferred later. In some counties, it is an article in the generality of leases, that all, or much of the dung,

of

of a farm should be spread on the grass ; but such
covenants are contrary to the spirit of good husbandry.

DIG UP CARROTS.

About the end of this month the carrot crop
should be dug up : some persons leave it till November ; but, in case of wet weather, they suffer. They
may be taken up either with three-pronged forks or
with spades, if the land is not hard, which it will not
be, if the crop has been well cultivated : a little loosening of the earth with the tool, and at the same time
drawing up the carrot by the top, will take them up
very quickly. They should be left spread over the
field till dry, which will be in a day or two : then
thrown into heaps and carted home ; which moving
will clear the dirt from them. Unload them in a
barn or some out-house, and let the tops be chopped
off, and given to the swine : then lay the roots where
they are to remain. Some pile them up in a heap,
and cover them with dry sand ; others cover them
with straw : they will keep very well, if packed close
together in any building : and if it be only a boarded
one, cover them with some straw, enough to keep
out the frost. There are many ways of preserving them ; one is, to pile them in circle, finished
conically, with just a scattering of tops left on
a few of the outside carrots, as shall form a thin
thatch of tops hanging down : not thickly, for the
tops will then ferment, and heat the roots : if these
few tops rot away, a thin scattering of straw should
be laid, and the pile be not more than five feet diameter. Others put them a ridge of earth, like potatoe *pies*. The general way in Suffolk is, to leave
them

them in the ground, and take them up as wanted ; but this is applicable only to a district where every man has a field, else they would be stolen. From 1s. 6d. to 2s. a load of 40 bushels, are given for taking up.

Respecting the application of the crop, much has already been said on that head. They are to be given to the team ; if without oats, two bushels per horse per diem ; and they will eat but little hay : they are of incomparable use in fattening oxen, and in feeding stock swine. Sows that have pigs may be kept on them, for they breed much milk. Cows eat them greedily, and they give no ill taste to the milk, cream, or butter. Their use, in short, is universal ; you can cultivate no plant that will answer more purposes.

PLOUGH UP POTATOES.

There is not the same reason for digging up this crop as for carrots : the plough among the latter is apt to cut, break, and bury them ; but not so with potatoes, for it turns them over, damaging scarcely any. First let a number of women preceded by a cart, pull up the tops, and throw the potatoes that adhere to them into baskets, and the stalks into the cart, which should convey them to the hog-yards, where they will presently be trampled into dung : then each plough taking its ground, attended by six or eight women, or more, if the crop is very large, each with a basket, divide the furrow, by setting up white sticks into as many parts as there are women, that each may pick her own share ; a range of bushel skeps, at a small distance, for the baskets being emptied, and three or four carts ready for men (who do nothing else), one to eight or ten women, to take the skeps to the carts. The furrow being picked, I used, many years ago, to work it by men with three-pronged forks, each with a woman, or boy, to pick up the roots ; but finding this expensive, I contrived a diagonal harrow in a shim beam, with 2 or 3 teeth, drawn by one horse, which tears the furrow in pieces, and lays bare the mass of the crop : the women then pick again ; and another common cross-harrowing, with a second ploughing and harrowing, all three attended with two women to each plough, will finish the business, and clean the roots all away ; so that I have found the pigs, when let in, make but very poor gleanings. The use of the little harrow saved me from 14s. to 20s. per acre in labour.

The best way of storing the roots, is in what are called potatoe *pies.* A trench, one foot deep and six feet wide, is dug, and the earth clean shovelled out, and laid on one side : this has a bedding of straw, and the one-horse carts shoot down the potatoes into the trench ; women pile them up about three feet high, in the shape of a house-roof ; straw is then carefully laid on six or eight inches thick, and covered with earth a foot thick, neatly smoothed by flat strokes of the spade. In this method I never lost any by the severest frosts ; but, in case of its freezing with un-common severity, another coat of straw over all, gives absolute security.

These pies, when opened, should each be quite cleared, or they are liable to depredation. To receive one at a time, besides also being at first filled for im-mediate use, I have a house that holds about 700 bushels,

bushels, formed of posts from fir plantations, with wattled sides, then a layer of straw, and against that earth, six feet thick at bottom and eighteen inches at top; the roof flat, and a stack of beans built upon it. This I have found frost-tight. The beans keep out the weather, and yet admit any steam which rises from the roots, which, if it did not escape, would rot them.

LAY UP THE FALLOWS.

This month must conclude the autumnal tillage on all stiff or moist lands; for in the following, they may probably be too wet: but on very light sandy soils, ploughing goes on all winter. Lay it down as an invariable rule, never to have a piece of stubble unploughed the end of November. It is of importance to leave the land for winter in such a manner, that the frosts may get into it.

But here our young farmer's attention must be particularly turned to the greatest of all modern improvements on strong land ; that of discarding as much as possible all spring tillage ; the summer fallows intended for barley and oats ; the bean, pea, and tare stubbles designed for the same crops ; and the white corn stubbles intended for any spring crop, must now be ploughed very carefully, being the future seed-earth ; no more ploughings being allowable on any account whatever. In the Calendars for February and March I have entered into some explanations of this system, which is of such importance, that too firm a resolution cannot be taken to carry it into execution. By means of it the crops are much greater, and the expences considerably reduced. The

k k 2　　　　　　　　leading

leading principle is this: if the land is so laid in autumn, on to ridges of that exact breadth which suits the tools (whatever they may be, whether harrows, scarifiers, scufflers, or drills), so that the horses which draw them may walk only in the furrows, the frosts will have left so fine and friable a surface, that any of these operations may be performed long before the land in the common system could be ploughed. The seed is securely in the ground before the old-fashioned farmer thinks of moving. If he ploughs, he turns down a dry crumbling surface, and brings up the stiff wet *clung* bottom ; if rain comes, then he is in the mire, and must wait for a season : if a drying sharp north-east wind comes, his furrows are converted to oblong stripes of a stony hardness. In one case he is plagued with mud, in the other with impenetrable clods : he was possessed of just the surface he wanted, and which once lost, is not often regained. This surface may be scuffled, and immediately drilled securely. If this husbandry be intended, it is of particular importance that the lands or stitches be laid out with great exactness. See the Calendar for the spring months.

A caution in the first forming of these stitches should be mentioned ; which is, the difficulty, without two or three ploughings, of bringing them from old breadths into a correct form. If the men are not skilful, they raise the centers too high ; and, in reversing, are apt to leave the outside furrows also too high : the stitch should be flat, or, if rounded, very slightly so ; without this attention, the seed in some drills will be deposited deeper than in others.

sow

SOW WHEAT.

All the cases of sowing this grain in September, are equally applicable to October, if the weather was too dry in that month. October is the principal month in the year for putting in wheat throughout the kingdom, and it is every where partly performed in this month, though some like to postpone it to November. The management is however bad, should the season suit in October.

SOW WINTER TARES.

There should be two sowings of tares in this month, as it is a material object to have a succession for soiling.

MANURE AND PLOUGH FOR BEANS.

A successful bean husbandry upon harsh and difficult soils, depends upon the exertions which are made in this month, or, in favourable weather, in November. As soon as the farmer has finished his wheat sowing (and before, if he has been delayed by drought), he should cart on the manure, all that is possible for beans. It is the wheat, barley, or oat stubbles, or layers which will come in course to receive it : if the wheat, the stubble must be mown and carted first ; the manure then carted and spread, and the land carefully ploughed into that form on which the crop is in the spring to be drilled or dibbled. If the former, the stitches must be of the exact breadth which suits the drill-machine ; if the latter, of that which is adapted to the scuffle and scarifier. The dung will lie safe, and the frosts will pulverize the surface, a main point for drilling, but not for dibbling. By means of effecting this before the bad weather comes,

he

he will be able, if the weather be open, to get in the crop in February, which is of much importance. Let him be assured that there is no crop which will pay him better for dung than this.

PLOUGH FOR PEASE.

On the same principle which governed the preceding observation in relation to beans, he must now plough the lands intended for pease; laying the stitches of the breadth which will suit the drill-machine, or for dibbling them on layers. But the manure should be applied to other crops.

PLOUGH FOR BARLEY AND OATS.

Whatever lands are intended for these crops (except such as are now under green winter ones), should be ploughed, as remarked in the preceding articles, in this season for the seed-earth, in order that no ploughs may be wanted to stir on wet land in the spring. Attention is to be paid very carefully to the breadth of the stitches, that the mode of putting in these crops may be duly prepared for. See the Calendar for March. No reasonable man, who has seen the effect of this system, can value the modern improvement at less than the rent of the land.

PLOUGH FOR MADDER.

This is the right time to give the first stirring to the land designed for madder. It requires great depth of ploughing. Loams, that in common conversation are called clays, will, with a proper quantity of dung, do for madder. The article of manuring is the soul of this culture; the plant delights to grow in a dunghill, so that you need not fear overdoing it.

Let

Let the farmer, however, determine to have nothing to do with this, or with any crop not in an
easy and safe mode of sale, unless he has previously
ascertained the certainty and price of the market.

DIGGING FOR LIQUORICE.

The best culture for this root, and which is common at Pontefract, is to dig for it four or five feet
deep.   This plant sends down only one tap-root,
like the carrot ; consequently the great profit of it,
is the length of the root, which is exactly proportioned to the depth of the tillage.   In this husbandry,
as in that of madder, the same land is preferable for
successive crops, as one digging serves both for the
old crop and the new.   For liquorice you must manure very richly : it will not answer well without this
attention.   Leave the land well water-furrowed for
the spring.

COURSES OF CROPS.

I esteem this to be the most important subject
that has been treated of by the modern writers of
husbandry, and that on which they have thrown far
more light than upon any other circumstance in
agriculture.   It is a very singular and remarkable
circumstance, that before the reign of his present
Majesty, notwithstanding the multitude of books
on Agriculture, there is not one author who had any
tolerable ideas upon this subject, or even annexed to
it any importance.   They recite courses good, bad,
and execrable, in the same tone, as matters not open
to praise or censure, and unconnected with any
principles that could throw light on the arrangement
of fields.   But, when once the idea was properly

started, its importance presently became obvious, so that thirty years have carried to great perfection the precepts which practice has afforded in this branch of rural economy. This subject will demand a principal attention from our young farmer, who should well consider the courses to which his soil is applicable.

General Principle.—It is now well known that some crops exhaust land much more than others : that some, notwithstanding they exhaust, return by being consumed on the farm, as much, or more, than they drew from the soil in their growth : that some admit profitable tillage and cleaning while growing ; and consequently, clean, instead of rendering the land foul with weeds ; while others, not admitting such tillage, and being exhausters, if combined in succession, will deteriorate the land and fill it with weeds. Practice tells us, that by a due arrangement of these crops in courses, land of almost any description may be kept perpetually clean and in heart.

It will be useful to detail some of the best courses adapted to the most striking varieties of soil.

First, including a fallow on strong and wet land :

1. Fallow,	1. Fallow,
2. Barley,	2. Barley,
3. Beans,	3. Clover,
4. Wheat,	4. Beans,
5. Tares,	5. Wheat,
6. Barley,	6. Cabbages,
7. Clover,	7. Oats,
8. Beans,	8. Tares,
9. Wheat.	9. Barley.

Excluding a fallow on good sound loam :

1. Turnips,	2. Barley,

3. Clover,

3. Clover,   7. Tares,
4. Wheat,   8. Barley,
5. Cabbages,   9. Beans,
6. Oats,   10. Wheat.

1. Turnips,   5. Beans,
2. Ruta baga,   6. Wheat,
3. Barley,   7. Beans,
4. Clover,   8. Wheat.

On good sand:
1. Turnips,   4. Barley,
2. Barley,   5. Clover,
3. Carrots,   6. Wheat.

On peat, and on soils long harassed by corn:
1. Cole-seed, or Turnips,   8. Grasses,
2. Ditto,   9. Ditto,
3. Oats,   10. Potatoes,
4. Ruta baga,   11. Barley,
5. Barley,   12. Tares, or Pease,
6. Grasses,   13. Barley and Grass.
7. Ditto,

On dry and calcareous soils:
1. Turnips,   5. Turnips,
2. Ditto,   6. Barley,
3. Barley,   7. Pease,
4. Sainfoin for ten years,   8. Wheat.
    and upwards; then pared
    and burnt for

THE DRILL HUSBANDRY.

Upon settling in his farm, our young farmer has, among many other objects that require his attention, to determine in what degree and for what crops he will adopt the drill husbandry. It has long been known that this system is applicable, without inconvenience, to sandy soils and to dry loams, which may be safely laid flat, and accordingly, on such it made
a great

a great and rapid progress in Norfolk ; but it travelled
no further in that county. A great revolution which
has taken place in the wet land district of Suffolk, has
introduced it with equal success on the strong soils of
that county.

This great change is the banishment of the plough,
to as great a degree as possible from heavy soils in the
spring ; all barley, oats, pease, and beans, that can by
any means be thus managed, are put in on an autumnal ploughing, which has thrown the stitches very
carefully ploughed to the exact breadth which suits
either one movement of the drill, or a *bout* of that
tool ; according to the system the farmer is in, some
preferring one and some the other. The frosts give
a considerable friability to the surface, so that the
farmer can go on very early in the spring, and after
one scarifying and harrowing, drill the corn without
a horse's foot treading any where except in the stitch
furrows.

The advantages of this system are beyond conception. In the common husbandry of giving two
or three spring ploughings, or even one, that friable
surface, the gift of the atmosphere, is turned down,
and in eight seasons out of ten lost, to be had no
more. Successive rain and sharp N. E. winds give a
succession of mire and clods, to the material delay,
expence and vexation of the farmer. His crops suffer greatly, and he is generally *in the afternoon* of
spring operations.

The improvement is applicable to the broad-cast
system, as well as to the drill ; but as it was introduced, I believe, by those who had been in the habit
of

of drilling wheat, they applied it to drilling barley and oats. It removed at once the main objection to this part of the system, and has been pursued with very great success by the best farmers on the strong lands of Suffolk.

Turnips *fed late* will in some seasons cause an exception, and render one ploughing necessary.

The main objection to drilling being by this circumstance done away, there remains no reason for any farmer, on a soil where the drill machine can freely move, rejecting the husbandry; and he certainly ought to apply it in such cases as it is better adapted to, than dibbling, which, however, never made any progress for barley.

Upon soils of such tenacity as deny effect to hollow drains, the system of drilling must be different; such soils must be laid on round high arched ridges, and to drill these to advantage will demand a machine made for the purpose.

RIDGES, STITCHES, OR LANDS FOR DRILLING.

Drilling, if performed on ridges, demands those of various breadths, according to the system intended to be pursued. Some farmers prefer such as admit but one stroke or movement of the drill machine; others prefer a *bout*, or two movements. Suppose the machine sows six rows at one foot; that one foot be allowed for the ridge-furrow; and that one movement is preferred: in this case, the ridge must be six feet wide, always measuring from centre to centre of the furrows; but if eighteen inches be allowed for the furrows, which will make better work, then the ridges must be six feet six inches. Two movements of the

machine

machine will demand in the former case eleven feet, and in the latter, thirteen feet six inches; and thus, in all other distances, the measure is ascertained on the same principles : but where, from the dryness of the soil, the ridges are ploughed to a great breadth, or the land quite flat, as in Kent, without any lands or ridges, the more common method is, to drill across the path of the plough by means of a marker attached to the axle-tree of the machine, and moved at turning on the headland. The most correct work I have seen in this way was not by leading the horse, but by a boy riding him, and keeping the mark always between the horses ears.

Another system of drilling has been recommended, which is that of double rows at nine inches, on a three-feet ridge, which may do for beans ; but I never heard of its being attempted in Suffolk for barley or oats, though that ridge is very common there; and all I conversed with on the subject condemned it in opinion ; but for cabbages in single rows, to be drilled in April, these ridges are of the proper breadth.

STANDING SHEEP-FOLD.

This being the month for mowing wheat-stubbles, it is proper to mention the great advantages which attend one application of the haulm thus procured : it is that of forming one or more standing sheep-folds for winter folding, especially in the lambing season. This is the best, and indeed, the only admissible system of folding on farms not particularly open.

COLLECT LEAVES.

In woodland countries it is of great advantage to rake up all leaves that can be procured at a moderate expence,

expence, and cart them to the yards, and standing folds, for littering and making them into dung : I do it at 3d. per one-horse cart-load. They do not rot easily, but that is no objection to them ; they are a sponge to be saturated with urine, and if not touched previously to carting on to the land, will convey to the field much of what might otherwise be lost ; and they are extremely useful in aiding the main object of bedding the yards.

DEPTH OF PLOUGHING.

Our young farmer, on entering his farm at a season when the ploughs will be all at work for various purposes, will necessarily have the question of depth come often across his mind ; and it is a subject that will demand no trifling attention. In some of our well cultivated counties, the shallowness of the ploughing is remarkable ; when almost every other point of management is very spirited and complete, a deficiency in this may not be at once perceived in the crops; but I have no doubt but failures are often caused by it, though attributed to other circumstances. It is a subject too ample, fully to discuss in a work of this nature ; but the following hints may have their use.

1. An additional depth should first be gained in autumn, that a successive change of seasons may take effect in atmospheric influences before any seed is ventured in the raw stratum brought up.

2. The quality of that stratum should be examined; it is sometimes steril by reason of an acid, discoverable by boiling in water, and putting that water to the test of blue infusions.

3. Animal

3. Animal and vegetable manures cannot be *buried:* at whatever depth they are deposited, their constant tendency is to rise to the atmosphere.

4. Fossil manures are extremely liable to be buried, having a constant tendency downwards. Chalk, marle, and clay, are sufficiently soluble, or so miscible with water as to sink in a regular mass, and are sometimes found much below the path of the plough.

5. In soils of a poor *hungry* quality, there should be some proportion observed between the depth of ploughing and the quantity of manure usually spread; but this does not hold good upon better soils.

6. Soils are rarely found that ought not to be ploughed, in common, six inches deep; many ought to be stirred eight inches, and some ten.

7. One deep ploughing (to the full depth) should be given once in 12, 18, or 24 months; if this be secured, shallow tillage by scaling, scarifying, scuffling, shimming, or broad-sharing, is in many cases preferable to deep working oftener, and especially for wheat, which loves a firm bottom.

These hints are enough to make a farmer *think,* which is no inconsiderable point gained.

GATHER APPLES AND PEARS.

These crops are now ready to gather: they should be taken from the trees in a dry season. Some persons keep them some time in heaps to sweat, before they are deposited where they are to remain. The safest place is a bricked vault, with broad shelves around, in order that they may not lay too deeply disposed. They should be examined for about a month, and moved and wiped if any moisture adheres. This

is

is easily done, if one space is left unoccupied when the cellar is filled.

PUT RAMS TO EWES.

Of all the systems of barbarity in relation to sheep, there is none more prominent than the management, almost every where common twenty years ago, of turning a number of rams promiscuously into a flock of some hundreds of ewes. Where breeding is pursued on enlightened principles, much attention is given to the choice both of ewes and rams, in selecting the lots of the former (50 to 60 in each) and in assigning the latter to the respective parcels. I have been present with that excellent farmer, the late Duke of Bedford, when he attended to this business; he had every ram, with the lambs got by him the preceding year, in distinct pens, that he might not only examine the ram himself, but also his progeny, before he determined what ewes to draw off for him; and the conduct is perfectly reasonable: such attention, united with a careful selection of cull lambs, must keep a flock in a state of progressive improvement, proportioned to the accuracy of judgment, eye, and hand of the farmer who practices it

NOVEMBER.

THRESHING.

As soon as the cattle are taken into the yards, if they are to be fed with straw, the threshers must be set to work, to supply the lean beasts, and they must be kept regularly to it.

FENCES.

This is the first month for hedging and ditching : October is too soon. After you have once brought your fences into good order, which should always be effected within the three first years of a lease, the best way is, to divide the length of hedging into 12 parts, and to make it a rule to do one twelfth every year afterwards; by which means the whole will always be kept in good order. The best method for all old hedges, consisting not wholly of white thorn, is the plashing, in which so much of the hedge is made of live wood, that it holds up and lasts far longer than hedges made all of dead wood, which is the practice of some countries; they are rotten, broken down, and gone, before the quick wood gets up to form a fence; whereas, in the plashing method, by leaving as many hedge-stakes alive as possible, and by laying down much growing wood, the hedge is constantly impenetrable.

BORDERS.

The borders in many counties, where the inclosures are small, take up a tenth or a twelfth part of whole

whole arable farms ; but, in all enclosed counties, they occupy a great space. It is highly expedient that such land, as it cannot be applied to the profit of the fields themselves, be reduced as much as possible ; that is, be no wider than requisite for a horse to turn at the end of the furrow in ploughing ; but in- many farms this is no rule, and borders overrun with rubbish, such as thorns, brambles, thistles, and other trumpery, spread into the fields, to a width that usurps a quantity of land which ought to be appropriated to more profitable purposes.

FOLDING.

If folding be the system pursued, and this month proves wet, you must leave off for the arable lands and begin with the dry grass fields. Many farmers stop about this time for the winter ; but that is bad management : the idea that winter folding is of but little use is a mistake. Winter is a proper season for manuring grass lands which you could not fold in summer ; nor does winter folding on very dry grass land do such harm to the sheep as arable folding, and the benefit to the land is great. You may manure mossy ground often, before you destroy that weed ; but the treading of the sheep at the same time that the dung and urine are dropped, completely destroys it : and this manuring is more adapted to turf, than dung to be spread on the surface, which is troublesome to get in.

WATER MEADOWS.

In this month you may begin to winter water the meadows and pastures, wherever it can be done ; and be assured that no improvement will pay better : a

winter's watering will answer in the hay, fully equal
to a common manuring of the best stuff you can lay
on the land; and the expence, in some situations, is
trifling.   The lower parts of a farm are generally in
grass; the farmer should attend to his ditches, so
that the water from all the higher parts of the farm
may have an unobstructed course to a ditch a little
above the bottom, from which it may be let at pleasure over the meadows, observing that it only runs
over them, and does not stagnate.

BURNET

It is a common error with the cultivators of burnet,
to let cattle go into the fields at this time of the year;
but it is bad management, and contrary to the intention of the culture.   Keep it throughout autumn and
winter from any cattle : it will then be ready in the
spring, when most wanted, for sheep.

WALLING.

In the dry stony countries, walls are the common
fence, and, when well made, are impenetrable, and
extremely durable.   This is the proper season to begin building them : they are made of whatever stone
is most plentiful; either lime-stone, which is generally in quarries, rag-stone, or grit-stone.   The best
are lime and grit; because generally most plentiful,
and at the same time much the easiest cut; but
whin-stone cannot be used to profit for this or any
other work, as it is so hard that it will not cut without
great difficulty.   But some grit-stones cut with so much
ease, that you may build walls of it without mortar,
as true as with, and will, if well laid, last nearly as
long.   In the enclosures of wastes, it is proper, by all
means,

means, to begin the walling in this month, which is so soon after the hiring time, Michaelmas, and continue the work all winter.

DIG MANURES.

All this month the carts should be employed in carrying marle, chalk, clay, or ditch-earth, upon soils that are light enough to admit carting on through winter.

CUT ANT-HILLS.

This is the proper season for destroying ant hills. Many ploughs have been invented for cutting them off level with the surface of the field, ready to be carted away ; and if that is the way you take with them, such machines are of great use ; for they will certainly do the work of many men.

MADDER.

Look well to the land deep ploughed in the preceding month for madder, to observe if it lays perfectly dry : if the water hangs at all in the furrows, or the water-furrows, let them be immediately cleansed, so as to run off without the least obstruction ; for it is very pernicious to any land to be soaking in stagnant water : instead of receiving benefit by autumnal tillage, it is much injured.

PEASE.

Upon dry soils, that plough well in winter, the end of this month is a good time to sow the hardy hogpea, which will remain uninjured by frosts, and be much earlier in the following year, than those sown in the spring.

SHEEP.

The lean stock sheep will yet be kept in the re-
L l 2   mains

mains of the summer grass, and on the sheep-walks;
but the fat stock must now be at turnips or cabbages.
Remember that fatting cattle, of whatever sort, should
have as much meat as they like; but should, at the
same time, be prevented from making any waste.
Giving fat sheep the turnips or cabbages is a dubious
point; many farmers urging strenuously the necessity
of saving carriage, by letting the sheep feed them off
where they grow, provided the land be dry enough;
but others are of a contrary opinion, and carry the
turnips to a grass field, where they give them to the
sheep as required, and without so much waste as is
made in the other case. Upon these systems I shall
remark, that, if the land is dry, you may feed off
without waste; because the soil is so clean, that there
is no soiling by dirt or poaching; and, by bringing a
stock of lean sheep to eat up the leavings of the fat
ones, there will not be the least loss: but this
point, of lean stock following the fat ones, is too
much neglected by many farmers, who only run over
their fat sheep, and consequently spoil a proportion
of their crops. It is good management, in many
farmers, to have a sheep-rack filled with hay always in
the turnip field that is fed by fatting sheep: others
give them bran or barley-meal, oil-cake, or pollard,
or malt-dust, in troughs; the dryness of which are a
corrective to the moisture of the turnips, and will
contribute well to the more speedy fattening of them.
I do not, however, mention these assistants as being
absolutely necessary; because I know that thousands
of sheep are fatted on turnips, without any such
help. Another article of dry food, which agrees excellently

cellently with turnips, is cut chaff: this makes very good dry meat for sheep that are fatting on turnips or cabbages.

THE TEAMS.

This is an idle month for the teams in many farms; but should not be so with good husbandmen : for, as I have often remarked, they must be constantly well fed, and employed. There are many works that may be executed in this month after ploughing is finished : on light dry soils, the marle, chalk, or clay carts should not stop : they may work from the first day to the last. In wetter soils, you may cart any sort of manure on to grass lands, provided you use small carts.

DRAINING.

In this month you may begin the work of hollow-draining, which, on wet lands, is the *sine qua non* of husbandry. It is in vain to think of farming them to any profit, without this improvement. Manuring before this is done, is but expending money for 5 per cent. advantage, where 50 ought to be the return. Lay your land dry before you attempt other improvements : the first step is cutting deep and large ditches around the wet fields ; then you gain a requisite fall to take the water clean away from the drains.

If the soil is very wet, it will be necessary to cut the drains near each other ; for instance, about a perch, a perch and a half, or two perches asunder ; by which means it will be laid in most soils in dry and wholesome order. Fill them with whatever materials you can get the easiest, bush-faggots, stones, straw, &c. &c. No improvement in agriculture is greater

than what is effected by these drains, nor any that
will sooner repay the expences. In many parts it is
well known, that the first arable crop will repay the
whole expence, which is a profit not to be reaped in
any other article, to which a man can attend.

ELKINGTON's DRAINING.

A very necessary attention is to be paid by every
occupier of wet land, to the cause of the moisture
which injures him : if, as common in many parts of
Essex, Suffolk, and Norfolk, which are, in general,
countries not marked by strong inequalities of surface, the wetness proceeds from the texture of the
soil, especially the under stratum, and not from
springs, the system of hollow-draining applied to the
whole surface is the best cure the evil may admit ;
but, in many districts, the case is different ; one
spring breaking out on the slope of a hill will damage much land below, and appear in so many places
irregularly, as to assume the appearance of many distinct springs, or a general wetness of surface. The
common system of hollow work, in such cases, may
fail entirely, though the expence may be greater than
that of another system discovered, or practised, or
published first by Mr. Elkington, under the patron-
age of the Board of Agriculture. It would be impossible, in the limits of a Calendar, fully to explain
this system ; but the principle of it is to discover
what may be called the mother spring, and to cut it
off by one deep drain passing across, but above the
spot where it breaks out. The boring at the bottom
of this deep cut has, sometimes, considerable effect,
not only on the spring immediately in contemplation,
                                                        but

but on others also that become visible at a distance, and has, in some cases, operated effectually to a great distance, and even on the opposite side of a hill more than a mile from the spot. The principal use of mentioning this system here is, to caution a young farmer to consider well the circumstances of the wetness which injures him, in order that he may ascertain the cause, and the best method of applying a remedy. If he duly and attentively examines his farm, and its relative situation in respect of hills or mountains in his immediate vicinity, and marks all the places where springs break out, he can hardly fail of going to work with more skill and more appropriated efforts, than if such a system of drainage had not been in his mind.

WOODS.

Now begins the business of wood-cutting. In some parts of the kingdom, this is a profitable part of husbandry ; but, in many others, it pays very indifferent returns. If there is a long carriage on the wood, it answers rarely well enough to induce a spirited farmer to apply his attention and money to it : arable and grass land will pay better ; and supposing one hundred or two hundred pounds, or more, of his stock, applied to hiring the wood, he may in general be assured, that the same sum, thrown into his farm in an increase of improvement by draining or manuring, will pay him better interest. But, if it is expedient to keep woods, it is much worth attention to apply them to the best use. Old experienced farmers are always attentive enough in this ; but young ones, and gentlemen just beginning their husbandry, are

L l 4 apt

apt to be too careless.   Labourers will ever persuade
them to what pays themselves best.   Hop-poles,
hoop-stuff, hurdles, short faggots, long ones, bushes,
stakes, and edders : each of these articles is, in some
places, more profitable than any of the rest; and I
believe, on an average, those will be found most
beneficial, for which the purchasers come and take
them away.   Carriage, on so cheap and bulky a commodity as wood, is a very great deduction from the
product.

COVER TURNIPS AGAINST FROST.

We are indebted to the Rev. Mr. Munnings, near
Dereham, in Norfolk, for a method of securing turnips
against frost, which deserves attention.   He drills at
two feet equi-distant on the flat; and, in a dry season,
towards the middle or end of November, he covers
them so by a deep ploughing as to secure them to a
great degree.   I must, however, remark upon it,
that it seems a husbandry better adapted to the Northumberland system of drilling, often mentioned in
this Calendar, than to that of flat work.   This mode
of drilling is upon the crowns of ridges.   If the turnips at this season are drawn, and two rows set close
together in one furrow, and then the ridges split, they
will be more effectually secured from the frost than
possible upon flat work.

STEAMING ROOTS.

This application of fire to the preparation of roots
for feeding stock, is a practice of the present age, and
it is thought a very favourable one, and largely practised by many very intelligent cultivators.   The best
apparatus which I have viewed is that of Mr. Stares,
of

of Hampshire, of which there is an explanation and a
plate in the *Annals of Agriculture,* vol. xxviii. p. 228.
The great effects of feeding cows with steamed turnips mixed with cut chaff, has been detailed by the
Rev. Henry Close, in the Communications to the
Board of Agriculture ; and this application of potatoes in feeding horses has been practised upon a very
large scale by H. C. Curwen, Esq. M. P.

FATTING BEASTS.

Of the food raised on the farm, the best for this business is parsnips ; next carrots; then come cabbages,
potatoes, and turnips. If a farmer has a due provision of these plants, with good hay for cutting into
chaff with straw, he will not find corn and oil-cake
profitable, unless beef promises to be very high, and
corn and cake very cheap. Whatever the food, it
cannot be too often repeated, that small quantities
are to be given at a time ; that troughs, cribs, &c.
are to be kept very clean, and that litter must þe so
plentiful, that the beasts have clean hides, and warm
beds.

BREEDING HORSES.

Our young farmer, on entering his farm, will have
to determine upon the system of keeping mares for
breeding, or using geldings or mares without increase.
As a general question, I should presume he would, if
he be well advised, pursue the latter system. Breeding demands a larger number to be kept, and a management much more careful and attentive, and more
obedient servants, than the more common conduct.
The hazards, also, are not inconsiderable: I have attempted, in various districts, to analyze the benefit
derived

derived from breeding ; but have not, in any, been so
convinced of the profit as to deem it proper to pay any
regular attention to the article in this Calendar.
Where it is the general practice of every farmer, the
servants and labourers acquire a certain degree of skill
and care, very useful in the business, and which may
render it not disadvantageous with tolerable *luck* ; but,
in such cases, there is not much need of Calendars to
remind ; in other situations, I cannot advise a young
farmer to breed ; he will find it more safe and profitable to avoid it.

GARDEN.

A good and well-cultivated kitchen-garden is a capital object to assist in house-keeping.   I have inserted
in the *Annals of Agriculture*, vol. xxxix. pp. 228, 304,
the scheme of a circular one for being kept under the
plough, which may be safely recommended as a plan
well calculated to save expence.

PLANT FRUIT TREES.

This is the proper season for planting fruit-trees.
A good orchard is a valuable article upon every farm,
as they well understand in Kent.   The mode of performing it, and the whole management belong to the
Calendars for gardeners rather than for farmers.

FISH-PONDS.

This is a proper season for making fish-ponds. The
object, in certain situations, may be advantageous to
a farmer who occupies his own land : it is in all agreeable, and a very comfortable assistance to house-
keeping.   The best means of doing it is by a head or
bank across a gentle vale, with a puddled wall in it,
and a sluice at the lowest part.   If the declivity of
the

the land be gentle a-head of six or eight feet depth, will in some cases float many acres with water. The expence, including every thing, may be reckoned at from 1s. to 1s. 6d. per cubical yard of the head. The late Mr. Bakewell thought, that water well managed would pay better than any grass: it must of course depend on the price and demand for fish.

SALT FOR SHEEP.

If the land of a farm be wet or moist, or otherwise unfavourable to sheep, the evil may be considerably remedied by the practice of giving salt in shallow troughs: they should have as much as they will eat, the quantity being very small, though they are exceedingly fond of it, little as they take.

DECEMBER.

DECEMBER.

THRESHING.

THE threshers must be kept constantly at work throughout this month, that the cattle feeding on straw may have a regular supply. Many farmers, who keep large stocks of lean or dry cattle, are attentive to threshing out their worst straw first, and the best last, proceeding upon the same gradation through the winter, that every change of straw may be for the better. This is a just conduct, and cannot fail of having good effects on the cattle, who, it is well known, often fall away in their looks on a change of straw that is the least for the worse. The wheat should, upon these principles, be threshed first, as that makes the worst fodder; next the oats, then the barley, and lastly the barley or oats that had much clover mown with them; for, in wet seasons, the clover rises so high, that the straw is almost as good as hay. There is but little trouble in attending at harvest, to lay the corn so as you may begin with what you please, and the advantages to the straw-fed cattle certainly are great.

The threshers should always be chosen from the labourers with some care : they should be honest, or the farmer will suffer much, if he does not watch them narrowly, as they have many opportunities of stealing corn.

A thresh-

A threshing machine is an object of such importance to every arable farmer, that no intelligent one will be without it.

FARM-YARD.

Attend, without ceasing, to the littering of the yards, stalls, stables, cow-houses, hog-sties, &c. With a little management, all the urine might be preserved: the drains that carry off the overflowings of the yard, should lead to a small well, with a pump fixed in it : this pump should have a light trough, turning on a pivot, to receive the liquid, and a heap of turf or marle be kept within reach of the trough : it should convey the liquid over the whole, which, being carted on to the land, would prove an excellent manure.

PLOUGH UP LAYS.

It is by this time wet enough to begin to break up grass-lands, a work that should not be done while the land is dry : for it will not then turn up in clean, well-cut furrows. Ploughing grass lands is a very good piece of husbandry, when they are worn out and over-run with moss and other rubbish, or hide-bound. To keep land under such unprofitable turf is bad management : it should, by all means, be broken up, and kept in a course of tillage for three or four years, and then laid down again ; by which conduct four times the profit will arise, that could be gained from keeping it in lay.

SHEEP.

This month your forward ewes may be expected to lamb, when you should be attentive to keep them much better than they have been in common through

the

the autumn : they should have plenty of turnips or
cabbages, as fast as they lamb ; for cattle that have
young require as good keeping as those that are fatting ; and if you let them have a rack of hay always
in the field, it will be much the better for them.
Draw the turnips or cabbages, and give them on a
dry grass field.   One great advantage of cabbages
over turnips, is the ease of cutting them, in case of
the hardest frosts, when turnips cannot be had.

In case of extreme bad weather, it will be advise-
able to bring your sheep under shelter. Most farmers are sensible of this, and drive them on such occasions into their hay-stack yard, which is not a bad
way ; but much inferior to giving them their hay in
racks, in a warm yard, with sheds around it for them
to feed under.   The use of such a yard is so great,
that I wonder they are not more common.   In
driving snows, sleet and rain, the injury sheep take
in the open fields is great.   Another circumstance,
which ought to have weight, is the raising plenty
of rich dung : by keeping your sheep in very bad
weather all day, and constantly of nights, in a
yard proportioned to their number, you fold them
perhaps in the most advantageous method of all
others ; for, if a layer of turf or marle be spread
over the bottom of the yard in autumn, and under all
the sheds, and the sheep are kept well littered with
straw, fern, or stubble, so as to be always perfectly
clean and dry, they will in the winter make a great
quantity of excellent manure.

SWINE.

This is the season for making the right profit by
hogs,

hogs, which is their dung. See that all the sows with
pigs are well littered, so as always to be perfectly
clean, with bright, healthy looking skins. Also your
fat hogs should be constantly littered up to their
bellies. If they are not kept perfectly clean, you
may depend on losing money, by not making so much
dung as might be.

FENCES.

Keep the hedgers and ditchers close to work all
this month, so that they may be ready for other
work in the spring. The three first winters of the
lease should get the fences into good order; afterwards divide them into twelve parts, and do one
every year, which will bring the whole to regular
cuttings.

DIG MANURES.

Upon light and very dry soils it will be proper to
keep the marle, chalk, or clay-carts at work : indeed
they should never stop ; for, when a man hires such
soils for improvement, the sooner they receive the
manure, the greater will be his profit : for in some
countries, landlords, after the first lease, either raise
the rents considerably, or turn the tenants out. It is
therefore highly incumbent on them to regain their
expence with profit, within the term of the first lease;
and that can only be done by marling very quickly at
first.

MANURE HOPS

Hops are by many planters manured in this month,
if the season be favourable.

THE LABORATORY

To have permitted, fifty years ago, such an article

as this to form part of a Farmer's Calendar, would have been thought an absurdity; but such an opinion will not, I trust, prevail at present. The intimate connexion between agriculture and chemistry is unquestioned. Let it not, however, be imagined that I propose a farmer should addict himself to a pursuit which is not only very captivating but also very expensive; I would merely have him able to analyze, in a rough way, his soils, and the fossil manures which may be found under them ; for this purpose the apparatus is not formidable : and for a laboratory, if he has a small blacksmith's shop and forge (which no large farm should be without), it will serve the purpose very well It is only providing a cupboard, under lock and key, to secure his glasses and the other articles necessary for these experiments. One caution, however, I must premise; if he has no forge, and converts some other room to this purpose, if it be attached to his house or offices, and a fire (from whatever cause) should consume them, the laboratory would vitiate his insurance at a fire-office, unless he enters it, and pays for the whole as *doubly hazardous,* in the language of insuring.

The apparatus necessary consists of the following particulars : A deal table with a drawer, which drawer should be sufficiently short of the full breadth to allow a range of holes in the upper surface for glass jars to stand, free from danger of falling.

Half a dozen glass cylindrical jars, nine or ten inches long, and three or four inches diameter ; with a few of the same shape, but smaller. One or two others, five or six inches diameter and eighteen inches long:

long : the contents in ounce measures marked on them with a diamond, beginning at top (when they stand inverted, or the mouths downwards) 1,.2, 3, &c. descending. These to receive and measure the air or gas expelled by heat.

Two or three old gun-barrels (the touch-hole closed), cut to the length of eighteen inches, and a small bent tube of iron, or of tin, finishing in an iron screw, for screwing to the end of the barrel already mentioned.

If a forge is not at hand, a cast-iron furnace, nine or ten inches diameter, with a circular hole to receive the gun-barrel, and a moveable dome cover to receive the end of a tin pipe six inches diameter, and 12 feet long, and moveable while up the chimney of the room. The fire to be of charcoal.

A trough or small tub of water, on legs, adapted in height to the elevation of the gun-barrel when in the furnace, with a perforated shelf in it, on which the jars to stand for receiving the air expelled.

A correct pair of scales and weights.

To try whether the gun-barrels or any retorts are really air-tight, an air-pump is very useful, as I have found that blowing in them when under water is not a criterion to be depended on.

An evaporating saucepan ; that is, a tin saucepan with a circular fixed frame of tin, four inches high, to receive a glass jar containing the earth to be dried by the boiling heat of the water, as it is necessary in comparisons for all specimens to be of the same degree of dryness.

Pint or quart phials with ground stoppers of sul-
M m 2					phuric

phuric acid ; muriatic acid ; carbonate of potash ; solution of potash ; ammonia (caustic) ; muriate of ammonia (the common solution of sal ammoniac).
And small phials of the substances mentioned in the
Appendix, as tests for the examination of water.

A few glass funnels ribbed, for filtrating with blot-
ting-paper.—A hydrostatic balance.

The whole of this apparatus may cost from ten to
twenty pounds.

The most material point in examining a soil, and it
is a point in which the authors I have read have
committed great errors, is that of taking the specimen.   I have always crossed a field in several directions, and taken about a tea spoonful in abundance
of places ; suppose an hundred, and thus taking
about a quart, reserved it for trials in glass phials
with ground stoppers.   The under stratum should be
examined, to know if it be retentive, permeable, or
calcareous.

All specimens may be kept a month before trying,
which will enable the farmer to compare various soils
with his own, under every similar circumstance.

In trials with the gun-barrel, he may put one
ounce in it, and then fill up with pounded flint boiled
in muriatic acid, which yields no air or gas.

The experiments should be double; in this dry, and
also in the humid way : upon the latter the following
passage from Dr. Fordyce's Elements of Agriculture,
will explain his method of analysis.

" Take one thousand grains of the dry soil, apply
to it half an ounce of muriatic acid, and four ounces
of water in a glass, stone ware, or porcelain vessel,
                                                  suffi-

sufficiently large; let them stand together till no more effervescence takes place; and if it was very considerable, pour in half an ounce more of the acid; let this stand also till the effervescence ceases; if any arose upon pouring it in, continue to add more acid in the same manner, until what was poured in last, produces little effervescence, which is often at the first, and generally at the second or third half ounce.

After the effervescence has ceased, put the whole in a filter; let the solution filtrate through; pour half a pint of water upon what remains in the filter, let that filtrate also in the same vessel; add to the solution thus filtrated, an ounce and a half of caustic volatile alkali for every ounce of acid used; if any precipitation take place, there is magnesia, earth of alum, or the calx of a metal (generally iron or copper) contained in the soil; after adding the volatile alkali, the whole is to be thrown into a filter again; after the filtration has taken place, pour into the liquor a solution of mild, fixed, vegetable alkali, in water; if there be any calcareous earth in the soil, a precipitation will take place; continue to add the solution of the alkali till no fresh precipitation ensues; throw the whole into a filter; let the liquor filtrate off; pour on by degrees a pint of water; let that filtrate off also; dry what remains in the filter; it is the calcareous earth.

To know the proportion of sand and clay.

Take what remains in the filter after the first solution in the foregoing operation, and by elutriation separate the sand from the clay, dry and weigh them; if there be any pyrites it will appear in the sand.

In

In the above processes, the principal things to be attended to, are,

Whether there be any metallic, or aluminous salts, as these are absolute poisons, and therefore are to be decomposed by quick lime.

Whether there be such a proportion of neutral or earthy salts as to be hurtful, in which case, the solution in *process* (second) will taste salt : a soil containing them in so large a proportion, will hardly ever admit of culture for grain.

Whether there be calcareous earth, and in what proportion, as that ascertains the propriety of applying any manure containing it, and the quantity of that manure.

What the proportion of sand and clay is, which ascertains the propriety of adding sand or clay.

Whether there be pyrites, as that shews why and when a soil will be long of being brought into cultivation.

Pyrites are best destroyed by fallowing, and afterwards applying lime."

When Mr. Professor Davy, of the Board of Agriculture, shall publish his excellent lectures, a more exact analysis will be explained.

If in this method of analyzing a soil the proportion of calcareous earth be large, the trial of the gunbarrel will give a quantity of gas proportioned to such quantity of calcareous earth, for which allowance must be made ; but if that quantity be small, the ounce measures of air or gas gained from an ounce of soil, and its degree of inflammability, will be the easiest test of the fertility of the specimen tried.

To

To aid a young beginner, I shall venture to recommend his reading the papers on this subject in the 6th, 7th, 8th, and 12th volumes of the *Annals of Agriculture*.

Another method of examining soils deserving attention is, by weighing them hydrostatically, as their fertility will generally be in proportion to their specific levity, if the expression may be permitted. Fabroni was, I believe, the first who recommended this test. He gives the following trials :

Various soils, weighed hydrostatically, have given the following result, the barometer being at 27-7, and *his* thermometer at 13 :

1. The fertile soil of a wood,	1,530
2. A kindly soil,	1,582
3. Green marle,	1,591
4. Fertile earth of a friable staple,	2,100
5. Volcanic earth, which does well for vines,	2,111
6. Friable reddish earth,	2,131
7. Strong land for wheat, vetches, &c.	2,160
8. Earth of a mountain, where they cultivate olives, barley, &c.	2,200
9. Sandy sterile land,	2,120

The long evenings of December will give a farmer time for acquiring these branches of chemical knowledge.

FARM ACCOUNTS.

In the month of October this subject was touched upon, but in the greater leisure at present, our farmer may be more likely to be able to give the requisite attention to a point which demands much consideration.

There is not a single step in the life of a farmer that does not prove the advantage of his keeping re-

gular

gular accounts; and yet there is not one in a thousand who does this. This is among the many instances in which the unenlightened situation of the practisers of the art is the evident reason for the backwardness in which the art itself is found by any man who searches for the principles deduced from practice, which ought to give it the regularity of a cultivated science.

A few rough memoranda, or figures, to yield a gross account of the general receipt or payment, are usually the greatest exertions that common farmers, who pretend to keep accounts, make in this line.

The advantages of clear accounts are obvious in every other pursuit in life; and to conduct those of a merchant by the Italian method of double entry, has been made an essential branch of education for the classes intended for commerce. Men engaged in large speculations, who are not regular in their accounts, are always supposed by the prudent part of the world to be in a dangerous situation; nor is there a greater reproach to a merchant, short of actual bankruptcy.

But agriculture is destined to be, in all its detail, an exception to every thing else. Men engage in it without previous education, or even study and inquiry, and they conduct large concerns in it without those accounts known to be necessary in every other pursuit. With the lowest and most uneducated farmers this is pardonable: but what excuse have gentlemen for such a conduct?

It should be remembered that experimental agriculture, or even those ideas more or less detailed which

which we meet with in conversation, must depend for
their justness very much on accuracy of accounts.
For a supposition deduced from general observation
on a farm, and grossly conceived, must fall exceedingly short for correctness, of the regular detail of
exact accounts.

The general fact is, however, admitted; and accordingly it is common to hear gentlemen speak of
their accounts.　But, unfortunately, they are usually
kept in such a manner as to prove rather the means
of fortifying prejudices, than removing errors; all
those questions of nicety, where the contrasts are not
exceedingly strong, relative to the comparative profit
of different soils, of different courses, of different
applications of the same soil, of different modes of
culture, &c. depend on accounts.　Keep your accounts in the mode of one man, grass is more profitable than tillage; keep them in a different method,
and the contrary shall be the result.　The variety in
the mode of keeping these accounts is very great,
even among gentlemen of considerable attention,
carefulness, and accuracy.

This comes from the great and undoubted difficulties which rise in many forms, whenever an attempt is made at positive accuracy.　They are not
imaginary, but real difficulties, and such as will demand a considerable attention to obviate.　I have
reflected on the subject for many years, and they are
few in which I have been satisfied with any approach
towards accuracy.　For while huore are distinctions
which must every where be kept up, there are many
minutiæ that must be sacrificed, in order to render
the

the account tolerably easy to keep, without an attention that a man in an active line of life cannot give.
To keep to this medium is the great difficulty.

The nature of the farm must, in some instances,
regulate the mode of the accounts.   Suppose a man
has the evil of an open field one, with scraps and bits
of land scattered amongst his neighbours : in such a
case it is imposssible for him to keep an account for
every field ; and yet this is one of the most indispensable points that in general must be adhered to ; for
he who does not know what every field has paid him,
is deficient in the very foundation of expetience.   In
this light all little fields on a large farm are nuisances:
they derange accounts entirely, if the greatest attention be not paid, and they are as inconvenient in cultivation, and attended with as much loss in headlands,
and borders, as they are ruinous to any exactness of
account.

But as many persons keep accounts without attending to this point, I would observe, that when all the
wheat, all the barley, all the oats, &c. are respectively
thrown together, some very essential objects of experience depend on guesses, which ought to be ascertained correctly.   Has fallow, or clover, or beans,
paid best, as preparations for wheat ?   How is that
question to be answered, if all are huddled together in
one barn or stack, and meet in the same account ? *The
farmer can guess nearly.* He may : but go to a chemist,
ask him if his science was pushed to the present perfection by accepting such guesses, instead of experiment ?
besides, they are in their nature quite uncertain ; and
when a comparison is formed by two guesses, a very
                                                        little

little error in each will amount to so much in both, as to overturn all authority. Another point is, a man's guess being influenced by a favourite theory: a rigid friend to fallows, when he draws, by guess, a comparison between them and beans, will be apt, in the nature of things, to be partial: he should not put himself in the situation: he who would abhor the idea of falsifying a fact that is before him, might guess, at least, without sufficient accuracy.

If the fields be not very small, the inconvenience of keeping crops separate is little. Stacking corn is better understood and executed in the Isle of Wight, than in most other parts of the kingdom: a great stack is rarely seen there: a farmer who has 500 acres of corn has only small ones. With such, accounts are kept separate with great facility. At least, if there be difficulties in it, there are others we shall meet with abundantly greater.

To sow one field with several crops at the same time, part wheat, part clover, &c. is very bad and inconvenient management, and ought to be avoided, were accounts out of the question. If they cannot be shunned, these must necessarily be more complex.

The first object in keeping accounts is to ascertain the expences, in order to divide them accordingly.

Rent, Tithe, and Parish Taxes.—These articles demand three accounts, to be kept separate; but they are all to be arranged on the same principle. The amount of the two last, when known, which is at the end of the year, must, like the rent, be divided over every field for which an account is kept: this is

very

very easy, when the measure of the fields is known. I need not observe, that the farmer, in dividing the rent, should do it as exactly and as fairly as possible, and that the two other articles should be proportioned to the rent.

But here occurs one difficulty, which is, I confess, puzzling : it is the difference between the gross and the neat measure of the fields of an inclosed farm. The hedges, ditches, and borders, take up, in many farms, a considerable portion of the field ; from one-eighth to one-twelfth, and in some, even more * ; now if these be reckoned and accounted for as a part of the field, then the acreable produce is affected, and even the profit of the husbandry, by a circumstance not essentially connected with it ; and if two fields be compared in their husbandry, that may be most advantageous which has least border, and for that reason, which would derange a comparison entirely. I know but one method of getting rid of this difficulty, which is to measure the neat contents where the plough goes in an arable field, and where the scythes cut in a grass one, and then, deducting the total of those measures from the gross contents of the farm, throw the difference into one account by itself, under the title of fences and borders, to which account must be charged the proportion of rent, tithe, and parish taxes. If wood be cut or grubbed from these borders, or grass mown from them, the value of the wood or hay to be credited. The expence of the fences to be charged, and the balance of the whole

---

* Margins of grass are common round the fields in Suffolk.

for

for it may every where be expected to prove a losing
account, considered as the expence of *fences*, and
acreably divided over the whole farm, like rent, tithe,
or parish taxes. The only person who ever had an
attention to this accuracy was Mr. Baker, the experi-
menter to the Dublin Society. He published a map
of his farm, with the gross and neat contents of every
field. For want of observing the precaution, many
experiments have been made, and many conclusions
drawn, which are mere errors.

Sundry expences may be the title of an account
which must have place on every farm. Whatever
payments concern the farm in general, and not any
field or object in particular, and is not included in the
preceding articles, must be entered under this title.
Instances are : a bailiff's salary ; payments to rat or
mole catcher ; mending roads ; expences at markets,
&c.

Wear and Tear includes all payments to blacksmith,
carpenter, wheelwright, harness-maker, &c. But in
the division of this article, there must be a variation
from the preceding ; they are divided over the whole
farm, but these must be proportioned differently :
the arable lands will absorb the greatest part of these
expences ; mowing grass, very little ; and feeding-
ground still less. But to avoid any arbitrary estimation when a rule can be established, the proper mode
of dividing this expence per acre, will be by making
the expence of the teams a rule for it : to find how
much per cent. or in the pound, of the team-account,
this expence of wear and tear amounts to, and charge
it accordingly.

The

The *team* account is that which is in general more mistaken than any other on a farm. Nothing is more common, than every day to see accounts in which ploughing is charged at 4s. an acre, or at 5s. or at 10s. or whatever may be the hiring price of the country : but few words are necessary to shew that this is entirely fallacious : it is probably much under the real expence. Every practical farmer must know, that the way to have cheap tillage is to keep the teams well employed : when a man's own work is done, his team stands still if he do not employ it for his neighbours ; to do which, he will work for them below the value, and yet find some advantage in it. In consequence of such a conduct being common, to say that such is the price of tillage, can never be accurate. It has by no means that best accuracy of *price* ; because you cannot buy your commodity when you want it ; and he who depended on the market for all the work of his farm, would soon find the state of his fields calling for a very different system.

The means of ascertaining the real expence of all team-work is very obvious, but depends totally and absolutely on accurate accounts. So much per week in summer for their green food : so much hay and oats eaten ; so much for shoeing and farrier ; so much for the actual decline of value ; and so much in labour for attendance, give the real expence of the team. In order to divide this total expence among the work executed, a day-book is necessary : which a man may keep himself, or trust to his bailiff, as he pleases ; it must contain the work of the teams and men every day in the year, specifying the field or business they are

are employed in.   At the end of the year the amount of
expence is proportionally divided among the work,
and the clearest truth and correctness are necessarily
the result.

I ought to observe, that this accuracy is very desirable for ascertaining various circumstances.   The
comparative profit of grass and arable land depends
much on it.   Some persons, from too lightly estimating the expence of teams, think arable the most
profitable; and others, whose calculation of those
charges runs perhaps too high, give too much in the
counter opinion.   I can easily conceive, that many
strenuous advocates for fallows might lose a little of
their warmth, if they knew what the expence of
ploughing an acre of land really was on their farms.
Such instances might be multiplied : they are indeed
obvious to every man capable of uniting the theory
with the practice of a business.

The article of manure is much more complex, and,
upon the whole, the most difficult account there is
for a farmer to keep.   It must be arranged under the
title *Farm-yard*; and it connects with so many objects,
that no little care is necessary to keep it ; and with
the greatest attention some doubts will still remain.

Suppose the system to be that of carting a stratum
of marle over the yard before foddering begins : that
expence is to be ascertained at once without any difficulty ; but how is the straw to be charged ? Cattle
may be put out to straw in this country at 1s.
or 1s. 6d. per week.   At these prices a ton will pay
about

about 7s. or 8s. ; but, while the cattle may be thus supported, the farmer may buy straw, with a view to the dung, at 20s. or 30s. a ton. This contrast is difficult to settle. The price per week is arbitrary, though actual : men take them at those rates, because they have none, or not enough, of their own ; and it is not ascertained what value cattle will really pay for the straw ; which may be more, or may be less. The whole is uncertain.

But with the straw of one's own crop, there is a double difficulty ; because there must be two valuations instead of one. We must reckon so much an acre, or load, for it, and so much a week for the cattle that eat it ; but both suppositions. Among counter objections, we must chuse the least. The best method, perhaps, is to charge the *farm-yard* account with the price of the straw, at which it could be sold, deducting the expence of carrying it out ; and to credit the same account with the price per week of keeping the cattle; which price is charged to the debtor side of the *cattle* account, as a part of the expences of keeping them. Whatever labour is bestowed on the dung, in shovelling and cleaning yards, throwing up the urine, turning over, &c. is charged of course to it. When the whole is carted on the land, the total expence is divided by the number of cubic yards, and the price per yard ascertained. It is charged to the account of the fields on which it is spread; and though the whole advantage is by no means exhausted

by

by one crop, yet the whole expence must be charged to the crop that receives it, or the accounts would be kept open so long as to create confusion.

The time of balancing the books every year, should be that of entering the farm : this is most usual at Michaelmas ; but the crop of the year is not then disposed of : to avoid valuations, which ought never to be relied on, when certainty can possibly be gained, the old year's accounts are to be kept open long after the new year ones are begun ; that is, till the corn is all threshed and sold, till the fatting beasts are gone, and till all those circumstances are decided which relate to the preceding year. This is essential to exact accounts, and can by no means be dispensed with. In this case, valuations may be nearly rejected, but there are others in which no management can exclude them : these are, in *live stock* not bought and sold within the year ; and *implements*.

A man may stock his farm with cows at 10l. each ; but if he suppose them, some years after, to be worth the same sum, he will grossly deceive himself. He must value them every year, and also the young stock which he rears with a view to keep up the number, or for sale: and the rule by which he should make the valuation, ought to be the price they would sell for at the moment. The same management must direct him with succession beasts, bought or bred for fatten-ing ; and also with a flock of sheep. On which last head I must observe, that the want of keeping such accounts as I am describing, is alone the reason for a difference relating to the profit of sheep. Can any thing be a clearer proof of the barbarity of accounts as they are kept at present by flock-masters, than the surprizing question once in agitation among them, whether they gain or lose by their flocks : a question that has arisen from Mr. Macro's paper on that subject, published in the *Annals of Agriculture.* Such uncertainty could not obtain, if farmers kept regular accounts. The description of the profits of a flock not being properly a calculation, but an account, it ought to be transcribed from a man's private books ; unfortunately, they are kept in such a manner, that difficulties multiply at every step in the endeavour to understand them.

Implements must all be valued every year, and the balance, being the expence, carried to the wear and tear account, of which it makes a part.

One of the most complex and difficult accounts, if not the most so of all, is that of grass-lands, fed. It involves itself with cattle of all kinds, with hay, with the team, &c. ; and in such a manner as to make an accurate separation very difficult. How is the value of the food to be calculated ? If 3s. a week for a cow or a bullock, or 6d. for a sheep be charged, it is merely arbitrary : such estimates are fallacious. They imply profits, but allow nothing for losses. On the other hand, if the actual profit or loss on the live stock be made the product ; in that case, the grass land must be made a mere cattle account : there are obvious objections to this ; but it is, upon the whole, less objectionable than a valuation per week, which must, in the nature of it, lead to error. On this principle, the account may be kept in the following manner :

One account opened for *mowing ground,* to which the

the rent, tithe, taxes, and all expences, in one total for every field mown, are carried : and the credit of it to consist of the value at the market price (carriage deducted) of the hay mown, as delivered to the team, fatting beasts, cows, sheep, &c. which several accounts are debited with their respective consumptions.

But the fields which are mown have also an aftergrass, which is fed; the account of the week's stock which are supported by it, ascertains the value in the manner presently to be mentioned.

The account of *feeding ground* comes next : all the total debits of the fielde must be carried to it. The credit side to consist of the food of the teams charged at the price per week, suppose 3s. 6d., and of that of any cattle taken in to joist. These articles may be arranged ; but those which result from profit on stock kept are not so easy.

There is farther, a *sheep* account; a *dairy* one; and another for *fatting beasts*. In these are to be charged all the expences peculiar to those articles : shepherd's wages ; market expences, &c. to the sheep : fuel, straw-yard, &c. to the cows : and the purchase money of lean stock to the fatting beasts.

Further : the fatting beasts are put to turnips ; the cows have turnips, the sheep have turnips ; how is this to be accounted for ? It creates a new difficulty; but we must examine the best mode of clearing it.

If the cattle account be charged with the prime cost of the turnips, that is, with the expence of cultivating them, it will by no means be fair, for the expence is usually greater than the value; and a man may in a turnip country buy them cheaper than he can cultivate them: he submits in some cases to a known loss, because he knows he shall be more than repaid in the barley that follows; but to transfer this loss to the cattle would be unfair. One way of proceeding is, to value the turnips at what they would sell for, and to debit the cattle accounts with their respective consumptions. But there are two prices of turnips; one, for carrying the crop on to another man's land; the other, for eating them in the field. The latter ought to be the rate chosen on this occasion, charging the cattle with the labour and expences of carriage. But the actual profit is a better guide.

Here, therefore, at the end of the year, five or six, or more, unsettled accounts are open, not one of which can be closed but by reference to each other. Hence arises the great complexity of the farmer's accounts; but, amidst this apparent confusion, order must be made to arise, or our labour is vain.

The reader will see that the main question on which this arrangement depends, is this—shall the profit or loss on live stock be assigned *to the stock,* by a valuation per week; or, *to the land they feed on* by a division per week of the actual profit or loss arising?

Suppose that two hundred pounds profit would be the balance of the sheep account if food be not charged; shall this sum stand as *profit,* and the fields fed charged necessarily with *loss*; or shall that balance be distributed proportionably among the fields which have supported the flock? The balance of the account, 200l. amounts to 6d. per head per week for 52 weeks. They have been fed 15 weeks in *grass-lands fed,* 10 in in *grass-lands mown,* 12 in *Great Staines* (a clover crop), 5 in *Ardera* (ruta baga), and 10 *Jermyn* (turnips). It is easy to divide the total among them ; and if he wishes further accuracy, he may vary the price per week, according to the scarcity or real importance of the several sorts of food : but still keeping to the real total. This method of dividing the profit among the crops is far preferable to assigning the 200l. as profit to sheep.

This remark applies equally to all the other live stock accounts.

The farmer sees clearly what he makes by the different kind of stock, by turning to their respective accounts ; but none of them appear in the profit and loss account ; there they are absorbed in the accounts of the distinct arable fields which produced food; and in the two others of *grass fed* and *grass mown* ; or in the two last thrown together in one of *grass-land.*

That there is a complexity in this mode of arranging the accounts of live stock, is beyond doubt ; but after the greatest attention that I have been able to give it, I see no mode of simplifying it. Submit to the rules here laid down, and you have the satisfaction of all the accuracy that is attainable ; but in any other method it will remain unknown, whether the profit or the loss belong to the land, or to the stock that feeds upon it.

I am clear this method will be rejected by those who only read this paper in a common, transitory manner, without studiously examining all the points on which the arrangement depends ; but, to such as will reflect on what they read, and give the due atten-

tion,

tion, I have little doubt but the method will appear satisfactory.

When so much profit is actually made, to divide it by a weekly account to the fields that fed the stock, is making an easy calculation, with full data before you : but to charge the stock with so much per week for feeding certain fields, when you do not know but the account of stock may be loss, not profit, is calculating without any better data than mere supposition.

Such are, I apprehend, the principal difficulties in keeping the accounts of a farm. I do not offer the mode as one that obviates all objections.   I do not conceive it possible to obviate all : but I think that fewer sources of inaccuracy will be found in it than in any other.

WOODS.

The woodmen are at work through this month.   In Worcestershire, &c. the sale of woods is very easy. " Those belonging to the Earl of Coventry are extensive, and are divided into fourteen equal parts, one fourteenth of which is annually felled ; this fourteenth is again subdivided into small parcels or lots of 40 yards by 20 ; which subdivisions are made by cutting right lines through the wood, just wide enough to admit a person to pass, who examines and values each parcel according to its growth and quality, numbering the lots in a book prepared for the purpose, with the price affixed to each ; this being done, a day is appointed for the sale, and persons chiefly of the neighbouring villages attend to purchase ; amongst whom, the poor form no inconsiderable part, and for whose accommodation the wood is thus divided.

There

There is one circumstance attending the sale, which, to a person unacquainted with the nature of the business, may appear extraordinary; that of disposing of the wood, and receiving the money, without the purchaser having any knowledge of the lot he pays for; this is done to prevent any dispute that might arise from several persons fixing upon the same lot. The purchaser describing to what uses he intends converting his wood, is placed by the person who disposes of the parcels, in that part which appears most suitable to his purposes, and the woodward having the name of each person prefixed to his lot, is prevented from making mistakes; this entirely answers the desired intent, and murmurings are seldom heard at succeeding sales : sometimes in the month of January the woodward begins cutting the underwood, taking care to leave a sufficient quantity of young thriving plants, either oak or ash, for the purpose of preserving a succession in the stock : the wood consists of two sorts of stores, which are called tellers ; the oldest are called black barks, and are of 42 years growth ; these are felled in the barking season, for the purpose of procuring the bark, and are then carried off with the faggots by the purchaser of the lot ; the next are called white barks, and are 28 years growth, and remain standing for stores, with a proper quantity of 14 years growth, till the wood is again felled.

Some of the woods in Herefordshire are stored in the same way, and some are felled at 20 years growth; some are cut at 15, when the wood is completely felled, and the poles used as hop-poles, which in that county is deemed a profitable mode; those of 20 years are regularly stored, and the poles converted into hoops, spokes, lath, hurdles, cord, wood for charcoal, and various other purposes.

Twining.   *Wm. Phelps.*"

STRAW-FED CATTLE.

" I met with an idea that cattle may be satiated with straw ; or, in other words, may be served with it in too great plenty. It has been observed, that after a dry summer, when straw is scarce, and the cattle have it dealt out to them regularly, they do better than when, after a plentiful year, it is thrown before them in profusion from the threshing floor ; not through the superior quality of the straw in a scarce year, as these effects have been observed to be produced from the same straw. This subject is by no means uninteresting to those who winter large quantities of cattle ; I have observed in Yorkshire, where cattle are kept tied up, and of course are regularly fed, that they in general do better at straw, than cattle in the south of England, where they go loose among a much greater plenty ; but whether it proceed from the warmth, from their resting better, from the breed of cattle, or from their being regularly fed and *eating with an appetite*, I will not pretend to decide."—*Marshall.*

LITTER FARM-YARD, &c.

Littering all sorts of cattle, &c. is never to be omitted at this season. The quantity of manure made is an essential object : the following notes will shew certain proportions of dung to straw.

Mr. Moody.—Forty-five fat oxen, in fatting, littered with

with 20 waggon loads of stubble, raised 200 loads, each three tons, of rotten dung, worth 7s. 6d. a load.

Every load of hay and litter given to beasts fatting on oil-cake, yield seven loads of dung, each $1\frac{1}{2}$ ton, exclusive of the weight of the cake.

On a comparison between the oil-cake dung and common farm-yard dung, 12 loads an acre of the former much exceeded 24 of the latter.

Mr. Arbuthnot.—One hundred and thirty-four sheep and thirty lambs, penned six weeks in a standing fold, and littered with five loads and 40 truss of straw, made 28 large loads of dung. Fed morning and evening in the fold with turnips. Ate two acres of turnips.

					£.		
Value, dung,	-	-	-	-	£. 10	0	0
Straw, at 20s.	-	-	-	‸	5	15	0
			Profit,	-	£. 4	5	0
Per acre for turnips,	-	-	-		£. 2	2	6
And per score per week,	-		-		£: 0	1	$9\frac{1}{4}$

William White.—Thirty-six cows and four horses tied up, ate 50 tons of hay, and had 20 acres of straw for litter : they made 200 loads of dung, in rotten order for the land.

The experiments of Mr. Moody and Mr. Arbuthnot prove how well it answers to buy litter with a view to the dung : in feeding oxen with oil-cake, one load of straw makes seven of dung, each $1\frac{1}{2}$ ton; and with feeding sheep with turnips, one trussed load made more than four and a half large loads, worth 7s. 6d. each.　With Mr. White, 20 acres of straw,

suppose

suppose 30 loads, made 200 of rotten dung in littering cows, which are six and a half for one?
whence it appears, that litter may safely be purchased
at a very high price, rather than be without it.   An argument which should be convincing with those
who have it in their wheat stubbles, and yet will
not be at the trouble of chopping and carting it
home.

POULTRY.

Throughout this month poultry is on full sale.   I
made the following memoranda at Mr. Boys's in
Kent :

Mrs. Boys, who is as intelligent in her walk of
management as her husband is in his, conducts her
poultry with greater success than any person I have
met with.   While I was at Betshanger, a higler's cart
carried off above 12 dozen of fowls for one draft :
inquiring what could be the process that commanded
such plenty, I found it so simple as to be explained
in a moment—the labourers wives and families who
live on Mr. Boys's farm, do the whole ; he supplies
them with what offal corn is necessary, and they return Mrs. Boys the grown fowls, ready for market, at
3d. each, 6d. for turkies and geese, and 3d. for ducks ;
and her account, well kept, states a profit of 20l. a
year, after all expences are paid, and the family well
supplied ; have also all the eggs without any payment.
It answers as well to the people as it does to the farmer.   A fat turkey, 21lb. alive, is 14lb dead   The
climate and soil here both agree with poultry   for
here is a farmer, of the name of Kelly, who rears
and sells 140 turkies per annum.

FATTING

FATTING BEASTS

Demand constant attention, as already so often noted ; the effect of acid food in fatting swine has been long well understood, and it is remarkable that I found it applied in France to fatting bullocks.

To fatten a pair of good oxen at la Ville Aubrun, would take 45 cart loads of *raves,* a sort of turnip, cut in pieces, and 20 quintals of hay : when the raves are done, they give the flour of rye or other corn, with water enough added to form a paste ; this they leave four or five days to become sour, and then they dilute it with water, thicken it with cut· chaff, and give it to the oxen thrice a day : when fed with raves the oxen do not want to drink.

At Bassie they finish with flour of rye, mixed as before mentioned : they assert that the oxen like it the better for being sour, and that it answers better in fatting them.   They eat about a boisseau a day (weighing 22lb.) and never give this acid liquor without chopped hay.   It is proper here to remark, that in coming from Paris, we have met a great many droves of these oxen, to the amount, I guess, of from 12 to 1500, and that they were, with few exceptions, very fat ; and considering the season, May, the most difficult of the year, they were fatter than oxen are commonly seen in England in the spring. I handled many scores of them, and found them an excellent breed, and very well fattened.

Limoges.—After the raves, give rye paste, as described above, but with the addition of a *leven (levain)* to the paste, to quicken the fermentation, and make it quite sour : at first the oxen will not drink it, but they

they are starved to it ; usually take it the second day, and after they have begun, like it much, and never leave a drop.

Usarch.—Fatten their oxen with raves, as above, and then with rye flour, made into a paste with leven and given sour, as before described.

Between Brive and Cressensac they fatten with maize, but, in order to render it tender, pour boiling water upon it, cover it up close, and give it to the cattle the same day ; and, in this method, it is a most excellent fattener, both of oxen and poultry. But, in order to make them fatten sooner and better, they give them, every night, and sometimes of a morning, a ball of pork grease, as large as an apple : they say this is both physic and food, and makes them thrive the better.

The fact of hog's grease being given was confirmed at Souilliac : it is given to increase the appetite, and answers so well, that the beasts perfectly devour their food after it, and their coats become smooth and shining. The most fattening food they know for a bullock is walnut oil-cake. All here give salt plentifully, both to cattle and sheep, being but 1d. per lb. But this practice is, more or less, universal through the whole kingdom.

In Flanders, from Valenciennes to Orchies, for fattening beasts and for cows, they dissolve linseed cake in hot water, and the animal drinks, not eats it, having various other food given at the same time, as hay, bran, &c. ; for there is no point they adhere to more than always to give variety of foods to a fattening beast.

APPENDIX.

APPENDIX.

USEFUL TABLES.

No. I.

Equality in the Weight of Cattle, between Scores, Stones, and Hundred-Weights. By the Right Honourable the Lord Somerville.

Scores.	Stones, at 14lb.		Stones, at 8lb.		Hundred, 112lb.		
	St.	lb.	St.	lb.	Cwt.	qrs.	lb.
20 equal	28	8	50	0	3	2	8
25 —	35	10	62	4	4	1	24
30 —	42	12	75	0	5	1	12
35 —	50	0	87	4	6	1	0
40 —	57	2	100	0	7	0	16
45 —	64	4	112	4	8	0	4
50 —	71	6	125	0	8	3	20
55 —	78	8	137	4	9	3	8
60 —	85	10	150	0	10	2	24
65 —	92	12	162	4	11	2	12
70 —	100	0	175	0	12	2	0
75 —	107	2	187	4	13	1	16
80 —	114	4	200	0	14	1	4

No. II.

SALTS. BY DR. FORDYCE.

Vitriolic acid unites with,

    1. Fixed vegetable alkali, forming vitriolated tartar.

    2. Fixed fossil alkali, forming Glauber's salts.

    3. Calcareous earth, forming selenite.

    4. Magnesia, forming magnesian Glauber's salts.

    5. Clay or earth of alum, forming alum.

Nitrous acid unites with,

    1. Fixed vegetable alkali, forming nitre.

    2. Volatile alkali, forming nitrous ammoniac.

    3. Calcareous earth.

    4. Magnesia.

Muriatic acid unites with,

    1. Fixed fossil alkali, forming sea-salt.

    2. Volatile alkali, forming sal ammoniac.

Muriatic

Muriatic acid, unites with,

    3. Calcareous earth, forming fixed ammoniac

    4. Magnesia.

    5. Earth of alum.

Fixed vegetable alkali,

Combined with air
- Salt of tartar.
- Ditto wormwood.
- Pot-ash.
- Pearl-ash.
- Fixed nitre.

Free from air -
- Caustic fixed volatile alkali.
- Common caustic.
- Soap leys.

——————— unites with,

    1. Vitriolic acid, forming vitriolic tartar.

    2. Nitric acid, forming common nitre.

    3. Muriatic acid, forming digested salt of Sylvius.

        When caustic, it dissolves,

           1. Oil, forming soap.

           2. Animal and vegetable substances.

Fixed fossil alkali,

Combined with air
- Natron.
- Soda.
- Sal soda.
- Barilla.
- Kelp.

Free from air -
- Soap leys.
- Caustic fixed fossil alkali

——————— unites with,

    1. Vitriolic acid, forming Glauber's salts.

    2. Muriatic acid, forming common sea salt.

        When caustic, it dissolves,

           1. Oils, forming Castile soap.

           2. Animal and vegetable substances.

Volatile alkali, unites with,

    1. Nitric acid, forming nitrated ammoniac.

    2. Muriatic acid, forming common sal ammoniac.

    3. Phosphoric acid.

No. II

No. III.

COMPARISON OF THERMOMETERS.

REAUM.	FAHRN.	REAUM.	FAHRN.	REAUM.	FAHRN.
0 —	32	15 —	66	30 —	$99\frac{3}{4}$
1 —	$34\frac{1}{2}$	16 —	$68\frac{1}{4}$	31 —	102
2 —	$36\frac{1}{2}$	17 —	$70\frac{1}{2}$	32 —	$104\frac{1}{4}$
3 —	$38\frac{3}{4}$	18 —	$72\frac{3}{4}$	33 —	$106\frac{1}{2}$
4 —	41	19 —	75	34 —	$108\frac{3}{4}$
5 —	$43\frac{1}{4}$	20 —	$77\frac{1}{4}$	35 —	111
6 —	$45\frac{1}{2}$	21 —	$79\frac{1}{2}$	36 —	$113\frac{1}{4}$
7 —	$47\frac{3}{4}$	22 —	$81\frac{3}{4}$	37 —	$115\frac{1}{2}$
8 —	50	23 —	84	38 —	$117\frac{3}{4}$
9 —	$52\frac{1}{4}$	24 —	$86\frac{1}{4}$	39 —	120
10 —	$54\frac{1}{2}$	25 —	$88\frac{1}{2}$	40 —	$122\frac{1}{4}$
11 —	$56\frac{3}{4}$	26 —	$90\frac{3}{4}$	41 —	$124\frac{1}{2}$
12 —	$59\frac{1}{4}$	27 —	93	42 —	$126\frac{3}{4}$
13 —	$61\frac{1}{2}$	28 —	$95\frac{1}{4}$	43 —	129
14 —	$63\frac{3}{4}$	29 —	$97\frac{1}{2}$	44 —	$131\frac{1}{4}$

No. IV.

COMPARISON OF ACRES—FROM PAUCTON.

France arpent 100 perches 22 pieds, - - 1,0000

Paris 100 ditto, - - - - 0,6694

England acre, - - - - 0,7929

Ancona rubbio, - - - - 2,541

Bergame pertica, - - - 0,12867

Bologna tornatura, - - - 0,3947

Brescia piò, - - - - 0,6381

Calenberg acre, - - - 0,5165

Crema pertica, - - - - 0,14812

Cremona pertica, - - 0,15608

Denmark tonde-hart-korn, - - 2,159

Spain fanega, - - - - 0,6720

Ferrara biolca, - - - - 1,2614

Florence stioro, - - - - 0,11461

Francfort-on-the-Maine acre, - - 0,3955

Gotha acre, - - - - 0,3967

Inspruck janch, - - - - 0,8472

Livurnia stiora, - - - - 0,1094

Mantua biolca, - - - - 0,6059

Milan pertica, - - - - 0,1472

Modena biolca, - - - - 0,8169

Muscovy

Muscovy décétine,	-	-	-	-	2,907
Naples moggio,	-	-	-	-	0,6546
Padua campo,	-	-	-	-	1,0866
Parma biolca,	-	-	-	-	0,5967
Placentia pertica,	-	-	-		0,1494
Rhine arpent,	-	-	-	-	0,3336
—— morgen Rhinland,	-	-	-		1,668
Rome quartuccio,	-	-	-		0,11308
—— rubbio,	-	-	-	-	3,619
—— pezzo,	-	-	-	-	0,5170
Rovigo campo,	-	-	-	-	1,2597
Saxony morgen,	-	-	-	-	1,0542
Trente pio,	-	-	-	-	0,6810
Trevisa campo,	-	-	-	-	1,0201
Turin giornata,	-	-	-	-	0,7440
Venice passo quadrato,	-	-	-	-	0,000588
Verona vaneza,	-	-	-	-	0,02454
—— campo,	-	-	-	-	0,5889
Vicenzia campo,	-	-	-	-	0,7100
Zurich zuchart,	-	-	-	-	0,4883

No. V.
WEIGHT.

France livre poids de marc,	-	-	-	1,000
—— quintal,	-	-	-	100,0
—— tonneau,	-	-	-	2000
England pound Troy,	-	-	-	0,7618
—— pound avoirdupois,	-	-	-	0,9264
—— stone, 14 lb.	-	-	-	12,970
—— hundred, 112 lb.	-	-		103,76
Germany pound,	-	-	-	0,7320
—— marc of Cologne,	-	-		0,4777
Amsterdam pound of 2 marcs,	-	-		1,0046
—— stone,	-	-	-	10,046
Barcelona pound,	-	-	-	0,6278
Bremen pound,	-	-	-	1,0043
—— stone,	-	-	-	10,043

No. VI.

No. VI.

Measures of Length of several Countries.

	Inch.	Decim.
English foot, - - - -	12	
Paris foot, - - - - -	12	816
Venetian foot, - - -	13	944
Rhinland foot, - - -	12	396
Strasburgh foot, - - -	11	424
Norimbergh foot, - -	12	
Dantzic foot, - - - - -	11	328
Danish foot, - - - -	12	504
Swedish foot, - - - - -	11	733
Derahor cubit of Cairo, - - -	21	888
Persian arish, - - - -	38	364
Greater Turkish pike, - - -	26	4
Lesser ditto, - - - -	25	572
Braccio at Florence, - - -	22	956
——— ditto, for woollen, at Sienna, - -	14	904
——— ditto, for linen, at ditto, - -	23	688
Vera at Almaria and Gibraltar, - - -	33	12
Palmo di Archetti at Rome, - -	8	784
Canna di Arcetti, - - -	87	84
Palmo di Braccio Mercantia - -	8	346
Genoa palm, - - - -	9	78
Bolognian foot, - - - -	15	
Antwerp ell, - - - -	27	396
Amsterdam ditto, - - - -	27	216
Leyden ditto, - - - -	27	12
Paris drapers' ditto, - - -	47	148
Ditto mercers' ditto, - - -	47	244
Roman foot, - - - -	11	604
Greek foot, - - - -	12	0875

No. VII.

No. VII.

SEVERAL USEFUL TABLES FOR PLANTERS.

A Table, to shew how many Plants, or Trees, may be planted on an Acre of Land at different Distances.

In an acre are,

4 Roods, each rood 40 rods, poles, or perches.

160 Rods, 16 feet and an half each.

4840 Square yards, 9 feet each.

43560 Square feet, 144 inches.

174240 Squares of 6 inches each, 36 inches.

6272640 Inches, or squares of 1 inch each.

A Table, to shew many Plants may be raised on a Rod of Land at different Distances.

In a rod are $272\frac{1}{4}$ square feet, or 39,204 square inches: a rod will contain,

Trees or plants.					Number of inches asunder.			Square feet to each.
2450 and	4 inches over.		-		4 by	4	-	16
1960	-	-		-	5 —	4	-	20
1633 and	12 over.		-	-	6 —	4	-	24
1089		-		-	6 —	6	-	36
816 —	36 over.		-	-	8 —	6	-	48
612 —	36		-	-	8 —	8	-	64
490 —	4	-		-	10 —	8	-	80
392 —	4	-		-	10 —	10	-	100
272 —	36	-		-	12 —	12	-	144
261 —	54	-		-	15 —	10	-	150

An acre will contain,

108 and 360 feet over, at		-	20	feet asunder, or		400
160	-	-	$16\frac{1}{2}$		-		$272\frac{1}{4}$
134 — 144 feet over		-	18		-		324
302 — 72		-		-	12	-	144
435 — 60		-		-	10	-	100
680 — 40		-		-	8	-	64
888 — 48		-		-	7	-	49
1089	-		-	8 by 5		-	40
1210	-		-	6		- -	36

Trees

Trees or plants.					Number of inches asunder.		Square feet to each.
1361 and	8	-	-		8 by 4	-	32
1452	-	-	-		6 — 5		30
1555 —	20	-	-	-	7 — 4	-	28
1815	-	-	-	-	6 — 4	-	24
2178	-	-	-	-	5 — 4	-	20
2722 —	8	-	-	-	4 — 4	-	16
2904	-	-	-	-	5 — 3	-	15
3630	-	-	-	-	4 — 3	-	12
4840	-	-	-	-	3 — 3	-	9
5445	-	-	-	-	4 — 2	-	8
7261	-	-	-	-	3 — 2	-	6
8712	-	-	-	-	$2\frac{1}{2}$— 2	-	5
10890	-	-	-	-	2 — 2	-	4
19305	-	-	-	-	$1\frac{1}{2}$— $1\frac{1}{2}$		$2\frac{1}{4}$
21780	-	-	-	-	2 — 1	-	2
43560	-	-	-		1	-	1

A Table for the more readily calculating the Value of several Crops on an Acre of Land.

			£.	s.	d.
19360 plants, at - $0\frac{1}{4}$d. each. ⎫					
9680 - $0\frac{1}{2}$d. ⎪					
4840 - 1d. ⎪					
2420 - 2d. ⎬ -			20	13	4
1210 - 4d. ⎪					
605 - 8d. ⎭					
7000 plants, at 2d. each,	-	-	62	6	8
5200 ditto,	-	-	43	6	8
2200 ditto,	-	-	18	6	8
9980 plants, ⎫	-		40	6	8
6970 ⎪	-		31	0	10
6534 ⎪	-		27	4	6
5400 ⎪	-		22	15	0
5445 ⎬ at 1d. each,			22	13	9
4356 ⎪	-		18	3	0
3630 ⎪	-		15	2	6
1000 ⎪	-		4	3	4
160 ⎭	-		0	13	4
15000 plants ⎫	-		31	5	0
7000 ⎪	-		15	11	8
6534 ⎬ at $1\frac{1}{4}$d. each,			13	12	3
6660 ⎪	-		13	17	8
5000 ⎭	-		10	8	4

A Table

A Table of the Specific Gravity of several sorts of Wood.

	Specific Gravity.				Weight of a Cube Foot.	
					lb.	oz.
Thorn,	87	-	-	-	54	6
Crab-tree,	85	-	-	-	53	2
Quince-tree,	83	-	-	-	51	14
Mahogany,	82	-	-	-	51	4
Plum-tree,	80	-	-	-	50	0
Holly,	80	-	-	-	50	0
Ash,	76	-	-	-	47	8
Barberry,	76	-	-	-	47	8
Nut-tree,	76	-	-	-	47	8
English oak,	75	-	-	-	46	14
Beech,	74	-	-	-	46	4
Elder,	73	-	-	-	45	10
Pear-tree,	73	-	-	-	45	10
Mulberry,	71	-	-	-	44	6
Walnut,	69	-	-	-	43	2
Yew,	67		-	-	41	14
Maple,	66	-	-	-	41	4
Yellow deal	63	-	-	-	39	6
Cherry,	61	-	-	-	38	2
Norway oak,	60	-	-	-	37	8
Sallow,	59	-	-	-	36	14
Sycamore,	59	-	-	-	36	14
Elm,	50	-	-	-	31	4

N. B. All the woods were very good of the sort, except the elm,
and all very dry; the measure is English, and the weight avoirdupois.

No. VIII.

No. VIII.

A Table of Expence.

By the Day. s. d.	By the Week. £. s. d.	By the Month*. £. s. d.	By the Year £. s. d.
0 1	0 0 7	0 2 4	1 10 5
0 2	0 1 2	0 4 8	3 0 10
0 3	0 1 9	0 7 0	4 11 3
0 4	0 2 4	0 9 4	6 1 8
0 5	0 2 11	0 11 8	7 12 1
0 6	0 3 6	0 14 0	9 2 6
0 7	0 4 1	0 16 4	10 12 11
0 8	0 4 8	0 18 8	12 3 4
0 9	0 5 3	1 1 0	13 13 9
0 10	0 5 10	1 3 4	15 4 2
0 11	0 6 5	1 5 8	16 14 7
1 0	0 7 0	1 8 0	18 5 0
2 0	0 14 0	2 16 0	36 10 0
3 0	1 1 0	4 4 0	54 15 0
4 0	1 8 0	5 12 0	73 0 0
5 0	1 15 0	7 0 0	91 5 0
6 0	2 2 0	8 8 0	109 10 0
7 0	2 9 0	9 16 0	127 15 0
8 0	2 16 0	11 4 0	146 0 0
9 0	3 3 0	12 12 0	164 5 0
10 0	3 10 0	14 0 0	182 10 0
11 0	3 17 0	15 8 0	200 15 0
12 0	4 4 0	16 16 0	219 0 0
13 0	4 11 0	18 4 0	237 5 0
14 0	4 18 0	19 12 0	255 10 0
15 0	5 5 0	21 0 0	273 15 0
16 0	5 12 0	22 8 0	292 0 0
17 0	5 19 0	23 16 0	310 5 0
18 0	6 6 0	25 4 0	328 10 0
19 0	6 13 0	26 12 0	346 15 0
20 0	7 0 0	28 0 0	365 0 0

* The month is 28 days.

A Table of Expence.

By the Year.		By the Month*.		By the Week.		By the Day.
£. s. d.		£. s. d.		£. s. d.		£. s. d.
1 0 0	—	0 1 6½	—	0 0 4½	—	0 0 0¾
2 0 0	—	0 3 0¾	—	0 0 9¼	—	0 0 1¼
3 0 0	—	0 4 7¼	—	0 1 1¾	—	0 0 2
4 0 0	—	0 6 1¾	—	0 1 6½	—	0 0 2¾
5 0 0	—	0 7 8	—	0 1 11	—	0 0 3¼
6 0 0	—	0 9 2½	—	0 2 3½	—	0 0 4
7 0 0	—	0 10 9	—	0 2 8¼	—	0 0 4½
8 0 0	—	0 12 3¼	—	0 3 0¾	—	0 0 5¼
9 0 0	—	0 13 9¾	—	0 3 5½	—	0 0 6
10 0 0	—	0 15 4	—	0 3 10	—	0 0 6½
11 0 0	—	0 16 10½	—	0 4 2¾	—	0 0 7¼
12 0 0	—	0 18 5	—	0 4 7¼	—	0 0 8
13 0 0	—	0 19 11¼	—	0 4 11¾	—	0 0 8½
14 0 0	—	1 1 5¾	—	0 5 4½	—	0 0 9¼
15 0 0	—	1 3 0¼	—	0 5 9	—	0 0 9¾
16 0 0	—	1 4 6½	—	0 6 1¾	—	0 0 10½
17 0 0	—	1 6 1	—	0 6 6½	—	0 0 11¼
18 0 0	—	1 7 7½	—	0 6 0¾	—	0 0 11¾
19 0 0	—	1 9 1¾	—	0 7 3½	—	0 1 0½
20 0 0	—	1 10 8¼	—	0 7 8	—	0 1 1¼
30 0 0	—	2 6 0¼	—	0 11 6	—	0 1 7¾
40 0 0	—	3 1 4½	—	0 15 4	—	0 2 2¼
50 0 0	—	3 16 8½	—	0 19 2¼	—	0 2 9
60 0 0	—	4 12 0¾	—	1 3 0¼	—	0 3 3½
70 0 0	—	5 7 4¾	—	1 6 10¼	—	0 3 10
80 0 0	—	6 2 9	—	1 10 8¼	—	0 4 4½
90 0 0	—	6 18 1	—	1 14 6¼	—	0 4 11¼
100 0 0	—	7 13 5	—	1 18 4¼	—	0 5 5¾
200 0 0	—	15 6 10¼	—	3 16 8½	—	0 10 11½
300 0 0	—	23 0 3¼	—	5 15 0¾	—	0 16 5¼
400 0 0	—	30 13 8½	—	7 13 5	—	1 1 11
500 0 0	—	38 7 1½	—	9 11 9¼	—	1 7 4¾
1000 0 0	—	76 14 3	—	10 3 6¾	—	2 14 9½

* The month is 28 days.

No. IX.

No. IX.
TESTS FOR THE ANALYSIS OF WATER.
Litmus infusion—delicate for acids.

Acid of sugar—lime.

Tincture of turmeric and Brazil wood—alkalis.

Tincture of galls by spirit of wine—iron.

Phlogisticated alkalis—excellent for iron.

Lime water—carbonic acid in water.

Nitrated silver dissolved in dist. water—complete for muriatic acid.

No. X.
FOR THE STRONG GULLION.
Rhubarb, in powder, 1 oz.; long pepper, in ditto, half an ounce; peppermint-water, 12 oz.; compound spirit of juniper berries, 4 oz.; liquid laudanum, $1\frac{1}{2}$ oz. Mix well and shake. The above one dose.

FOR A PINCH ON A HORSE's WITHERS.
Mercurial ointment; then rye-meal poultice and brandy, if skin not broken.

DISORDER FROM HEAT IN A HOT CLIMATE.
One drachm camphor; dessert spoonful of brandy; half an ounce of sugar. Rub the camphor and brandy together, then add the sugar: when well mixed, add one pint of boiling water by degrees; cover up, keep till cold, and take a quarter or half a pint, or all, in a day, after James's powder. If a tendency to putrefaction, 2 dr. sweet sp. of vitriol to the pint.

James's powder—Pulvis antimonialis, according to last Dispensatory.—Take, going to bed, 4 or 5 grains. If necessary, as far as 16 grains a day, at three or four times, may be taken.

POWDER FOR RHEUMATISM, WHEN FIRST COMING.
Pulvis ipecacuanha compositus, of the last London Dispensatory. In common cases, with no violent pain, 10 gr. at going to bed; in great pain, 20 gr.: wash down with diluting liquor.

No. XI.

No. XI.

BURDON's HORSE OINTMENT.

Into a clean pipkin that holds about a quart, put the bigness of a pullet's egg of yellow resin; when it is melted over a middling fire, add the same quantity of bees-wax : when that is melted, put in half a pound of hog's-lard; when it is dissolved, put in two ounces of honey; when that is dissolved, put in half a pound of common turpentine: keep it gently boiling, stirring it with a stick all the time: when the turpentine is dissolved, put in two ounces of verdigrease, finely powdered; but, before you put in the verdigrease, you must take off the pipkin (élse it will rise into the fire in a moment); set it on again, and give it two or three wambles, and strain it through a coarse sieve into a clean vessel, for use, and throw away the dregs.

This is an extraordinary ointment for a wound or bruise in flesh or hoof, broken knees, galled backs, bites, cracked heels, mallenderse, or, when you geld a horse, to heal, and keep the flies away. Nothing takes fire out of a burn or scald in human flesh so soon : I have had personal experience of it. I had it out of De Grey; but finding it apt to heal a wound at the top before the bottom was sound, I improved it, by adding an ounce of verdigrease.

No. XII.

The New Covenants in letting the Farms of T. W. Coke, Esq. M. P.

Supposing a farm to contain 540 acres arable land :

" Shall and will at all times keep and leave 90 acres, part of the arable land laid to grass, of one or more years laying; also 90 acres grass, of two or more years laying; each to be laid down with a crop of corn after turnips, and to continue laid two years at least; the time of laying to be computed from the harvest next after sowing the said seeds, and upon breaking up the same. After January 1, 1804, may be permitted to sow 45 acres (part thereof annually) with pease or tares for seed, to be twice well hoed; other part thereof with tares, for green food; buck-wheat, or any leguminous plant, for ploughing in as manure, or summer-tilling any portion of the remainder.

" Shall not sow any of the lands with two successive crops of corn, grain, pulse, rape, or turnips, for seed (except the above-mentioned
pea

pea and tare stubble), without the leave or consent of the said

his heirs or assigns being first had and obtained in writing.

" The land intended to be sown with pease, should not be till four years and a half after the commencement of the lease, upon the supposition that the *new tenant* may not be so situated as to have the turnips (covenanted to be left by old lease) completely clean.

" Lands for turnips, four clean earths at least.

" The turnips covenanted to be left in the last year, ninety acres to be mucked so far as the same will extend, and to be paid for by valuation, at the same time a due regard to be had to the cleanness of the land upon which they grow.

" Sheep, cattle, and all the other live stock, to be lodged upon some part of the premises when consuming the produce of the farm.

" Straw, chaff, and colder to be left without allowance.

" Incoming tenant to carry out the crop of corn, not exceeding the distance of ten miles, gratis.

" Rent payable forty days before Michaelmas (wherever a threshing machine is, or shall be erected) if demanded, by notice in writing being left at the farm-house to that purpose."

The husbandry alluded to, but not clearly expressed, is,

1. Turnips,	-	-	90 acres.
2. Barley,	-	-	90
3. Seeds,	-	-	90
4. Ditto,	-	-	90
5. { Pease,	-	-	45
Tares,	-	-	45
6. Wheat,	-	-	90
			540

Forbidding two crops of white corn running, not a bad covenant, simply; yet open to the following course:

1. Turnips,
2. Barley,
3. Clover,
4. Wheat, which would fill the land with
5. Pease, weeds.
6. Wheat,
7. Clover,
8. Wheat,

No. XIII.

No. XIII.

A CATALOGUE OF FARMING IMPLEMENTS.

	£.	s.	d.
Asbey's threshing-mill, fixed, - -	105	0	0
Do. moveable, - -	170	0	0
A winnowing machine, - -	10	10	0
Cooke's drill, and the corresponding tools, -	40	0	0
A scuffler, - - -	10	10	0
A stout plough with wheels, -	10	10	0
A do. smaller, - -	7	7	0
A Norfolk plough, -	4	4	0
The best swing-plough, - -	4	4	0
Harrows, -	10	0	0
Do, - - - -	5	0	0
Do. - - -	3	0	0
A fallowing machine, the Norfolk heavy drill-roller,	20	0	0
Great iron roller for grass, - -	35	0	0
Smaller do. - -	25	0	0
Roller for arable land, - -	8	0	0
Barley roller, - -	4	0	0
Northumberland turnip-drill, -	2	12	6
Garden drill and hoe, - -	4	0	0
Sir Joseph Bank's hay-stack apparatus, -	20	0	0
Fen paring plough for burning, - -	4	4	0
Amos's bean-drill, - -	4	10	0
Kentish turn-wrest plough, - -	9	9	0
The mole drain-plough with wheel and roller,	6	0	0
One-horse carts with ladders, the only machine for carriage (except the following article), which should be on any farm; each - -	12	12	0
Small three-wheeled cart, - 10 to	12	0	0
Machine for weighing cattle alive,	20	0	0
Salmon's cage and steel-yard for weighing sheep and hogs,	8	0	0
———— wheel-plough, - -	3	0	0
———— machine for weighing cattle, hay, &c.	100	0	0
———— chaff-cutter, - -	12	12	0
———— machine for ascertaining draught, -	3	3	0
Sowing machine for broad-cast turnips, -	0	10	0

Set

	£.	s.	d.
Set of hollow-draining spades, - -	1	1	0
Ant-hill plough, - - -	4	0	0
Machine for bruising beans and oats, -	5	0	0
Double mould-board plough with expanding wings,	10	0	0
Smaller do. - - -	5	0	0
Berkshire shim, - - -	2	10	0
Do with three hoes, - - -	3	10	0
Isle of Thanet shim, - -	10	0	0
Ducket's skim-coulter, - -	1	1	0
———— hand-hoes, - -	0	10	0
Miner, - - -	3	3	0
The Hon. George Villiers' moveable sheep-house,	20	0	0
Apparatus for steaming roots, -	20	0	0
Machine for breaking oil-cake, - -	4	0	0
Turnip-slicer, - -	2	10	0
Potatoe-cutter, - - -	2	0	0
Potatoe harrow, - - -	2	10	0
Jointed borer, - - -	21	0	0
Hay-drag, - - -	4	0	0
Horse dew-rake, - - -	5	0	0
Wheel hay-rake, - - -	1	1	0
Mr. Bentinck's machine for drawing up trees by the root, - - -	100	0	0

No. XIV.

No. XIV.

SYNOPSIS OF THE BREEDS OF SHEEP. BY MR. CULLEY.

				Weight of fleece.	Prices per lb. 1791.			Wethers, per qr.	Years old when killed.
1 Dishley,	No horns,	White face and legs,	Combing wool,	8 lb.	£. 0	0	10	25 lb.	2
2 Lincoln,	Do.	Do.	Do.	11	0	0	10	25	3
3 Teeswater,	Do.	Do.	Do.	9	0	0	10	30	2
4 Dartmoor,	Do.	Do.	Do.	9	0	0	8	30	2½
5 Exmoor,	Horned,	Do.	Do.	6	0	0	8	16	2½
6 Dorset,	Do.	Do.	Carding wool,	3½	0	1	2	18	3½
7 Hereford,	No horns,	Do.	Fine do.	2	0	2	9	14	4½
8 South Down,	Do.	Grey faces and legs,	Do.	2½	0	2	0	18	2
9 Norfolk,	Horns,	Black faces and legs,	Do.	2	0	1	0	18	3½
10 Heath,	Do.	Do.	Course combing,	3½	0	0	6	15	4½
11 Herdwick,	No horns,	Speckled do.	Carding,	2	0	0	6	10	4½
12 Cheviot,	Do.	White faces and legs,	Do.	3	0	0	11	16	4½
13 Dunfaced,	Do.	Dun faces and legs,	Do.	1½	0	3	0	7	4½
14 Shetland,	Do.	Colours various,	Fine cottony,	1½	0	3	0	8	4½
15 Romney Marsh,	No horns,	White faces and legs,	Combing,	9	0	0	10	25	2½
16 Spanish,	Horns,	Do.	Carding,	3½	0	4	0		2¼

No. XV

No. XV.

DIVISION OF SOILS.

Soil.	Synonima.	Crops.
I. Clay.	Strong land. Strong loam. Stiff land. Stiff loam. Lime-stone clay. *Marmy* clay.	Cabbages. Beans. Wheat. Clover. W. tares. Oats.
II. Sand.	Red. White. Black. Yellow. Heathy. Gravelly.	Turnips. Potatoes. Carrots. Barley. Buckwheat. Trefoil. Ray, &c.
III. Loam.	Sandy. Gravelly. Hasel. Lime-stone. Stone *brash*.	Potatoes. Turnips. Barley. Oats. Pease. Clover. Beans. Cabbages. Hemp and flax.
IV. Chalk.	Chalky loam. Downs.	Sainfoin. Barley. Pease. Turnips.
V. Peat.	Moss. Bog, red and black. Fen.	Potatoes. Rape. Cabbages. Grass.

No. XVI.

No. XVI.
THE FARMER's LIBRARY

Hitt on Barren Lands, 8vo.

Twamley on Dairying, 8vo.

Lisle's Husbandry, 2 vol. 8vo.

Fordyce's Elements of Agriculture, 8vo.

Billing on Carrots, 8vo.

Culley on Live Stock, 8vo.

Boswell on Irrigation, 8vo.

Wright on Ditto, 8vo.

Blythe's Improver improved, 4to.

Baker's Experiments, 4 vol. 8vo.

Kirwan on Manures, 8vo.

Dossie's Memoirs of Agriculture, 3 vol. 8vo.

Transactions of the Society of Arts, 20 vol. 8vo.

Bath Society's Papers, 9 vol. 8vo.

Wight's Survey of Scotland, 6 vol. 8vo.

Darwin's Phytologia, 4to.

Gyllenborg's Elements of Agriculture, by Mills, 12mo.

Home's Principles of Agriculture and Vegetation, 8vo.

Marshall's Works, 16 vol. 8vo.

Lord Kaimes's Gentleman Farmer, 8vo.

Reprinted County Reports, by the Board of Agriculture, 20 vol. 8vo

Communications to the Board of Agriculture, 3 vol. 4to.

Curtis on Grasses, 8vo.

Swayne's Gramina Pascua, folio.

On Potatoes, Published by the Board of Agriculture, 4to.

Johnson on Draining, 8vo.

Martyn's edition of Miller, folio,

Bailey's Advancement of Arts, 4to.

Anderson's Essays on Agriculture, 3 vol. 8vo.

Bannister's Synopsis, 8vo.

Hunter's Georgical Essays, 4 vol. 8vo.

Stickney on Grubs, 8vo.

Bartlet's Farriery, 12mo *.

---

* Works of merit that may be here omitted, will be inserted should the work be again reprinted.

No. XVII.

No. XVII.
OLD AND NEW CHEMISTRY.

Old Names.	New Names.
Alkaline air.	Ammoniacal gas.
Alkali, caustic fixed.	Potash.
———— mild fixed.	Carbonate of potash.
———— fixed vegetable.	Ditto.
———— caustic marine.	Soda.
———— ditto mild.	Carbonate of soda.
———— phlogisticated.	Prussiate of potash.
———— mild volatile.	Carbonate of ammonia.
———— caustic volatile.	Ammonia.
Alum.	Sulphate of alumine.
Argil.	Alumine.
Barilla.	Carbonate of soda.
Blue, Prussian.	Prussiate of iron.
Butter of antimony.	Sublimated muriate of antimony.
Calces metallic.	Metallic oxides.
Chalk.	Carbonate of lime.
Charcoal.	Carbon.
Calcareous earth.	Lime.
Earth of alum.	Alumine.
Fixed air.	Carbonic acid gas.
Gas hepatic.	Sulphurated hydrogen gas.
—— inflammable.	Hydrogen gas.
—— inflammable burning with a blue flame.	Carbonated hydrogen gas.
—— mephitic.	Carbonic acid gas.
—— phlogisticated.	Nitrogen gas.
Gypsum.	Sulphate of lime,
Heat latent.	Caloric.
Magnesia, caustic.	Magnesia.
———— alba.	Carbonate of magnesia,
Natron,	Carbonate of soda.
Nitre.	Nitrate of potash.
—— cubic.	Nitrate of soda.
Oil of vitriol	Sulphuric acid.

Principle

Principle acidifying.   Oxygen.
Pytrites.   Sulphure of iron.
Rust of iron.   Oxide of iron.
Sal ammoniac.   Muriate of ammonia.
Salt, common.   —— of soda.
—— of sylvius.   —— of potash.
—— Glauber's.   Sulphate of soda.
—— Epsom.   —— of magnesia.
—— petre.   Nitrate of potash.
—— vegetable.   Tartarite of potash.
Selenite.   Sulphate of lime.
Spirit of salt.   Muriatic acid.
Tartar.   Tartarite of potash.
—— vitriolated.   Sulphate of potash.
Verdegris.   Green oxide of copper.
Vinegar.   Acetous acid.
Vitriol, blue.   Sulphate of copper.
—— green   —— of iron.

No. XVIII.

No. XVIII.

TABLE OF COMPARATIVE PRICES OF WOOL.

By the pound of 16 ounces.			By the stone of 12 lb.			By the tod of 28 lb.			By the tod of 32 lb.			By the pack of 240 lb.		
£.	s.	d.	£.	s.	d.	£.	s.	d.	£.	s.	d.	£.	s.	d.
At 0	0	0 per lb.												
0	0	9*	0	9	0	1	1	0	1	4	0	9	0	0
0	0	10	0	10	0	1	3	4	1	6	8	10	0	0
0	0	11	0	11	0	1	5	8	1	9	4	11	0	0
0	1	0	0	12	0	1	8	0	1	12	0	12	0	0
0	1	1	0	13	0	1	10	4	1	14	8	13	0	0
0	1	2	0	14	0	1	12	8	1	17	4	14	0	0
0	1	3	0	15	0	1	15	0	2	0	0	15	0	0
0	1	4	0	16	0	1	17	4	2	2	8	16	0	0
0	1	5	0	17	0	1	19	8	2	5	4	17	0	0
0	1	6	0	18	0	2	2	0	2	8	0	18	0	0
0	1	7	0	19	0	2	4	4	2	10	8	19	0	0
0	1	8	1	0	0	2	6	8	2	13	4	20	0	0
0	1	9	1	1	0	2	9	0	2	16	0	21	0	0
0	1	10	1	2	0	2	11	4	2	18	8	22	0	0
0	1	11	1	3	0	2	13	8	3	1	4	23	0	0
0	2	0	1	4	0	2	16	0	3	4	0	24	0	0
0	2	1	1	5	0	2	18	4	3	6	8	25	0	0
0	2	2	1	6	0	3	0	8	3	9	4	26	0	0
0	2	3	1	7	0	3	3	0	3	12	0	27	0	0
0	2	4	1	8	0	3	5	4	3	14	8	28	0	0
0	2	5	1	9	0	3	7	8	3	17	4	29	0	0
0	2	6	1	10	0	3	10	0	4	0	8	30	0	0

* Rules for calculating the value of each quantity, when the price per lb. is not any number of even pence.

1. If the increase be one farthing, add to the sum found in the table, opposite to the given number of pence, 3d. for the stone; 7d. for the tod of 28 lb.; 8d. for the tod of 32 lb.; and 5s. for the pack. Thus, if the price be 9¼d. per lb. the value of the stone will be 9s. 3d.; of the tod of 28 lb. it will be 1l. 1s. 7d.; of the tod of 32 lb. it will be 1l. 4s. 8d.; and of the pack, it will be 9l. 5s.

2. If the increase be one halfpenny, add to the sum found in the table opposite to the given number of pence, 6d. for the stone; 1s. 2d. for the tod of 28 lb. 1s. 4d. for the tod of 32 lb. and 10s. for the pack. Thus, if the price be 1s. 3½d. per lb., the value of the stone will be 15s. 6d.; of the tod of 28 lb. will be 1l. 16s. 2d.; of the tod of 32 lb will be 2l. 1s. 4d.; and of the pack, it will be 15l. 10s.

3. If the increase be three farthings, add to the sum found in the table opposite to the given number of pence, 9d. for the stone; 1s. 9d. for the tod of 28 lb.; 2s. for the tod of 32 lb.; and 15s. 8d. for the pack. Thus, if the price be 2s. 0¾d. per lb., the value of the stone will be 1l. 4s. 9d.; of the tod of 28 lb., it will be 2l. 17s. 9d.; of the tod of 32 lb., it will be 3l. 16s.; and of the pack, it will be 24l. 15s.

INDEX.

A

 —— the

——but

— qnestion

— question whether on doing this, the whole of the pasture grounds should be at once thrown open to them, or divided off successively ; advantages and disadvantages of each method, 251 to 253.

— attentions in the point of stocking grass-lands, 253.

management of young cattle in October, 493.

may be served with Straw (as a food) in too great plenty, 550.

See also the articles *Calves* ; *Cows* ; *Farm-yard* ; *Fattening beasts* ; and *Litter*.

Chaff. See *Cut-chaff.*

Chaff-house. See *Threshing-mill.*

Chalking ; method of performing this business, in Hertfordshire, 42.

—sinking the pit, 42.

—raising the chalk, 43.

—operations at the bottom of the pit, 43.

—quantity of land to each pit ; expence, 44.

deepest chalk the best ; flints to be picked out, 44.

successful practice of Mr. John Hill, of Coddicot ; once chalking produces abundant crops for ten years, 44, 45.

distinction between the different sorts of chalk ; and suitable soils, 370.

dig throughout July and August ; use the small three-wheeled cart, 417, 418, 439.

dig in November, 515.

— in December, 527.

Chamomile, its soil, and planting ; the cultivation of it troublesome and not advantageous, 152.

Cheese. See *Dairy.*

Cheshire cheese ; general method of making, 285 to 288.

Chicory ; peculiar advantage of the cultivation of this plant, 124, 150.

objection to it ; answered, 125.

methods of sowing, broad-cast or drilling ; quantity of seed ; harrowing after sowing, 125, 6, 204.

Churns, apparatus to those used in Cheshire, 285.

Clay. See *Marling.*

Close feeding pasturage with sheep ; important advantages of an attention to this point, 269 to 271.

Clover : great benefit of this crop; various methods of sowing it, 117.

— advantage of each, 118.

value of the produce, in different views, 118.

it is peculiarly liable to fail in districts where it has been long grown ; course on a farm in Surrey, which succeeded well with clover every third year, 119.

— in such circumstances, Tares may also be dibbled into the vacant spots with advantage, 243.

quantity of seed, 119.

White clover alone, for seed, 120.

time of mowing clover ; this the most profitable application of the crop ; direction in making the hay, 346.

— Wheat

—· begin

H

general

L

— consi‑

— Mr.

Mountain

— advan-

— potatoes

— attention

Threshing :

TO THE BINDER.

Place the Plate of Irrigation opposite page 306.
And that of Watered Meadow opposite page 308.

Printed by B. M'Millan, }
Bow-Street, Covent-Garden. }

Valua